Essentials of Mesenchymal Stem Cell Biology and Its Clinical Translation

Robert Chunhua Zhao
Editor

Essentials of Mesenchymal Stem Cell Biology and Its Clinical Translation

Springer

Editor
Robert Chunhua Zhao
Center of Excellence in Tissue Engineering
Institute of Basic Medical Sciences and School of Basic Medicine
Chinese Academy of Medical Sciences and Peking
Union Medical College
Beijing, China, People's Republic

ISBN 978-94-007-6715-7 ISBN 978-94-007-6716-4 (eBook)
DOI 10.1007/978-94-007-6716-4
Springer Dordrecht Heidelberg New York London

Library of Congress Control Number: 2013940097

Printed on acid-free paper

Springer is part of Springer Science+Business Media (www.springer.com)

Preface

Once you open this book, we are somewhat connected to stem cell science, and it will take you walking into the amazing world of stem cells.

You may have read books or attended classes about stem cells; you may have even reported important scientific results related to stem cells. This book will lead you to a specific type of stem cells – mesenchymal stem cells (MSCs), which have attracted the attention of both scientists and physicians due to their unique biological properties and promise for disease treatment. This book will be valuable to you as it bridges the gap between basic research and therapeutic approaches on stem cell clinical translation.

A decade ago, scientists obtained human embryonic stem cell (ESC) and began to reveal that adult stem cells could generate differentiated cells beyond their own tissue boundaries, which was termed developmental plasticity; yet development of therapeutic approaches with stem cells is still in its infancy. Day by day, the field of stem cells develops at rapid pace, and the transition of stem cells from basic research to clinical application is making enormous progress. More than ever, stem cell biologists and physicians are joining in this field to better understand the molecular mechanisms and develop novel therapeutic paradigm. As stem cell research is sophisticated and the translation of basic research to clinical application faces great challenges, it is important to have leading expertise in this field to update the most recent information and share their views and perspectives. To this end, we would bring out this book, *Essentials of Mesenchymal Stem Cell Biology and its Clinical Translation*. It first addressed and discussed current advances and concepts pertaining to MSC biology, covering topics such as MSC secretome, homing, signaling pathways, miRNAs, and manipulation with biomaterials and so on. Especially, we introduce the hypothesis that post-embryonic pluripotent stem cells exist as a small subset of cells in MSCs. As MSC plays a key role in immunomodulation, we explored the clinical application of MSCs in a variety of diseases, taking into account cardiovascular diseases, liver diseases, graft-versus-host diseases and diabetes. International regulations and guidelines governing stem-cell-based products are also brought in here. Overall, this book covers a broad range of topics about MSCs during their transition from bench side to bedside. The chapters of the book are all written by experts in their respective disciplines, which allow each of them to be a "stand-alone" entity although there is continuity of style from chapter to chapter

Last year MSCs as the first stem cell drug were lauched into the market, and currently there are more than 270 clinical trials registered in the public clinical trials database (http://clinicaltrials.gov), 66 of which are conducted in China. Chinese government exercises the most strict and stringent rule on stem cell products. In 2004, Flk1^{+} MSCs in our laboratory became the first stem-cell-product that received official approval for clinical trial from the Chinese State Food and Drug Administration (SFDA). Since then our studies demonstrate that Flk1+ MSCs represent a safe and effective treatment for several disorders. These encouraging results promoted me to organize a book to share the fascinating stem cell knowledge and technology with those who are interested in MSCs, and now the book is finally complete.

I wish to extend my gratitude to the staff of our publisher, Springer, for providing great support for this book. I want to express my appreciation to all the authors for their excellent contributions and dedication to scholarly pursuits. With their pioneering work and devoted efforts, this book could be brought to fruition. They are the true heroes in the backstage, although I am the one standing under the spotlight. I would also like to thank Dr. Shihua Wang in my stem cell center for her efforts in chapter collecting and assistance in editing. Lastly, as always, the goal of this book is to educate, stimulate and serve as a resource. I hope that you, as a reader, will enjoy this scientific stem cell book.

Beijing, China Robert Chunhua Zhao

Contents

Part II Clinical Translation

Part III International Regulations and Guidelines Governing Stem Cell Based Products

Part I
Basic Research/Mechanisms

A Historical Overview and Concepts of Mesenchymal Stem Cells

Shihua Wang and Robert Chunhua Zhao

Abstract Mesenchymal stem cells have generated great interest among researchers and physicians due to their unique biological characteristics and potential clinical applications. Here, we first give a brief introduction to mesenchymal stem cells, from their discovery to their definition, sources and types. During embryonic development, MSCs arise from two major sources: neural crest and mesoderm. We discuss these two developmental origins. Additionally, we propose for the first time the concept of a hierarchical system of MSCs and draw the conclusion that post-embryonic subtotipotent stem cells are cells that are leftover from embryonic development and are at the top of the hierarchy, serving as a source of MSCs. Then, we describe various concepts related to MSCs, such as their plasticity, immunomodulatory functions, homing and secretion of bioactive molecules. These concepts constitute an important part of the biological properties of MSCs, and a thorough understanding of these concepts can help researchers gain better insight into MSCs. Finally, we provide an overview of the recent clinical findings related to MSC therapeutic effects. MSC-based clinical trials have been conducted for at least 12 types of pathological conditions, with many completed trials demonstrating their safety and efficacy.

S. Wang • R.C. Zhao (✉)
Center of Excellence in Tissue Engineering, Institute of Basic Medical Sciences and School of Basic Medicine, Chinese Academy of Medical Sciences and Peking Union Medical College, 5# Dongdansantiao, 100005 Beijing, China, People's Republic
e-mail: chunhuaz@public.tpt.tj.cn; zhaoch16@hotmail.com

R.C. Zhao (ed.), *Essentials of Mesenchymal Stem Cell Biology and Its Clinical Translation*, DOI 10.1007/978-94-007-6716-4_1,

Keywords MSC • Developmental origin • Plasticity • Homeing • Immunomodulatory functions • Clinical application

Introduction

Stem cells have the capacity to self-renew and to give rise to cells of various lineages. Thus, they represent an important paradigm of cell-based therapy for a variety of diseases. Broadly speaking, there are two main types of stem cells, embryonic and non-embryonic. Embryonic stem cells (ESCs) are derived from the inner cell mass of the blastocyst and can differentiate into the cells of all three germ layers. However, teratoma formation and ethical controversy hamper their research and clinical application. Contrastingly, non-embryonic stem cells, mostly adult stem cells, are already somewhat specialized and have limited differentiation potential. They can be isolated from various tissues and are currently the most commonly used seed cells in regenerative medicine. Recently, another type of non-embryonic stem cell, known as an induced pluripotent stem cell (iPSC), has emerged as a major breakthrough in regenerative biology. These cells are generated through the forced expression of a defined set of transcription factors, which reset the fate of somatic cells to an embryonic stem-cell-like state.

Cellular therapy has evolved quickly over the last decade both at the level of in vitro and in vivo preclinical research and in clinical trials. Embryonic stem cells and non-embryonic stem cells have both been explored as potential therapeutic strategies for a number of diseases. One type of adult stem cell, the mesenchymal stem cell, has generated a great amount of interest in the field of regenerative medicine due to its unique biological properties. MSCs were first discovered in 1968 by Friedenstein as an adherent fibroblast-like population in the bone marrow capable of differentiating into adipocytes, chondrocytes and osteocytes, both in vitro [1] and in vivo [2]. Caplan demonstrated that bone and cartilage turnover was mediated by MSCs, and the surrounding conditions were critical to inducing MSC differentiation [3]. They termed these cells "mesenchymal stem cells," and the term "MSC" became popular after the work of A.I. Caplan et al. in 1991. Later, the multilineage differentiation capability of MSCs was definitively demonstrated by Pittenger [4]. During the late 1990s, Kopen et al. then described the capacity of MSCs to transdifferentiate into ectoderm-derived tissue [5].

Definition, Sources and Types of Mesenchymal Stem Cells

The defining characteristics of MSCs are inconsistent among investigators. Many laboratories have developed methods to isolate and expand MSCs, which invariably have subtle, and occasionally quite significant, differences. To address this problem, in 2006, the Mesenchymal and Tissue Stem Cell Committee of International Society for Cellular Therapy (ISCT) proposed a set of standards to define human MSCs for both laboratory-based scientific investigations and for pre-clinical studies. First, MSCs must be plastic-adherent when maintained in standard culture conditions using tissue culture flasks. Second, ≥95 % of the MSC population must express CD105, CD73 and CD90, as measured by flow cytometry. Additionally, these cells must lack the expression (≤2 % positive) of CD45, CD34, CD14 or CD11b, CD79a or CD19 and HLA class II. Third, the cells must be able to differentiate into osteoblasts, adipocytes and chondroblasts under standard in vitro differentiating conditions [6].

MSCs have been identified in almost every tissue type, including placenta, umbilical cord blood, amniotic fluid, bone marrow, adipose tissue, and the liver. Most of the adult sources, including large volumes of normal bone marrow, are relatively difficult to access as a tissue source for the isolation of MSCs. In contrast, birth-associated tissues, including placenta, are readily and widely available. However, bone marrow remains the principal source of MSCs for most preclinical and clinical studies. It is estimated that MSCs represent only between approximately 0.01 and 0.001 % of the total nucleated cells within isolated bone marrow aspirates [4, 7]. Despite this low number, there remains a great interest in these cells, as they can be isolated easily from a small aspirate and culture-expanded through as many as 40 population doublings to significant numbers in approximately 8–10 weeks. MSCs from different sources have been studied, and each type has been reported to vary in its proliferative and multilineage potential [7]. Therefore, it is important to realize that the varied approaches used to culture-expand and select for MSCs make it difficult to directly compare experimental results. Moreover, some isolation schemes introduce epigenetic and genetic changes in cells that may dramatically affect their plasticity and therapeutic utility [8].

Developmental Origin of MSCs

Although the biological characteristics and therapeutic potential of MSCs have been extensively studied, the in vivo behavior and developmental origin of these cells remain largely unknown. During embryonic development, MSCs arise from two major sources: neural crest and mesoderm. The adult MSCs are commonly considered to be of mesodermal origin, whereas embryonic MSCs derive mainly from the neural crest. The neural crest is a transient embryonic tissue that originates at the neural folds during vertebrate development. Morikawa et al. found that the development of MSCs partially originate from the neural crest [9]. Takashima et al. showed that the earliest wave of MSCs in the embryonic trunk is generated from Sox1+ neuroepithelium, and they provided evidence that Sox1+ neuroepithelium gives rise to MSCs in part through a neural crest intermediate stage [10]. The mesoderm is considered to be another major source of mesenchymal cells giving rise to skeletal and connective tissues [11]. Using hESCs directed towards mesendodermal differentiation, Vodyanik et al. showed that mesoderm-derived MSCs arise from a common endothelial and mesenchymal cell precursor, the mesenchymoangioblast, which is a transient population of cells within the APLNR+ mesodermal subset that can be identified using an FGF2-dependent mesenchymal colony-forming cell (MS-CFC) assay in serum-free semisolid suspension culture. Recently, the Olsen group revealed that vascular endothelial cells can transform into MSCs by an ALK2 receptor-dependent mechanism. Expressing mutant ALK2 in human endothelial cells causes an endothelial-mesenchymal transition (endMT) and the acquisition of a multipotent stem cell-like phenotype [12]. This result indicates that endothelial cells could be an important source of MSCs in postnatal life. Conversely, the transition from MSCs to endothelial cells has also been described in several studies. These studies suggest a cycle of cell-fate transition from endothelium to MSCs and back to endothelium. Because multiple parallels could be drawn between the endMT described in adult tissues and that during hESC differentiation, one may wonder whether bipotential cells with endothelial and MSC potential similar to embryonic mesenchymoangioblasts are present and constitute an important element of the EndMT circuit in adults [13]. The number of MSCs of neuroepithelial origin in the adult bone marrow decreases rapidly, which suggests that in post-natal life, the relative importance of MSCs derived from other developmental lineages decreases due to the increasing importance of mesodermal MSCs. We isolated $Flk1^{+}CD31^{-}CD34^{-}$ stem cells, which are MSCs from human fetal bone marrow, and found that they could differentiate into cells of the three germ layers, such as endothelial, hepatocyte-like, neural, and erythroid cells, at the single-cell level [14, 15]. Based on this result, we hypothesized that post-embryonic subtotipotent stem cells exist, and this hypothesis was later confirmed by other scientists (Table 1).

Here, for the first time, we propose the existence of a hierarchical system of MSCs (Fig. 1), which is composed of all mesenchymal stem cells from post-embryonic subtotipotent stem cells to MSCs progenitors. Post-embryonic subtotipotent stem cells are left-over cells during embryonic development and are on the top of the hierarchy. MSC system is a combination of cells that are derived from different stages of embryonic development, possess different differentiation potential and ultimately give rise to cells that share a similar set of phenotypic markers. The concept of MSC system entirely explains the three important biological characteristics of MSC: stem cell properties of MSCs, MSCs as components of tissue microenvironment and immunomodulatory functions of MSCs.

MSC Plasticity

As previously demonstrated, MSCs can differentiate into cells of mesenchymal lineages, such as osteoblasts, chondrocytes and adipocytes, under culture conditions containing specific growth factors and chemical agents. Furthermore, the important signaling pathways underlying these differentiation processes have been studied extensively. In addition to the abovementioned mesenchymal lineages, MSCs have been reported to give rise to cells of other lineages. Kopen et al. were the first

Table 1 Studies confirming the subtotipotent stem cell hypothesis

Tissue	Cell types produced	Reference
Term placental membranes	All embryonic germ layers, including alveolar type II cells	[16]
Wharton's jelly of umbilical cord	Ectoderm-, mesoderm- and endoderm-derived cells, including insulin-producing cells	[17]
Amniotic fluid	All embryonic germ layers, including neuronal lineage cells secreting the neurotransmitter L-glutamate or expressing G-protein-gated inwardly rectifying potassium channels, hepatic lineage cells producing urea, and osteogenic lineage cells forming tissue-engineered bone	[18]
Placenta and bone marrow	Adipocytes and osteoblast-like cells (mesoderm), glucagon- and insulin-expressing pancreatic-like cells (endoderm), as well as cells expressing the neuronal markers neuron-specific enolase, glutamic acid decarboxylase-67 (GAD), or class III beta-tubulin, and the astrocyte marker glial fibrillary acidic protein (ectoderm)	[19]
Human term placenta	All three germ layers in vitro – endoderm (liver, pancreas), mesoderm (cardiomyocyte), and ectoderm (neural cells)	[20]
placental cord blood	In vitro – osteoblasts, chondroblasts, adipocytes, and hematopoietic and neural cells, including astrocytes and neurons that express neurofilament, sodium channel protein, and various neurotransmitter phenotypes. In vivo – mesodermal and endodermal lineages demonstrated in animal models	[21]
Adult bone marrow	Cells with visceral mesoderm, neuroectoderm and endoderm characteristics in vitro	[22]

researchers to demonstrate that MSCs injected into the central nervous systems of newborn mice migrate throughout the brain and adopt morphological and phenotypic characteristics of astrocytes and neurons [5]. Spees et al. reported that coculture with heat-shocked small airway epithelial cells induced human MSCs to differentiate into epithelial-like cells, as evidenced by their expression of keratins 17, 18, and 19, the Clara cell marker CC26, and the formation of adherens junctions with neighboring epithelial cells [23].

These reports raised a number of critical issues and created controversy regarding the theories of MSC plasticity, which claimed that many factors may influence cell fate, such as fusion in vivo, criteria for differentiation and selection by rare cell populations. Alvarez-Dolado et al. were the first researchers to demonstrate that bone-marrow MSCs fuse spontaneously with neural progenitors in vitro. Furthermore, bone marrow transplantation demonstrates that BMDCs fuse in vivo with hepatocytes in the liver, Purkinje neurons in the brain and cardiac muscle in the heart, resulting in the formation of multinucleated cells [24]. As to the criteria for differentiation, it is difficult to conclude a differentiation process from the expression of a number of markers without the expression of the key transcription factors [25].

We are the first group to demonstrate that Flk1+-MSCs (Flk1+CD44+CD29+CD105+CD166+ CD34-CD31-Lin-) can give rise to multilineage cells of the three

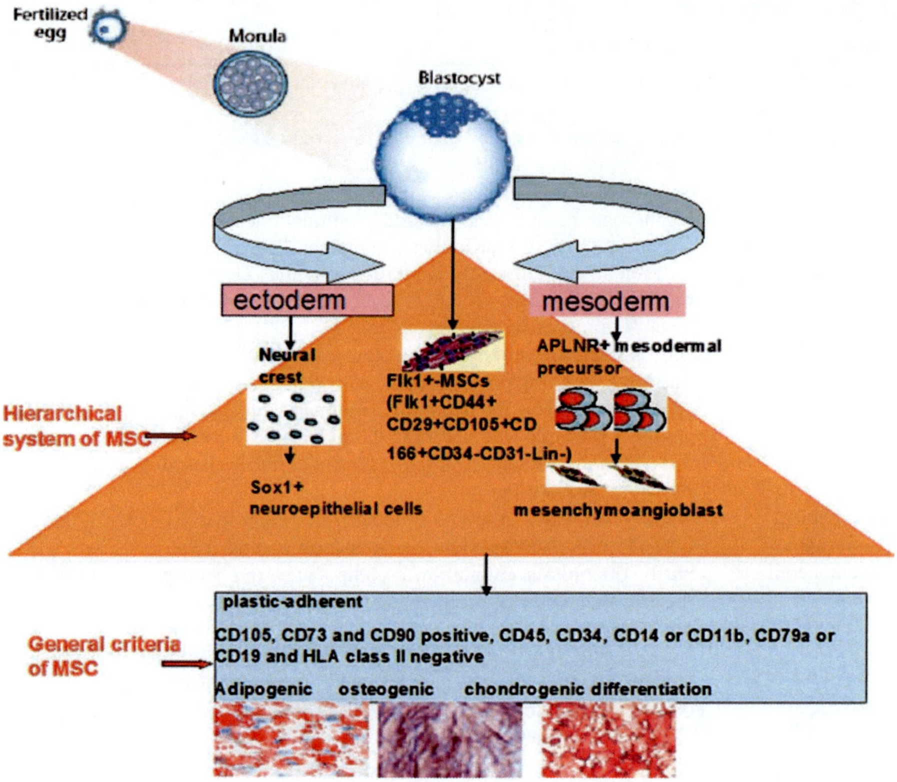

Fig. 1 A schematic description of the hierarchical system for mesenchymal stem cells. MSC system is a combination of cells that are derived from different stages of embryonic development, possess different differentiation potential and ultimately give rise to cells that share a similar set of phenotypic markers

germ layers at the clone level. To explore the mechanisms underlying the multilineage state and lineage specification of Flk1+-MSCs, we performed a genome-wide investigation of H3K4me3 and H3K27me3 profiles in these cells by ChIP-seq (n = 3) and compared these results with those obtained in embryonic stem cells (ESCs), hematopoietic stem cells (HSCs) and hematopoietic progenitor cells (HPCs). The pluripotent-associated gene, Klf4, was modified by the activating H3K4me3 histone modification; Sall4, Sox2, and Foxd3 were found to be bivalent; and Oct4 (Pou5f1) and Nanog exhibited either a repressive state or no modification in Flk1+-MSCs. However, all the above-mentioned genes were marked by H3K4me3 in ESCs and were either modified by H3K27me3 or carried no modification in HSCs and HPCs. We speculate that distinct histone modifications of pluripotency-associated genes might be partly responsible for the phenomenon that, among the four stem cell types, only ESCs give rise to teratomas in vivo. We next evaluated the histone methylation status of genes associated with lineage specification. As our analysis moved from ESCs to Flk1+-MSCs, HSCs, and finally, to HPCs, there was an increasing frequency of active modifications on hematopoietic lineage-related genes and a decreasing frequency of modifications on genes related to other lineages. These findings suggest that the histone modification patterns of differentiation-associated genes are closely related to a stem cell's multipotential state and can be used to predict its differentiation potential.

Immunomodulatory Properties of MSCs

MSCs lack immunogenicity because they express low levels of major histocompatibility complex-I (MHC-I) molecules and do not express MHC-II molecules or costimulatory molecules such as CD80, CD86, or CD40 [26]. This unique property allows for the transplantation of allogeneic MSCs. Another important reason for the large number of clinical studies using MSCs is their immunomodulatory functions. MSCs can also modulate the functions of the immune system by interacting with a wide range of immune cells, including T lymphocytes, B lymphocytes, and dendritic cells. The immunomodulatory properties of MSCs were initially reported in T-cell proliferation assays using one of a variety of stimuli, including mitogens, CD3/CD28, and alloantigens; these are settings in which the ability of MSCs to suppress T-cell proliferation can readily be determined [27–29]. MSCs regulate the proliferation, activation, and maturation of B lymphocytes in vitro in a dose-dependent and time-limited manner [30], and they can facilitate the immunosuppressive effect of cyclosporin A on T lymphocytes through Jagged-1-mediated inhibition of NF-κB signaling [31]. We first reported that MSCs could inhibit the upregulation of CD1a, CD40, CD80, CD86, and HLA-DR during DC differentiation and prevent an increase of CD40, CD86, and CD83 expression during DC maturation [32]. We also demonstrated that in the presence of MSCs, the percentage of cells with a cDC phenotype is significantly reduced, whereas the percentage of pDC phenotypes increases, further suggesting that MSCs can significantly influence DC development [33]. MSCs could drive maDCs to differentiate into a novel Jagged-2-dependent regulatory DC population and escape their apoptotic fate [34]. The immunomodulatory properties of MSCs in vivo have also become an exciting focus for investigators in terms of examining their potential implications in a variety of disease models such as diabetes, cardiovascular diseases, and liver diseases.

MSC Homing

Homing is the process by which cells migrate to, and engraft in, the tissue in which they exert their local, functional effects. MSC homing is defined as the arrest of MSCs within the vasculature of a tissue followed by transmigration across the endothelium. Such a nonmechanistic definition is appropriate, given the current absence of a definitive MSC homing mechanism, unlike the well-characterized leukocyte adhesion cascade that defines leukocyte homing [35]. The homing of MSC after systemic or local infusion has been studied in animal models in a variety of experimental settings. A growing number of studies of various pathologic conditions have demonstrated that MSCs selectively home to sites of injury [36]. For example, with the use of the high sensitivity of a combined single-photon emission CT (SPECT)/CT scanner, the in vivo trafficking of allogeneic MSCs co-labeled with a radiotracer and an MR contrast agent to acute myocardial infarction was dynamically determined. Focal and diffuse uptake of MSCs in the infarcted myocardium was visible in SPECT/CT images in the first 24 h after injection and persisted until 7 days after injection [37]. Ortiz et al. showed that MSC engraftment in lung tissue is enhanced in response to bleomycin exposure and ameliorates the fibrotic effects of the drug [38]. Although the homing of leukocytes to sites of inflammation is well studied, the mechanisms of MSC homing to sites of ischemia or injury are poorly understood. It is likely that increased inflammatory chemokine concentration at the site of inflammation is a major factor causing MSCs to preferentially migrate to these sites. Chemokines are released after tissue damage, and MSCs express the receptors for several chemokines. The migration capacity of MSCs was found to be under the control of a large range of receptor tyrosine kinase growth factors, such as platelet-derived growth factor (PDGF) and insulin-like growth factor 1 (IGF-1), and chemokines, such as CCR2, CCR3, CCR4 and CCL5, as assessed by in vitro migration assays [36].

MSC Secreting Bioactive Molecules

MSCs can secrete multiple bioactive molecules, including many known growth factors, cytokines and chemokines, that have profound effects on local cellular dynamics (Table 2). The administration of MSC-conditioned medium can recapitulate the beneficial effects of MSCs on tissue repair. For instance, data from Van Poll D et al. provide the first clear evidence that MSC-conditioned medium (MSC-CM) provides trophic support to the injured liver by inhibiting hepatocellular death and stimulating regeneration, potentially creating new avenues for the treatment of fulminant hepatic failure (FHF) [52]. Takahashi et al. demonstrated that various cytokines were produced by BM-MSCs, and these cytokines contributed to functional improvement of the infarcted heart by directly preserving the contractile capacity of the myocardium, inhibiting apoptosis of cardiomyocytes, and inducing therapeutic angiogenesis of the infarcted heart [53].

A protein-array analysis of MSC-CM detected 69 of 174 assayed proteins, and most of these detected molecules were growth factors, cytokines, and chemokines with known anti-apoptotic and regeneration-stimulating effects [54]. These effects can be either direct or indirect (or both): direct by causing intracellular signaling, or indirect by causing another cell in the microenvironment to secrete the functionally active agent.

Table 2 Important bioactive molecules secreted by MSCs and their functions

Bioactive molecules	Functions
Prostaglandin-E2 (PGE2)	Anti-proliferative mediators [39] Anti-inflammation [40]
Interleukin-10 (IL-10)	Anti-inflammatory [41, 42]
Transforming growth factorβ-1 (TGFβ1), hepatocyte growth factor (HGF)	Suppress T-lymphocyte proliferation [43]
Interleukin-1 receptor antagonist	Anti-inflammatory [44]
human leukocyte antigen G isoform (HLA-G5)	Anti-proliferative for naive T-cells [45]
LL-37	Anti-microbial peptide and reduce inflammation [46]
Angiopoietin-1	Restore epithelial protein permeability [47]
MMP3, MMP9	Mediating neovascularization [48]
Keratinocyte growth factor	Alveolar epithelial fluid transport [49]
Endothelial growth factor (VEGF), basic fibroblast growth factor (bFGF), placental growth factor (PlGF), and monocyte chemoattractant protein-1 (MCP-1)	Enhance proliferation of endothelial cells and smooth muscle cells [50, 51]

Clinical Applications of MSCs

Although accumulating data have shown the therapeutic effects of MSCs in animal models of various diseases, we only focus on the clinical application of MSCs in this review. The first clinical trial using culture-expanded MSCs was conducted in 1995, and 15 patients were recipients of the autologous cells [55]. Since then, a number of clinical trials have been conducted to test the feasibility and efficacy of MSC therapy. By 2011/12/13, the public clinical trial database http://clinicaltrials.gov showed 206 clinical trials using MSCs for a wide range of therapeutic applications (Fig. 2). Most of these trials are in Phase I (safety studies), Phase II (proof of concept for efficacy in human patients), or a mixture of Phase I/II studies. Only a small number of these trials are in Phase III (comparing a newer treatment to the standard or best known treatment) or Phase II/III. In general, MSCs appear to be well-tolerated, with most trials reporting a lack of adverse effects in the medium term, although a few showed mild and transient peri-injection effects [56]. In addition, many completed clinical trials have demonstrated the efficacy of MSC infusion for diseases such as acute myocardial ischemia (AMI), stroke, liver cirrhosis, amyotrophic lateral sclerosis (ALS) and GVHD.

Conclusions and Future Prospects

MSCs hold the promise to fulfill unmet needs in regenerative medicine and have recently emerged as potential candidates for cell-based therapy because these cells can differentiate into a wide range of cells; produce a series of growth factors,

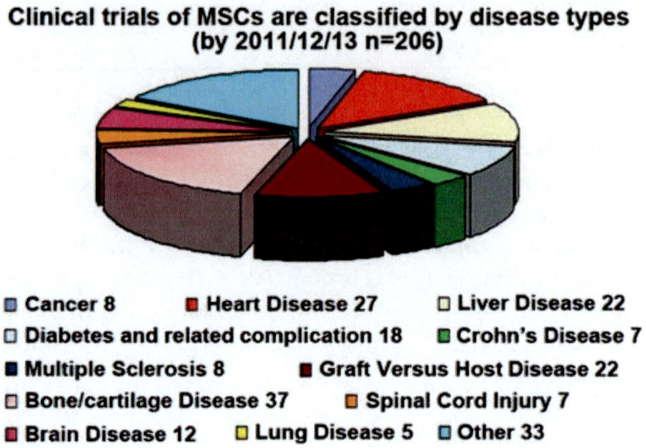

Fig. 2 The public clinical trial database http://clinicaltrials.gov showed 206 clinical trials using MSCs for a wide range of therapeutic applications

cytokines and signal molecules; and modulate the immune response in various ways. Despite tremendous progress having been made by both basic scientists and clinicians, future research in this field should continue to focus on elucidating the following issues. (1) The mechanisms underlying the multilineage differentiation of MSCs. The lineage specification of MSCs is tightly controlled by both genetic and epigenetic factors. Recently, microRNAs, a class of non-coding RNAs that regulate gene expression at the post-transcriptional level, have been demonstrated to play an important role in MSC differentiation. We found that microRNA-138 could inhibit the adipogenic differentiation of human MSCs through EID-1 [57]. Genetic and epigenetic factors interact, further complicating the mechanisms governing MSC differentiation. (2) How MSCs react to the environment and secrete bioactive molecules. (3) The mechanisms underlying MSC immunomodulatory function. (4) Determination of the possible adverse effects and complications that might arise with MSC transplantation. We believe that eventually, a novel and safe therapy utilizing MSCs will emerge and revolutionize the treatment and therapies for patients with severe diseases.

References

1. Friedenstein AJ, Petrakova KV, Kurolesova AI, Frolova GP. Heterotopic of bone marrow. Analysis of precursor cells for osteogenic and hematopoietic tissues. Transplantation. 1968;6:230–47.
2. Friedenstein AJ, Chailakhyan RK, Latsinik NV, Panasyuk AF, Keiliss-Borok IV. Stromal cells responsible for transferring the microenvironment of the hemopoietic tissues. Cloning in vitro and retransplantation in vivo. Transplantation. 1974;17:331–40.
3. Caplan AI. Mesenchymal stem cells. J Orthop Res. 1991;9:641–50.
4. Pittenger MF, Mackay AM, Beck SC, Jaiswal RK, Douglas R, et al. Multilineage potential of adult human mesenchymal stem cells. Science. 1999;284:143–7.
5. Kopen GC, Prockop DJ, Phinney DG. Marrow stromal cells migrate throughout forebrain and cerebellum, and they differentiate into astrocytes after injection into neonatal mouse brains. Proc Natl Acad Sci USA. 1999;96:10711–6.

6. Dominici M, Le Blanc K, Mueller I, Slaper-Cortenbach I, Marini F, et al. Minimal criteria for defining multipotent mesenchymal stromal cells. The International Society for Cellular Therapy position statement. Cytotherapy. 2006;8:315–7.
7. Salem HK, Thiemermann C. Mesenchymal stromal cells: current understanding and clinical status. Stem Cells. 2010;28:585–96.
8. Phinney DG, Prockop DJ. Concise review: mesenchymal stem/multipotent stromal cells: the state of transdifferentiation and modes of tissue repair – current views. Stem Cells. 2007;25:2896–902.
9. Morikawa S, Mabuchi Y, Niibe K, Suzuki S, Nagoshi N, et al. Development of mesenchymal stem cells partially originate from the neural crest. Biochem Biophys Res Commun. 2009;379:1114–9.
10. Takashima Y, Era T, Nakao K, Kondo S, Kasuga M, et al. Neuroepithelial cells supply an initial transient wave of MSC differentiation. Cell. 2007;129:1377–88.
11. Dennis JE, Charbord P. Origin and differentiation of human and murine stroma. Stem Cells. 2002;20:205–14.
12. Medici D, Shore EM, Lounev VY, Kaplan FS, Kalluri R, et al. Conversion of vascular endothelial cells into multipotent stem-like cells. Nat Med. 2010;16:1400–6.
13. Slukvin II, Vodyanik M. Endothelial origin of mesenchymal stem cells. Cell Cycle. 2011;10:1370–3.
14. Fang B, Shi M, Liao L, Yang S, Liu Y, et al. Multiorgan engraftment and multilineage differentiation by human fetal bone marrow Flk1+/CD31-/CD34- Progenitors. J Hematother Stem Cell Res. 2003;12:603–13.
15. Fang B, Liao L, Shi M, Yang S, Zhao RC. Multipotency of Flk1CD34 progenitors derived from human fetal bone marrow. J Lab Clin Med. 2004;143:230–40.
16. Macias MI, Grande J, Moreno A, Dominguez I, Bornstein R, et al. Isolation and characterization of true mesenchymal stem cells derived from human term decidua capable of multilineage differentiation into all 3 embryonic layer. Am J Obstet Gynecol. 2010;203:495.e9–23.
17. Anzalone R, Lo Iacono M, Loria T, Di Stefano A, Giannuzzi P, et al. Wharton's jelly mesenchymal stem cells as candidates for beta cells regeneration: extending the differentiative and immunomodulatory benefits of adult mesenchymal stem cells for the treatment of type 1 diabetes. Stem Cell Rev. 2011;7:342–63.
18. De Coppi P, Bartsch Jr G, Siddiqui MM, Xu T, Santos CC, et al. Isolation of amniotic stem cell lines with potential for therapy. Nat Biotechnol. 2007;25:100–6.
19. Battula VL, Bareiss PM, Treml S, Conrad S, Albert I, et al. Human placenta and bone marrow derived MSC cultured in serum-free, b-FGF-containing medium express cell surface frizzled-9 and SSEA-4 and give rise to multilineage differentiation. Differentiation. 2007;75:279–91.
20. Miki T, Lehmann T, Cai H, Stolz DB, Strom SC. Stem cell characteristics of amniotic epithelial cells. Stem Cells. 2005;23:1549–59.
21. Kogler G, Sensken S, Airey JA, Trapp T, Muschen M, et al. A new human somatic stem cell from placental cord blood with intrinsic pluripotent differentiation potential. J Exp Med. 2004;200:123–35.
22. Jiang Y, Jahagirdar BN, Reinhardt RL, Schwartz RE, Keene CD, et al. Pluripotency of mesenchymal stem cells derived from adult marrow. Nature. 2002;418:41–9.
23. Spees JL, Olson SD, Ylostalo J, Lynch PJ, Smith J, et al. Differentiation, cell fusion, and nuclear fusion during ex vivo repair of epithelium by human adult stem cells from bone marrow stroma. Proc Natl Acad Sci USA. 2003;100:2397–402.
24. Alvarez-Dolado M, Pardal R, Garcia-Verdugo JM, Fike JR, Lee HO, et al. Fusion of bone-marrow-derived cells with Purkinje neurons, cardiomyocytes and hepatocytes. Nature. 2003;425:968–73.
25. Charbord P. Bone marrow mesenchymal stem cells: historical overview and concepts. Hum Gene Ther. 2010;21:1045–56.
26. Guo M, Sun Z, Sun QY, Han Q, Yu CL, et al. A modified haploidentical nonmyeloablative transplantation without T cell depletion for high-risk acute leukemia: successful engraftment and mild GVHD. Biol Blood Marrow Transplant. 2009;15:930–7.

27. Abdi R, Fiorina P, Adra CN, Atkinson M, Sayegh MH. Immunomodulation by mesenchymal stem cells: a potential therapeutic strategy for type 1 diabetes. Diabetes. 2008;57:1759–67.
28. Bartholomew A, Sturgeon C, Siatskas M, Ferrer K, McIntosh K, et al. Mesenchymal stem cells suppress lymphocyte proliferation in vitro and prolong skin graft survival in vivo. Exp Hematol. 2002;30:42–8.
29. Le Blanc K, Tammik L, Sundberg B, Haynesworth SE, Ringden O. Mesenchymal stem cells inhibit and stimulate mixed lymphocyte cultures and mitogenic responses independently of the major histocompatibility complex. Scand J Immunol. 2003;57:11–20.
30. Aggarwal S, Pittenger MF. Human mesenchymal stem cells modulate allogeneic immune cell responses. Blood. 2005;105:1815–22.
31. Shi D, Liao L, Zhang B, Liu R, Dou X, et al. Human adipose tissue-derived mesenchymal stem cells facilitate the immunosuppressive effect of cyclosporin A on T lymphocytes through Jagged-1-mediated inhibition of NF-kappaB signaling. Exp Hematol. 2011;39(214–224):e211.
32. Zhang W, Ge W, Li C, You S, Liao L, et al. Effects of mesenchymal stem cells on differentiation, maturation, and function of human monocyte-derived dendritic cells. Stem Cells Dev. 2004;13:263–71.
33. Chen L, Zhang W, Yue H, Han Q, Chen B, et al. Effects of human mesenchymal stem cells on the differentiation of dendritic cells from CD34+ cells. Stem Cells Dev. 2007;16:719–31.
34. Zhang B, Liu R, Shi D, Liu X, Chen Y, et al. Mesenchymal stem cells induce mature dendritic cells into a novel Jagged-2-dependent regulatory dendritic cell population. Blood. 2009;113:46–57.
35. Karp JM, Leng Teo GS. Mesenchymal stem cell homing: the devil is in the details. Cell Stem Cell. 2009;4:206–16.
36. Yagi H, Soto-Gutierrez A, Parekkadan B, Kitagawa Y, Tompkins RG, et al. Mesenchymal stem cells: mechanisms of immunomodulation and homing. Cell Transplant. 2010;19:667–79.
37. Kraitchman DL, Tatsumi M, Gilson WD, Ishimori T, Kedziorek D, et al. Dynamic imaging of allogeneic mesenchymal stem cells trafficking to myocardial infarction. Circulation. 2005;112:1451–61.
38. Ortiz LA, Gambelli F, McBride C, Gaupp D, Baddoo M, et al. Mesenchymal stem cell engraftment in lung is enhanced in response to bleomycin exposure and ameliorates its fibrotic effects. Proc Natl Acad Sci USA. 2003;100:8407–11.
39. Bouffi C, Bony C, Courties G, Jorgensen C, Noel D. IL-6-dependent PGE2 secretion by mesenchymal stem cells inhibits local inflammation in experimental arthritis. PLoS One. 2010;5:e14247.
40. Foraker JE, Oh JY, Ylostalo JH, Lee RH, Watanabe J, et al. Cross-talk between human mesenchymal stem/progenitor cells (MSCs) and rat hippocampal slices in LPS-stimulated cocultures: the MSCs are activated to secrete prostaglandin E2. J Neurochem. 2011;119:1052–63.
41. Nemeth K, Leelahavanichkul A, Yuen PS, Mayer B, Parmelee A, et al. Bone marrow stromal cells attenuate sepsis via prostaglandin E(2)-dependent reprogramming of host macrophages to increase their interleukin-10 production. Nat Med. 2009;15:42–9.
42. Gupta N, Su X, Popov B, Lee JW, Serikov V, et al. Intrapulmonary delivery of bone marrow-derived mesenchymal stem cells improves survival and attenuates endotoxin-induced acute lung injury in mice. J Immunol. 2007;179:1855–63.
43. Di Nicola M, Carlo-Stella C, Magni M, Milanesi M, Longoni PD, et al. Human bone marrow stromal cells suppress T-lymphocyte proliferation induced by cellular or nonspecific mitogenic stimuli. Blood. 2002;99:3838–43.
44. Ortiz LA, Dutreil M, Fattman C, Pandey AC, Torres G, et al. Interleukin 1 receptor antagonist mediates the antiinflammatory and antifibrotic effect of mesenchymal stem cells during lung injury. Proc Natl Acad Sci USA. 2007;104:11002–7.
45. Selmani Z, Naji A, Zidi I, Favier B, Gaiffe E, et al. Human leukocyte antigen-G5 secretion by human mesenchymal stem cells is required to suppress T lymphocyte and natural killer function and to induce CD4 + CD25highFOXP3+ regulatory T cells. Stem Cells. 2008;26:212–22.

46. Krasnodembskaya A, Song Y, Fang X, Gupta N, Serikov V, et al. Antibacterial effect of human mesenchymal stem cells is mediated in part from secretion of the antimicrobial peptide LL-37. Stem Cells. 2010;28:2229–38.
47. Fang X, Neyrinck AP, Matthay MA, Lee JW. Allogeneic human mesenchymal stem cells restore epithelial protein permeability in cultured human alveolar type II cells by secretion of angiopoietin-1. J Biol Chem. 2010;285:26211–22.
48. Kim Y, Kim H, Cho H, Bae Y, Suh K, et al. Direct comparison of human mesenchymal stem cells derived from adipose tissues and bone marrow in mediating neovascularization in response to vascular ischemia. Cell Physiol Biochem. 2007;20:867–76.
49. Lee JW, Fang X, Gupta N, Serikov V, Matthay MA. Allogeneic human mesenchymal stem cells for treatment of *E. coli* endotoxin-induced acute lung injury in the ex vivo perfused human lung. Proc Natl Acad Sci USA. 2009;106:16357–62.
50. Kinnaird T, Stabile E, Burnett MS, Shou M, Lee CW, et al. Local delivery of marrow-derived stromal cells augments collateral perfusion through paracrine mechanisms. Circulation. 2004;109:1543–9.
51. Kinnaird T, Stabile E, Burnett MS, Lee CW, Barr S, et al. Marrow-derived stromal cells express genes encoding a broad spectrum of arteriogenic cytokines and promote in vitro and in vivo arteriogenesis through paracrine mechanisms. Circ Res. 2004;94:678–85.
52. van Poll D, Parekkadan B, Cho CH, Berthiaume F, Nahmias Y, et al. Mesenchymal stem cell-derived molecules directly modulate hepatocellular death and regeneration in vitro and in vivo. Hepatology. 2008;47:1634–43.
53. Liechty KW, MacKenzie TC, Shaaban AF, Radu A, Moseley AM, et al. Human mesenchymal stem cells engraft and demonstrate site-specific differentiation after in utero transplantation in sheep. Nat Med. 2000;6:1282–6.
54. Parekkadan B, van Poll D, Suganuma K, Carter EA, Berthiaume F, et al. Mesenchymal stem cell-derived molecules reverse fulminant hepatic failure. PLoS One. 2007;2:e941.
55. Lazarus HM, Haynesworth SE, Gerson SL, Rosenthal NS, Caplan AI. Ex vivo expansion and subsequent infusion of human bone marrow-derived stromal progenitor cells (mesenchymal progenitor cells): implications for therapeutic use. Bone Marrow Transplant. 1995;16:557–64.
56. Otto WR, Wright NA. Mesenchymal stem cells: from experiment to clinic. Fibrogenesis Tissue Repair. 2011;4:20.
57. Yang Z, Bian C, Zhou H, Huang S, Wang S, et al. MicroRNA hsa-miR-138 inhibits adipogenic differentiation of human adipose tissue-derived mesenchymal stem cells through adenovirus EID-1. Stem Cells Dev. 2011;20:259–67.

Biology of MSCs Isolated from Different Tissues

Simone Pacini

Abstract Mesenchymal stem cells (MSCs) have been firstly isolated from bone marrow (BM). The relatively ease of MSC collection from BM samples alongside their high frequency, make it a widely used source of MSCs. For many years, BM was considered the main source of MSCs for clinical application. Subsequently, MSCs have been isolated from various other sources and the adipose tissue seems one of the most promising alternatives due to safer collecting procedures, and also the considerably larger amounts of cells obtained. Adipose tissue-derived MSCs, as well as other tissues-derived cells, and BM-MSCs share many biological characteristics; however, there are some differences in their immunophenotype, differentiation potential, transcriptome, proteome, and immunomodulatory activity. Some of these differences may represent specific features related to the different tissue origins, while others are suggestive of the inherent heterogeneity of *in vitro* expanded populations. Moreover, lack of a widely accepted consensus about MSC isolating and culture procedures represent an important source of variability.

The general approach to investigate the presence of MSCs in a specific tissue consists of culturing processed samples in minimal media selecting MSC-like cell population by plastic adherence, and verifying the clonogenity, the multilineage differentiation potential and surface markers expression. Applying this method, many different tissues have shown to be a feasible source of MSCs in humans and in animals, contributing to consolidate the emerging concept that MSCs could reside virtually in all organs and tissues.

Here, data about MSC isolation from some adult or birth-associated tissues are presented, discussed and compared.

Keywords MSC • Biology • Bone marrow • Adipose tissue

S. Pacini (✉)
Department of Clinical and Experimental Medicine, University of Pisa,
Via Roma 56, 56124 Pisa, Italy
e-mail: simone.pacini@do.unipi.it

R.C. Zhao (ed.), *Essentials of Mesenchymal Stem Cell Biology and Its Clinical Translation*, DOI 10.1007/978-94-007-6716-4_2,

The discovery of multipotent mesenchymal stromal cells (MSCs) is usually attributed to the work of A.J. Friedenstein and coworkers in the late 1960s in which the authors observed that culturing human bone marrow (BM) cell suspensions, in plastic dishes, lead to progressive lost of the hemopoietic counterpart in favor of a proliferating adhered colonies of fribroblastoid cells able to differentiate into chondrocytes or osteoblasts, *in vitro* [1], and *in vivo* [2]. Authors firstly described these cells as *colony forming units of fibroblastoid cells* (CFU-Fs) referring to their ability to form large colonies on plastic surfaces.

By that time, T.M. Dexter and colleaues developing a culture system to study hemopoiesis *in vitro*, demonstrated that the hemopoietic stem cells (HSC) residing in the bone marrow were unable to adhere onto the culture flasks and were dependent on the estabilishment of a layer of adherent cells that were considered be representative of the bone marrow stromal compartment [3]. Later, the concept that CFU-Fs were derived from the bone marrow stroma was demonstarted and the term "bone marrow stromal cells" became used refering to this culture adherent cells [4]. The acronymous "MSC" became popular after the work of A.I. Caplan et al. in 1991 where the authors proposed that in adult BM, a population of stem cells could differentiate into a spectrum of different tissues originated from the mesodermal layer, during embryonic development [5]. They termed these cells as "mesenchymal stem cells" (MSCs). Later, the multilineage differentiation capability of MSCs was then definitively demonstrated, these cells shown a stable phenotype and could be easily expanded in culture retaining the ability to differentiate, *in vitro*, into osteoblasts, chondrocytes, adipocytes, tenocytes, myocytes and hematopietic supporting stromal cells [6].

From these seminal findings, MSCs obtained increasing interest by the scientific community and subsequent studies revealed the possibility to isolate MSCs from some other adult and fetal/neonatal tissues [7–10]. The original design of these studies consist of applying the established culture condition to isolate BM-MSCs to other cell populations derived from different tissues, in order to verify the possibility that MSCs could reside in other organs. A comparative and comprehensive study from da Silva et al. demonstrated, in mice, that long-term MSC culture could be established from a wide range of different adult tissues including fat, muscles, pancreas, vena cava, kidney glomerulus, aorta, brain and many others alongside bone marrow [11]. Notably, the cell populations obtained by da Silva and colleagues can be characterized for their phenotype, capability of adherent long-term culture and differentiation along mesenchymal cell lineages. Surprisingly, all the MSC lines, independently from the embryonic origins of the tissue tested, exhibited these features. These data suggest that MSCs could reside virtually in all organs and tissues. To date, three hypothesis could explain MSC tissue distribution: (1) MSCs are tissue-resident cells and can be collected from distinct tissues and organs, (2) MSCs reside in some tissues and circulate in blood or (3) MSCs are derived from the circulating blood. The presence of CFU-Fs in blood of adult mammals was shown at the beginning of the twentieth century [12]. Anyway, contamination by fragments of connective tissue could be explain the presence of MSCs in the collected sample and then invalidate the experiments. The existence of circulating MSCs remains a discussed subject [13, 14], but the da Silva group excluding the possibility that MSC culture were partially or entirely derived from peripheral blood, by intravascular perfusion of the animals before the organ collection. Nonetheless, the possibility that MSCs may circulate locally or systemically under non-physiological conditions i.e. tissue injury, is not excluded.

Features of MSC population obtained by different organs were very similar, excepted for mild differences in differentiation potential and surface markers profile that could be expression of the influence of the local environment from which they originated (niche). At the beginning of the century, some reports suggested that MSC could be derived from the vasculature. Thank to the seminal findings of Doherty et al. [15] and Bianco et al. [16] that reported origins of MSCs from perivascolar cells (pericytes). Thus, a new proposed model for MSC *in vivo* localization hypothesized that the MSC compartment extends through the whole post-natal organism as a result of its perivascular location.

Bone Marrow-Derived MSCs

MSCs have been firstly isolated from bone marrow (BM). The relatively ease of MSC collection from BM samples alongside their frequency of $1/10^4$–$1/10^5$ BM-derived mononuclear cells (BM-MNCs) make it still a widely used source of MSCs. Small animals BM samples are usually collected, after euthanasia, by flushing the BM out of long bones as femurs or tibiae. Human BM samples are commonly obtained by small volume aspiration (less than 4 ml, to avoid hemodilution) after puncture of iliac crest or sternum. Larger amount of human BM samples could be also harvested during orthopedic surgery as hip replacement or knees implants, where BM is easily accessible after the osteotomy.

Standard procedure isolating MSCs from bone marrow samples start from a discontinuous density gradient centrifugation (1.077 g/dL) for 20–30 min at 400 g. This procedure allows collecting, at the liquid interface, a cell fraction enriched in mononucleated cells (BM-MNCs). Once harvested and washed twice with phosphate-buffered saline (PBS), BM-MNCs are usually plated at a cell density that could vary from 2×10^5 to 10^6 cells/cm^2, in growth medium and then incubated at 37 °C under controlled atmosphere of 5 % v/v of CO_2. After 48–72 h, non-adherent cells are washed out and the growth medium is entirely replaced with fresh one. Standard GM include minimal basal media as DMEM or αMEM supplemented with L-glutamine and 10 % of fetal bovine serum (FBS). Cultures are then maintained until they reach at least 80 % of confluence (passage 0, P0). At this point, adherent cells are treated with trypsin and re-plated at cell densities high enough to allow cell survival, and low enough to maximize cell yield at each passage. Human MSC (hMSCs) expansion in culture is highly variable [17]. Different studies on expandability of hMSCs underline that many factors could influence the expansion rate as donor age, cell density, supplements, serum batch-to-batch variability as well as basal media itself. Nonetheless, it is widely accepted that hMSCs, cultured in standard conditions, are able to expand until about 30 population doublings, showing logarithmic growing curves for less than ten passages [18]. Some works have focused on the optimization of culture conditions to maximize hMSC production in culture [19, 20]. However, data indicating that prolonged expansion of hMSCs *in vitro* could lead to undesired genetic alteration of these cells [21], make it unfeasible for clinical applications.

In the last years, many efforts have been applied to the obtainment of genetically stable MSCs with higher proliferation and wider differentiation capability; therefore different culture techniques have been developed for this purpose. Nonetheless, the applying of different methodologies to isolate and expand cells bearing the MSC characteristics lead to a possible selection of specific cell population. Consequently, morpho-functional variability of cell preparations could be consequences of the specific culture condition that select, or simply promote particular subpopulations of BM-derived multipotent cells. To date, MSC-related stem cells, isolated from human bone marrow, include rapid self-renewing (RS) cells [22] marrow-*isolated adult multilieage inducible* (MIAMI) cells [23], *mesodermal progenitor cells* (MPCs) [24–27] and *Flk-1*$^{+}$*CD31*$^{-}$ *CD34*$^{-}$ -MSCs [28, 29]. The lack of a definitive study, comparing these populations and analyzing the different cell types when cultured under conditions described for the others, lead to the impossibility to clarify if they constitute intrinsically different entities or if they can be described in a hierarchy.

In contrast to hMSCs, murine MSCs (mMSCs), have been show to be able to expand beyond 100 population doublings [11]. On the other hand, mMSCs isolation results more time-consuming than hMSCs, especially in the first phases of culture. The isolation by plastic adherence of mMSCs from BM is complicated by the considerable high percentage of adherent cells of non-stromal origins, which are not washed out after 48–72 h of incubation. Thus, the standard and unmodified method based on MSCs propensity to adhere to the plastic substrate resulted unsuccessful in mice, where various hemopoietic and endothelial cell proliferate in adhesion and therefore constitute a large percentage of the plastic adherent population, even after several passages. A wide range of different methods have been proposed to eradicate the hemopoietic contamination of mMSCs culture, including positive and negative selection of specific BM subpopulation, cytokine exposure of mMSCs culture and also specific cytotoxic treatments. Nevertheless, none of these alternative methods have been widely accepted due to the reported modification of mMSCs biology as consequence of modified protocols. Actually, the most promising isolating methods to obtain mMSCs from mouse BM include (1) short plastic adherent selection of whole BM (3 h), (2) frequent media exchange (every 8 h for the first 72 h of culture) and (3) mild trypsinization (0.25 % trypsin/EDTA for 2 min) [30]. Appling this method a purified culture of mMSCs can be obtained 3 weeks after the initial plating.

Summarizing, MSCs have been isolated from BM of numerous species and generally the three critical steps allowing MSCs to be isolated from other BM cells are (1) the ability to adhere to plastic surfaces, (2) the high proliferating capability in minimal essential media and (3) the higher susceptibility to trypsin digestion compared to other BM cells as monocytes for instance.

Even if the scientific community established a widely accepted consensus about standard MSC isolating procedures, several studies revealed that MSCs display high level of heterogeneity in terms of cell morphology. Different terms were used to describe morphology of plastic-adherent cells: fibroblastoid [31], giant fat cells and blanket cells [32], spindle shaped flattened cells [33] and very small round cells [22]. Thus, mesenchymal cell morphology seems to be highly correlated to the culture conditions as supplements, seeding density, number of passages and culture time [34] and it is still unclear how these different morphologies could be related to cell functions.

No unique specific marker as been found for BM-derived MSCs, so far. The markers widely applied, in combination, to characterize a cultured population are usually expressed, or not expressed, by other cell lineages. For that reason, a definitive identification of a specific MSC phenotype is still lacking. Several publication demonstrate the reproducible expression of the most important MSC markers such as CD105 (Endoglin, SH2), CD73 (NT5E), CD90 (Thy-1), CD44 and CD166 (ALCAM) and the absence of hemopoietic markers CD34, CD14, CD11b and CD45 [6], as well as the MHC class II complex and the co-stimulatory molecules CD80 (B7-1), CD86 (B7-2) or CD40. The current criteria for human MSC characterization are mainly based on the positive expression of CD73, CD90, and CD105 [35, 36]; however the expression of none of these markers is shared by all other species. CD90 shows strong expression in the majority of species tested but is absent on MSCs in goats and sheep [37]. Nonetheless, the variability of expression of CD73, CD105 and CD90 in MSCs from some animal species could be ascribed to the use of anti-human antibodies, due to the lack of species-specific antibodies. A more accurate evaluation of antibody cross-reactivity would be required to confirm the true expression pattern of these molecules. In mice, MSC characterization is complicated by the expression of Sca-1 that is also expressed by hemopoietic compartment, and by the fact that preparations from different strains could express two alternative CD90 antigens (CD90.1/Thy1.1 or CD90.2/Thy1.2), as well as CD106 instead of CD105 [30]. Other molecules are suggested to be useful to identify BM-derived MSCs such as CD29, STRO-1, CD146, MSCA-1 and CD271, but despite of the markers cited above which show almost stable expressions in cultures, the positivity to these latest markers seems to be useful for a prospective isolation of MSCs while their expression is absent in culture or influenced by the culture time [38–40].

As described above, immunophenotype of MSCs is heterogeneous and dynamic. Thus, differentiation potential seems to be the more feasible and stringent criteria to characterize cultured bone marrow adherent cell population as MSCs. From the clarification of the nomenclature by ISCT in 2005, MSCs have to show multilineage differentiation capability under specific culture conditions and stimuli. As extensively discussed in the following chapter (Chap. 4), MSCs are able to differentiate into osteogenic, adipogenic and chondrogenic lineages. However, it was further observed that MSCs show high variability of differentiation potential, not only related to donors [17], but also within different clones from the same individual, where MSC clones could be characterized as mono-, bi- or tri-potent on the basis of their ability to differentiate into, respectively one, two or three of the mesenchymal lineages (osteogenic, chondrogenic and adipogenic lineages) [41]. Moreover, it was also clearly demonstrated that repeated passaging progressively reduce the multineage differentiation ability of the clones, introducing a further origin for the heterogeneity of the cell preparations [42]. Multipotency of BM-derived MSC cell preparations is not only restricted to osteogenic, adipogenic and chondorgenic potential but it is also demonstrated, *in vitro* and *in vivo*, that these cells are able to differentiate into further mesodermal cells such as tenocytes [43], miocytes and hemopoietic supporting stroma [6]. Beside that, BM-MSCs plasticity as been reported to sustain differentiation toward tissues and cell lineages that arise from non-mesodermal embryonic layer (trans-differentiation), *in vitro* and *in vivo*. Controversies about MSCs trans-differentiation has been extensively discussed and remain a topic issue of BM-MSCs biology [44].

Adipose Tissue-Derived MSCs

Adipose tissue-derived MSCs (AT-MSCs) were firstly isolated by Zuk and colleagues in 2001, from human liposuction aspirates [7]. In this original article the authors noted that hAT-MSCs express, alongside the typical spindle-shaped morphology, immunophenotype pretty similar to the MSCs isolate from bone marrow. hAT-MSCs express CD105, CD90, CD44, CD29 and also STRO-1, while lacking the expression of hemopoietic lineage markers, and show multilineage differentiation capability. Although AT-MSCs were only identified relatively recently, their ease of harvest give rise to considerable amount of studies focused on these multipotent cells. To date, adipose tissue is considered the most feasible source of MSCs, alternative to bone marrow, and for some aspect it is even to prefer to BM. In fact, in view of possible clinical application of MSCs, sampling adipose tissue results less painful and safe than bone marrow aspiration.

AT-MSCs could be harvested from liposuction aspirate or excised fat, and small amount of adipose tissue (100–200 ml) could be obtained under local anesthesia with less patient discomfort. Furthermore, 1 g of adipose tissue yields an average number of approximately 5×10^3 MSCs that is enormously higher (around 500-fold) compared to the same amount of bone marrow. Thus, adipose tissue could be considered as a rich source of MSCs, available in large quantities and that could allows harvesting of large amount of cells with reduced *in vitro* expansion. General protocol to isolate MSCs, from adipose tissue, includes collagenase digestion of the extracellular matrix for 30′ at 37 °C with gentle agitation [45]. Tryptic activity is then inhibited by addition of an equal volume of grow medium. After centrifugation mature adipocytes, that constitute less than 50 % of all cells, are separated from the other heterogeneous cell population that is generally termed *stromal vascular fraction* (SVF). In fact, mature lipid-laden and low-density adipocytes float into the surpernatant, whereas SVF forms the denser cellular pellet, which contains the MSC fraction. AT-MSCs are then isolated by plastic adhesion culturing SVF applying the same protocol for BM-derived mononuclear cells.

As already noted by Zuk et al. the immunophenotype of AT-MSCs and BM-MSCs are greater than 90 % identical, however later study underline some minor differences [46]. Some authors reported the expression of CD34 in freshly isolate AT-MSCs and although this expression gradually declines with successive passages, it may not be entirely lost conversely to MSCs from other sources. Furthermore, AT-MSCs showed expression of CD49d (Integrin α4), at different intensity, but lack the expression of CD106 (VCAM-1), while BM-MSCs express CD49f (Integrin α6) instead of CD49d and high level of CD106. Similarly, CD54 (ICAM-1) expression is reported to be high on AT-MSCs while BM-MSCs show a minimal expression of this marker. Nonetheless, the immunophenotypic differences between AT-MSCs and BM-MSCs are still debated, and controversial data are reported from different groups. There are data that distinct subsets with different immunophenotype, proliferation capability and differentiation potential exist in the heterogeneous population of MSCs isolated from the same source, and the predominance of a particular subset could be ascribed to the different isolating and culture procedures, as happen in BM-MSC preparations. It is also possible to hypothesize that the immunophenotypic differences between AT-MSCs and BM-MSCs, already described or still unidentified, may contribute to differential response to grow factors or differentiating agents of adipose-derived MSCs versus bone marrow-derived. This hypothesis could also explain the controversial data reported about differences in differentiation potential of AT-MSCs versus BM-MSCs. Some authors reported that AT-MSCs display pronounced, *in vitro*, adipogenic differentiation compared to BM-MSCs, and conversely decreased osteogenic and chondrogenic differentiation capability (reviewed in [46]). Nonetheless, some other studies suggest that the AT-MSC response to the various differentiating agents do not differ significantly from the BM-MSCs, and that differences reported could be ascribed to many other factors as gender and donor age as well as to the heterogeneity of cell preparations as discussed above.

MSCs Derived from Synovial Membrane Tissues

A thin layer of synovial membrane tissue lines the non-articular surfaces of diarthrodial joints and provides producing synovial fluid that fills the cavity around cartilage and tendon surfaces. In 1995, FitzGerald and Bresnihan described the cells, derived from synovial tissues, in two different categories [47]. Together with the bone marrow derived cells, expressing macrophage markers as CD68 and CD14, the Authors described fibroblast-like cells showing prominent expression of adhesion molecules as VCAM-1 and CD44 and associated to matrix proteins synthesis. Only during 2001, De Bari et al. successfully isolated cells, bearing MSC characteristics, from synovial membrane tissues [8]. General procedure obtaining synovium-derived MSCS (S-MSCs) includes shattering the sample into pieces, after washing with steril PBS, followed by collagenase digestion similarly to AT-MSCs, but prolonged for several hours (around 3 h). Cells harvested after blocking collagenase activity and washing with PBS, are then seeded in growth medium and selected by plastic adherence, similarly to the protocols for obtaining MSCs from other sources. Recent study reported an average number of about 20'000 S-MSCs could be obtained from 1 mg of collected synovial tissue, after 2 weeks of culture [48]. Some studies reported that the morphology, immunophenotype, colony frequency and differentiation capability of S-MSCs are similar to that of BM-MSCs (reviewed in [49]), even if low percentage (40–60 %) of CD90 expression is reported for freshly isolated S-MSCs and even lower on further culturing [50]. It is generally believed that S-MSCs retain higher chondrogenic potential in comparison to MSCs from other sources. This idea is supported by some experimental evidences including higher CD44 (hyaluronan receptor) expression as well as diphosphoglucose dehydrogenase (UDPGD) activity, involved in hyaluronan synthesis.

In any case, any discussion about differences of synovium-derived MSCs versus MSCs from other sources should be commented taking in considerations the heterogeneity of cell preparations. Similarly to other MSCs, S-MSCs population is influenced by many factors including donor variability and cell culture techniques. Moreover, the synovial membrane is a thin layer very closely correlated with different sub-synovial tissue as areolar, fibrous and fat tissues that could contaminate sampling of synovium tissue and at the end contribute to the heterogeneity of S-MSCs population.

Dental Tissues as Sources of MSC-Like Cells

Dental tissues are specialized tissues that do not show continuous remodeling as bony tissue. Nonetheless, it has been reported that progenitor cell populations, sharing most of the MSC characteristics, may be isolated from teeth [51]. Firstly, stem/progenitor cells were isolated from the human pulp tissue and defined as "post-natal dental pulp stem cells" (DPSCs) [52]. DPSCs isolated from enzymatic or non-enzymatic treatment of human dental pulp tissue are able to form CFU-Fs when cultured under conditions similar to BM- or AT-MSCs. These cells exhibit multilineage differentiation ability even if DPSCs seems to be more committed to odontogenic rather than osteogenic development, with specific dentin-like tissue formation. Compared to BM-MSCs, DPSCs show higher *in vitro* proliferation capability that could vary from 60 to 120 population doublings, before appearing of cell senescence signs. Interestingly, DPSCs has been reported secreting neurotrophins as BDNF, NGF and GDNF and exhibited neuroprotective activity [53].

It is noteworthy that dental mesenchyme is usually termed as "ectomesenchyme" due to its earlier interaction with the neural crest, during embryonic development. Thus, it has been hypothesized that the ectomesenchyme-derived dental cells may possess different characteristics akin to those of neural crest cells. In this prospective, successive isolation of MSC-like cells from human exfoliated deciduous teeth (SHEDs) seems of particular interest. In fact, as well as DPSCs, SHEDs showed the ability to differentiate toward adipogenic and osteogenic lineages but additionally, under neurogenic conditions, SHEDs lost the fibroblast-like morphology and showed multicytoplasmic processes while increasing the expression of neural markers as βIII-tubulin, GAD and NeuN [54]. Moreover, SHEDs has been reported showing even higher expansion potential compare to DPSCs, reaching around 140 population doublings, as well as shorter population doubling time. Further dental MSC-like populations have been isolated and characterized as stem cells from apical papilla (SCAP) [55] and dental follicle precursor cells (DFPCs) [56], however the precise relationship among these cell population have to be more extensively investigated.

Periodontal ligament has been also reported containing post-natal stem/progenitor cells. Seo et al. successfully isolated clonogenic adherent cells with multidifferentiation potential from periodontal ligaments (PDLSCs) [9]. These cells express, alongside typical MSC-related marker as STRO-1, a tendon specific transcription factor: scleraxis (Scx), detected neither in DPSCs nor in BM-MSCs.

Tendon-Derived Stem/Progenitor Cells

The report from Seo et al. work changed the traditional idea that considers ligaments and tendons to only contain tenocytes, responsible for the tissue homeostasis. After the isolation of PDLSCs, further findings suggested that there might be a special cell population inside tendons that possesses self-renewal and multi-lineage differentiation potentials. However it was only in 2007 that Bi et al. directly demonstrated the presence of multipotent cells inside tendons from humans and animals [57]. Tendon-derived stem/progenitor cells (TDSCs), despite the chosen terminology, showed biological properties overlapping the MSC characteristics, including clonogenicity, self-renewal and multi-lineage differentiation capacities even after extended expansion *in vitro* and *in vivo*. As is the case for other MSCs, no single marker could unambiguously identify TDSCs [58]. Although TDSCs express many of the same markers as BM-MSCs, the expression patterns were not identical. TSPCs highly express tendon-related factors, such as Scx, TNMD, Comp and tenascin C. Mouse TSPCs expressed CD90.2, a fibroblast marker, but not CD18, usually associated to mBM-MSCs. These data suggest that TSPCs are closely related to BMSCs, but not identical. Similarly to other tissues, it is hypothesized that the tendon niche, where TDSCs reside *in vivo*, could influence the biological features of this cell population. Furthermore, tendon microenvironment results pretty peculiar compared to other discussed above, it is extremely rich in extra-cellular matrix (ECM) components and contains substantially fewer cells than most of the other tissues. Consequently, it is possible to hypothesize a unique niche predominantly composed by ECM, regulating the TDSCs fate.

MSCs Derived from Birth-Associated Tissues

In addition to the different adult tissues, cells bearing MSC characteristics can be isolated from birth-associated tissues as *placenta*, *amnion*, *umbilical cord* and *cord blood* [10, 59]. Several studies suggested that neo-natal tissue-derived MSC might have additional capacities and superior biological properties.

MSCs from human placenta (PL-MSCSs) showed a higher proliferation and engraftment capacity compared to BM-MSCs [60, 61]. Nonetheless, discussing on PL-MSC biology, it is relevant to note that placental tissues can have fetal or maternal origin and the characterization of these two cell types as well as the study of the MSC functions and biology, should take in consideration of the different origins. For instance, placenta-derived MSCs from fetal tissues including amnion membrane (AM-MSCs), chorion membrane (CM-MSCs) and chorion villi (CV-MSCs) have shown to posses a more limited lifespan that MSCs isolated from the maternal part of the extraembryonic membranes or decidua (D-MSCs) [62], however higher than adult MSCs as BM-MSCs or AT-MSCs. Moreover, studies on PL-MSC differentiation capability provide more reproducible and convincing data about the potential to differentiate into cells from the three germ layers, than adult tissue-derived MSCs. Similar properties have been demonstrated for MSCs derived from amniotic fluid (AF-MSCs) [63].

With respect to isolation from umbilical cord, different parts have been demonstrated feasible source of MSCs. MSCs can be obtained from whole umbilical cord (UC-MSCs) [64], from Wharton's jelly (WJ-MSCs) [65] or from umbilical cord blood (CB-MSCs) [66]. Majore et al. observed an adherent cell layer outgrowth from small pieces of human UC directly cultured in αMEM supplemented with 15 % of human serum, after 10 days. These cells (UC-MSCs) showed typical MSC markers as CD105, CD73, CD90 and low level of HLA-I alongside adipogenic and chondrogenic differentiation potential. However, osteogenic induction resulted less efficient than AT-MSCs.

Wharton's Jelly derived mesenchymal stem cells (WJ-MSCs) are located between the subamnion and the perivascular region. Wharton's jelly is a mucous-connective tissue matrix composed of stromal cells, collagen fibers, proteoglycans and mainly by hyaluronic acid (HA). Samples of Wharton's jelly could be obtained cutting the umbilical cord longitudinally and exposing the matrix surrounding the vessels. Fragments of this tissue could be directly cultured in growth medium, giving rise to adherent cell layer robustly growing for several passages [67]. It has been demonstrated that WJ-MSCs show faster and higher expansion potential compared to BM-MSCs, partially due to the higher expression of telomerase activity [68]. Phenotype of WJ-MSCs is substantially identical to BM-MSCs, moreover as well as other UC-derived MSCs, these cells beside matching the ISCT differentiation criteria, seems to show wider differentiation ability toward non-connective tissue as hepatocytes [67], pancreatic [69] or neural cells [70].

Cord blood has also reported to be a feasible source of MSCs [71]. After removal of the placenta, blood was allowed to drain from the severed end of the cord into samples tubes containing heparin. Then, cord blood is processed by the same protocols usually applied for bone marrow aspirates and cultured in almost the same growth media. However, it has been reported that MSCs could be isolated from no more than 60 % of processed CB. The time between harvest and the beginning of culture seems to be critical for the success and should be shorter than 15 h. Moreover, the volume of CB and the total quantity of mononuclear cells of the collected samples, influence the probability of obtaining CB-MSCs growing cultures [72], as well as cryopreservation. Although data reported lower frequency of MSCs in CB than in bone marrow (1 per 10^8 cells *vs.* 1 per 10^5 cells), they showed a greater proliferative potential [73, 74]. The differentiation potential of CB-MSCs in different tissues is also broader. After enrichment by depletion, CB-MSCs have been found to differentiate not only toward mesodermal but also toward the endodermal and ectodermal cell lineages [75]. This experimental data are also supported by the identification of a CB-MSC sub-population termed *unrestricted somatic stem cells* (USSCs), which show enormous proliferative capability up to more than 20 passages and retain great differentiation potential after several weeks of culture, and toward cell lineages from the three germ layers [76]. Notably, percentage of USSCs in the cord blood has been reported to dramatically decrease during cryopreservation [77].

While the superior osteogenic differentiation potential of CB-MSCs *versus* BM-MSCs is well documented, controversial data were reported about the adipogenic potential of CB-MSCs. Some authors described CB-MSCs as less sensitive to the adipogenic differentiating agents or even not able to differentiate into adipocytes [78, 79]. These latest findings seems to be in accordance with the proposed model of MSC origins in which the microenvironment of the source tissue could influence the biology of isolated MSCs, throughout specific interactions between the *in vivo* putative cell and its "niche". In fact, it is notably that adult bone marrow is an adipose-rich tissue while fetal bone morrow shows an absent adipogenesis which is reported increasing in correlation with aging [80].

Additionally, some other tissues have been reported as a feasible source for MSCs as skeletal muscle [81, 82], lungs [83], thymus [84, 85], tonsils [86], parathyroid gland [87], fallopian tube [88], etc. The general approach to investigate the presence of MSCs in a specific tissue consists of culturing processed samples in minimal media selecting MSC-like cell population by plastic adherence, and consequently verifying the clonogenity, the multilineage differentiation potential and some non-specific surface markers expression, according to ISCT guidelines. Consequently, the parameters applied to define a cultured cell population as MSC population are still not sufficiently stringent, leading to define heterogeneous cell populations with the same terminology.

About the origins of MSC heterogeneity, it is also important to notice that, additionally to species-, donor- and tissues origins-related variability, MSCs show variability even among cell clones from the same culture [41]. Moreover biological properties have reported to vary also within the cells that form a colony itself [89], which show different differentiation potential apparently related to the topographic localization inside the colony. It has been demonstrated that cells from the inner regions differ from the cells at the margins of the colonies, in terms of morphology, differentiation potential and markers expression [90, 91]. Thus, it is clear that when we use the term "multipotent mesenchymal stromal cells" we refer to a highly heterogeneous population of cells, the composition of which is dramatically affected by isolating methods and culture conditions, and that is hard to unambiguously characterize due to the lack of specific and stringent criteria of definition. Several possible mechanisms, at the basis of the MSC heterogeneity, have been hypothesized in addition to the well-documented variability introduced by isolating methods and *in vitro* cultivation [92]. Stochastic events, occurring during expansion and differentiation, have to be discussed as a possible origin of MSC variability, alongside a possible heterogeneity of the *in vivo* cell population that give rise to MSC in culture, which could be influenced by the different biological properties of the tissue niche in which they reside. In this latest hypothesis, MSC heterogeneity and morpho-functional variability of cell preparations could be consequences of the characteristic of the tissue from which MSCs have been derived.

References

1. Friedenstein AJ, Petrakova KV, Kurolesova AI, Frolova GP. Heterotopic of bone marrow. Analysis of precursor cells for osteogenic and hematopoietic tissues. Transplantation. 1968;6:230–47.
2. Friedenstein AJ, Chailakhyan RK, Latsinik NV, Panasyuk AF, Keiliss-Borok IV. Stromal cells responsible for transferring the microenvironment of the hemopoietic tissues. Cloning in vitro and retransplantation in vivo. Transplantation. 1974;17:331–40.
3. Dexter TM, Allen TD, Lajtha LG. Conditions controlling the proliferation of haemopoietic stem cells *in vitro*. J Cell Physiol. 1977;91:335–44.
4. Lanotte M, Allen TD, Dexter TM. Histochemical and ultrastructural characteristics of a cell line from human bone-marrow stroma. J Cell Sci. 1981;50:281–97.
5. Caplan AI. Mesenchymal stem cells. J Orthop Res. 1991;9:641–50.
6. Pittenger MF, Mackay AM, Beck SC, Jaiswal RK, Douglas R, Mosca JD, Moorman MA, Simonetti DW, Craig S, Marshak DR. Multilineage potential of adult human mesenchymal stem cells. Science. 1999;284(5411):143–7.
7. Zuk PA, Zhu M, Mizuno H, Huang J, Futrell JW, Katz AJ, Benhaim P, Lorenz HP, Hedrick MH. Multilineage cells from human adipose tissue: implications for cell-based therapies. Tissue Eng. 2001;7:211–28.
8. De Bari C, Dell'Accio F, Tylzanowski P, Luyten FP. Multipotent mesenchymal stem cells from adult human synovial membrane. Arthritis Rheum. 2001;44(8):1928–42.
9. Seo BM, Miura M, Gronthos S, Bartold PM, Batoli S, Brahim J, Young M, Robey PG, Wang CY, Shi S. Investigation of multipotent postnatal stem cells from human periodontal ligament. Lancet. 2004;364(9429):149–55.
10. In't Anker PS, Scherjon SA, Kleijburg-van der Keur C, de Groot-Swings GM, Claas FH, Fibbe WE, Kanhai HH. Isolation of mesenchymal stem cells of fetal and maternal origin from human placenta. Stem Cells. 2004;22:1338–45.
11. da Silva Meirelles L, Chagastelles PC, Nardi NB. Mesenchymal stem cells reside in virtually all post-natal organs and tissues. J Cell Sci. 2006;119(Pt 11):2204–13.
12. He Q, Wan C, Li G. Multipotent mesenchymal stromal cells in blood. Stem Cells. 2007;25:69–77.
13. Roufosse CA, Direkze NC, Otto WR, Wright NA. Circulating mesenchymal stem cells. Int J Biochem Cell Biol. 2004;36:585–97.

14. Kuznetsov SA, Mankani MH, Leet AI, Ziran N, Gronthos S, Robey PG. Circulating connective tissue precursors: extreme rarity in humans and chondrogenic potential in guinea pigs. Stem Cells. 2007;25:1830–9.
15. Doherty MJ, Ashton BA, Walsh S, Beresford JN, Grant ME, Canfield AE. Vascular pericytes express osteogenic potential in vitro and *in vivo*. J Bone Miner Res. 1998;13:828–38.
16. Bianco P, Riminucci M, Gronthos S, Robey PG. Bone marrow stromal stem cells: nature, biology, and potential applications. Stem Cells. 2001;19:180–92.
17. Phinney DG, Kopen G, Righter W, Webster S, Tremain N, Prockop DJ. Donor variation in the growth properties and osteogenic potential of human marrow stromal cells. J Cell Biochem. 1999;75(3):424–36.
18. Bruder SP, Jaiswal N, Haynesworth SE. Growth kinetics, self-renewal, and the osteogenic potential of purified human mesenchymal stem cells durino extensive subcultivation and following cryopreservation. J Cell Biochem. 1997;64:278–94.
19. Sekiya I, Larson BL, Smith JR, Pochampally R, Cui JG, Prockop DJ. Expansion of human adult stem cells from bone marrow stroma: conditions that maximize the yields of early progenitors and evaluate their quality. Stem Cells. 2002;20:530–41.
20. Sotiropoulou PA, Perez SA, Salagianni M, Baxevanis CN, Papamichail M. Cell culture medium composition and translational adult bone marrow-derived stem cell research. Stem Cells. 2006;24:1409–10.
21. Rubio D, Garcia-Castro J, Martin MC, de la Fuente R, Cigudosa JC, Lloyd AC, Bernad A. Spontaneous human adult stem cell transformation. Cancer Res. 2005;65:3035–9.
22. Colter DC, Sekiya I, Prockop DJ. Identification of a subpopulation of rapidly self-renewing and multipotential adult stem cells in colonies of human marrow stromal cells. Proc Natl Acad Sci USA. 2001;98(14):7841–5.
23. D'Ippolito G, Diabira S, Howard GA, Menei P, Roos BA, Schiller PC. Marrow-isolated adult multilineage inducible (MIAMI) cells, a unique population of postnatal young and old human cells with extensive expansion and differentiation potential. J Cell Sci. 2004;117(Pt 14):2971–81.
24. Petrini M, Pacini S, Trombi L, Fazzi R, Montali M, Ikehara S, Abraham NG. Identification and purification of mesodermal progenitor cells from human adult bone marrow. Stem Cells Dev. 2009;18(6):857–66.
25. Pacini S, Carnicelli V, Trombi L, Montali M, Fazzi R, Lazzarini E, Giannotti S, Petrini M. Costitutive expression of pluripotency-associated genes in mesodermal progenitor cells (MPCs). PLoS One. 2010;5(3):e9861.
26. Trombi L, Pacini S, Montali M, Fazzi R, Chiellini F, Ikehara S, Petrini M. Selective culture of mesodermal progenitor cells. Stem Cells Dev. 2009;18(8):1227–34.
27. Fazzi R, Pacini S, Carnicelli V, Trombi L, Montali M, Lazzarini E, Petrini M. Mesodermal progenitor cells (MPCs) differentiate into mesenchymal stromal cells (MSCs) by activation of Wnt5/calmodulin signalling pathway. PLoS One. 2011;6(9):e25600.
28. Fang B, Liao L, Shi M, Yang S, Zhao RC. Multipotency of Flk1+ CD34- progenitors derived from human fetal bone marrow. J Lab Clin Med. 2004;143(4):230–40.
29. Liu L, Sun Z, Chen B, Han Q, Liao L, Jia M, Cao Y, Ma J, Sun Q, Guo M, Liu Z, Ai H, Zhao RC. Ex vivo expansion and *in vivo* infusion of bone marrow-derived Flk-1+CD31-CD34- mesenchymal stem cells: feasibility and safety from monkey to human. Stem Cells Dev. 2006;15(3):349–57.
30. Soleimani M, Nadri S. A protocol for isolation and culture of mesenchymal stem cells from mouse bone marrow. Nat Protoc. 2009;4(1):102–6.
31. Werts ED, DeGowin RL, Knapp SK, Gibson DP. Characterization of marrow stromal (fibroblastoid) cells and their association with erythropoiesis. Exp Hematol. 1980;8(4):423–33.
32. Allen TD, Dexter TM. Long term bone marrow cultures: an ultrastructural review. Scan Electron Microsc. 1983;1983(Pt 4):1851–66.
33. Kuznetsov SA, Krebsbach PH, Satomura K, Kerr J, Riminucci M, Benayahu D, Robey PG. Single-colony derived strains of human marrow stromal fibroblasts form bone after transplantation in vivo. J Bone Miner Res. 1997;12(9):1335–47.

34. Wagner W, Ho AD. Mesenchymal stem cell preparations – comparing apples and oranges. Stem Cell Rev. 2007;3(4):239–48.
35. Horwitz EM, Le BK, Dominici M, et al. Clarification of the nomenclature for MSC: the International Society for Cellular Therapy position statement. Cytotherapy. 2005;7:393–5.
36. Dominici M, Le Blanc K, Mueller I, et al. Minimal criteria for defining multipotent mesenchymal stromal cells. The International Society for Cellular Therapy position statement. Cytotherapy. 2006;8:315–7.
37. Boxall SA, Jones E. Markers for characterization of bone marrow multipotential stromal cells. Stem Cells Int. 2012;2012:975871.
38. Aggarwal S, Pittenger MF. Human mesenchymal stem cells modulate allogeneic immune cell responses. Blood. 2005;105:1815–22.
39. Schaefer R, Dominici M, Muller I, Dazzi F, Bieback K, Godthardt K, Le Blanc K, Meisel R, Pochampally R, Richter R, Skutella T, Steinhoff G, Mitterberger M, Wendel H, Wiskirchen J, Handgretinger R, Northoff H. Progress in characterization, preparation and clinical applications of non-hematopoietic stem cells. Cytotherapy. 2007;9:397–405.
40. Sacchetti B, Funari A, Michienzi S, Di Cesare S, Piersanti S, Saggio I, Tagliafico E, Ferrari S, Gehron Robey P, Riminucci M, Bianco P. Self-renewing osteoprogenitors in bone marrow sinusoids can organize a hematopoietic microenvironment. Cell. 2007;131:324–36.
41. Russell KC, Phinney DG, Lacey MR, Barrilleaux BL, Meyertholen KE, O'Connor KC. In vitro high-capacity assay to quantify the clonal heterogeneity in trilineage potential of mesenchymal stem cells reveals a complex hierarchy of lineage commitment. Stem Cells. 2010;28(4):788–98.
42. Muraglia A, Cancedda R, Quarto R. Clonal mesenchymal progenitors from human bone marrow differentiate in vitro according to a hierarchical model. J Cell Sci. 2000;113(Pt 7):1161–6.
43. Pacini S, Spinabella S, Trombi L, Fazzi R, Galimberti S, Dini F, Carlucci F, Petrini M. Suspension of bone marrow-derived undifferentiated mesenchymal stromal cells for repair of superficial digital flexor tendon in race horses. Tissue Eng. 2007;13(12):2949–55.
44. Barzilay R, Melamed E, Offen D. Introducing transcription factors to multipotent mesenchymal stem cells: making transdifferentiation possible. Transl Psychiatry. 2011;13(1):e61.
45. Mosna F, Sensebé L, Krampera M. Human bone marrow and adipose tissue mesenchymal stem cells: a user's guide. Stem Cells Dev. 2010;19(10):1449–70.
46. Strioga M, Viswanathan S, Darinskas A, Slaby O, Michalek J. Same or not the same? Comparison of adipose tissue-derived versus bone marrow-derived mesenchymal stem and stromal cells. Stem Cells Dev. 2012;21(14):2724–52.
47. FitzGerald O, Bresnihan B. Synovial membrane cellularity and vascularity. Ann Rheum Dis. 1995;54(6):511–5.
48. Sakaguchi Y, Sekiya I, Yagishita K, Muneta T. Comparison of human stem cells derived from various mesenchymal tissues: superiority of synovium as a cell source. Arthritis Rheum. 2005;52(8):2521–9.
49. Fan J, Varshney RR, Ren L, Cai D, Wang DA. Synovium-derived mesenchymal stem cells: a new cell source for musculoskeletal regeneration. Tissue Eng Part B Rev. 2009;15(1):75–86.
50. Nagase T, Muneta T, Ju YJ, Hara K, Morito T, Koga H, Nimura A, Mochizuki T, Sekiya I. Analysis of the chondrogenic potential of human synovial stem cells according to harvest site and culture parameters in knees with medial compartment osteoarthritis. Arthritis Rheum. 2008;58(5):1389–98.
51. Huang GT, Gronthos S, Shi S. Mesenchymal stem cells derived from dental tissues vs. those from other sources: their biology and role in regenerative medicine. J Dent Res. 2009;88(9):792–806.
52. Gronthos S, Arthur A, Bartold PM, Shi S. A method to isolate and culture expand human dental pulp stem cells. Methods Mol Biol. 2011;698:107–21.
53. Nesti C, Pardini C, Barachini S, D'Alessandro D, Siciliano G, Murri L, Petrini M, Vaglini F. Human dental pulp stem cells protect mouse dopaminergic neurons against MPP+ or rotenone. Brain Res. 2011;1367:94–102.
54. Miura M, Gronthos S, Zhao M, Lu B, Fisher LW, Robey PG, Shi S. SHED: stem cells from human exfoliated deciduous teeth. Proc Natl Acad Sci USA. 2003;100(10):5807–12.

55. Sonoyama W, Liu Y, Yamaza T, Tuan RS, Wang S, Shi S, Huang GT. Characterization of the apical papilla and its residing stem cells from human immature permanent teeth: a pilot study. J Endod. 2008;34(2):166–71.
56. Morsczeck C, Götz W, Schierholz J, Zeilhofer F, Kühn U, Möhl C, Sippel C, Hoffmann KH. Isolation of precursor cells (PCs) from human dental follicle of wisdom teeth. Matrix Biol. 2005;24(2):155–65.
57. Bi Y, Ehirchiou D, Kilts TM, Inkson CA, Embree MC, Sonoyama W, Li L, Leet AI, Seo BM, Zhang L, Shi S, Young MF. Identification of tendon stem/progenitor cells and the role of the extracellular matrix in their niche. Nat Med. 2007;13(10):1219–27. Epub 2007 Sep 9.
58. Lui PP, Chan KM. Tendon-derived stem cells (TDSCs): from basic science to potential roles in tendon pathology and tissue engineering applications. Stem Cell Rev. 2011;7(4):883–97.
59. Hass R, Kasper C, Böhm S, Jacobs R. Different populations and sources of human mesenchymal stem cells (MSC): a comparison of adult and neonatal tissue-derived MSC. Cell Commun Signal. 2011;9:12.
60. Barlow S, Brooke G, Chatterjee K, Price G, Pelekanos R, Rossetti T, Doody M, Venter D, Pain S, Gilshenan K, Atkinson K. Comparison of human placenta- and bone marrow-derived multipotent mesenchymal stem cells. Stem Cells Dev. 2008;17(6):1095–107.
61. Brooke G, Tong H, Levesque JP, Atkinson K. Molecular trafficking mechanisms of multipotent mesenchymal stem cells derived from human bone marrow and placenta. Stem Cells Dev. 2008;17(5):929–40.
62. Soncini M, Vertua E, Gibelli L, Zorzi F, Denegri M, Albertini A, Wengler GS, Parolini O. Isolation and characterization of mesenchymal cells from human fetal membranes. J Tissue Eng Regen Med. 2007;1(4):296–305.
63. De Coppi P, Bartsch Jr G, Siddiqui MM, Xu T, Santos CC, Perin L, Mostoslavsky G, Serre AC, Snyder EY, Yoo JJ, Furth ME, Soker S, Atala A. Isolation of amniotic stem cell lines with potential for therapy. Nat Biotechnol. 2007;25(1):100–6.
64. Majore I, Moretti P, Stahl F, Hass R, Kasper C. Growth and differentiation properties of mesenchymal stromal cell populations derived from whole human umbilical cord. Stem Cell Rev. 2011;7(1):17–31.
65. Wang HS, Hung SC, Peng ST, Huang CC, Wei HM, Guo YJ, Fu YS, Lai MC, Chen CC. Mesenchymal stem cells in the Wharton's jelly of the human umbilical cord. Stem Cells. 2004;22(7):1330–7.
66. Broxmeyer HE, Srour E, Orschell C, Ingram DA, Cooper S, Plett PA, Mead LE, Yoder MC. Cord blood stem and progenitor cells. Methods Enzymol. 2006;419:439–73.
67. Anzalone R, Lo Iacono M, Corrao S, Magno F, Loria T, Cappello F, Zummo G, Farina F, La Rocca G. New emerging potentials for human Wharton's jelly mesenchymal stem cells: immunological features and hepatocyte-like differentiative capacity. Stem Cells Dev. 2010;19(4):423–38.
68. La Rocca G, Anzalone R, Corrao S, Magno F, Loria T, Lo Iacono M, Di Stefano A, Giannuzzi P, Marasà L, Cappello F, Zummo G, Farina F. Isolation and characterization of Oct-4+/HLA-G+ mesenchymal stem cells from human umbilical cord matrix: differentiation potential and detection of new markers. Histochem Cell Biol. 2009;131(2):267–82.
69. Anzalone R, Lo Iacono M, Loria T, Di Stefano A, Giannuzzi P, Farina F, La Rocca G. Wharton's jelly mesenchymal stem cells as candidates for beta cells regeneration: extending the differentiative and immunomodulatory benefits of adult mesenchymal stem cells for the treatment of type 1 diabetes. Stem Cell Rev. 2011;7(2):342–63.
70. Weiss ML, Medicetty S, Bledsoe AR, Rachakatla RS, Choi M, Merchav S, Luo Y, Rao MS, Velagaleti G, Troyer D. Human umbilical cord matrix stem cells: preliminary characterization and effect of transplantation in a rodent model of Parkinson's disease. Stem Cells. 2006;24:781–92.
71. Erices A, Conget P, Minguell JJ. Mesenchymal progenitor cells in human umbilical cord blood. Br J Haematol. 2000;109(1):235–42.
72. Bieback K, Kern S, Kluter H, Eichler H. Critical parameters for the isolation of mesenchymal stem cells from umbilical cord blood. Stem Cells. 2004;22:625–34.

73. Greschat S, Schira J, Küry P, Rosenbaum C, de Souza Silva MA, Kögler G, Wernet P, Müller HW. Unrestricted somatic stem cells from human umbilical cord blood can be differentiated into neurons with a dopaminergic phenotype. Stem Cells Dev. 2008;17(2):221–32.
74. Sensken S, Waclawczyk S, Knaupp AS, Trapp T, Enczmann J, Wernet P, Kogler G. In vitro differentiation of human cord blood-derived unrestricted somatic stem cells towards an endodermal pathway. Cytotherapy. 2007;9(4):362–78.
75. Lee OK, Kuo TK, Chen WM, Lee KD, Hsieh SL, Chen TH. Isolation of multipotent mesenchymal stem cells from umbilical cord blood. Blood. 2004;103(5):1669–75.
76. Kögler G, Sensken S, Airey JA, Trapp T, Muschen M, Feldhahn N, Liedtke S, Sorg RV, Fischer J, Rosenbaum C, Greschat S, Knipper A, Bender J, Degistirici O, Gao J, Caplan AI, Colletti EJ, Almeida-Porada G, Muller HW, Zanjani E, Wernet P. A new human somatic stem cell from placental cord blood with intrinsic pluripotent differentiation potential. J Exp Med. 2004;200:123–35.
77. Kögler G, Sensken S, Wernet P. Comparative generation and characterization of pluripotent unrestricted somatic stem cells with mesenchymal stem cells from human cord blood. Exp Hematol. 2006;34(11):1589–95.
78. Kern S, Eichler H, Stoeve J, Klüter H, Bieback K. Comparative analysis of mesenchymal stem cells from bone marrow, umbilical cord blood, or adipose tissue. Stem Cells. 2006;24(5):1294–301.
79. Barachini S, Trombi L, Danti S, D'Alessandro D, Battolla B, Legitimo A, Nesti C, Mucci I, D'Acunto M, Cascone MG, Lazzeri L, Mattii L, Consolini R, Petrini M. Morpho-functional characterization of human mesenchymal stem cells from umbilical cord blood for potential uses in regenerative medicine. Stem Cells Dev. 2009;18(2):293–305.
80. Moerman EJ, Teng K, Lipschitz DA, Lecka-Czernik B. Aging activates adipogenic and suppresses osteogenic programs in mesenchymal marrow stroma/stem cells: the role of PPAR-gamma2 transcription factor and TGF-beta/BMP signaling pathways. Aging Cell. 2004;3(6):379–89.
81. Alessandri G, Pagano S, Bez A, Benetti A, Pozzi S, Iannolo G, Baronio M, Invernici G, Caruso A, Muneretto C, Bisleri G, Parati E. Isolation and culture of human muscle-derived stem cells able to differentiate into myogenic and neurogenic cell lineages. Lancet. 2004;364(9448):1872–83.
82. Usas A, Huard J. Muscle-derived stem cells for tissue engineering and regenerative therapy. Biomaterials. 2007;28(36):5401–6.
83. Sabatini F, Petecchia L, Tavian M, Jodon de Villeroché V, Rossi GA, Brouty-Boyé D. Human bronchial fibroblasts exhibit a mesenchymal stem cell phenotype and multilineage differentiating potentialities. Lab Invest. 2005;85(8):962–71.
84. Siepe M, Thomsen AR, Duerkopp N, Krause U, Förster K, Hezel P, Beyersdorf F, Schlensak C, Südkamp NP, Bosse R, Niemeyer P. Human neonatal thymus-derived mesenchymal stromal cells: characterization, differentiation, and immunomodulatory properties. Tissue Eng Part A. 2009;15(7):1787–96.
85. Mouiseddine M, Mathieu N, Stefani J, Demarquay C, Bertho JM. Characterization and histological localization of multipotent mesenchymal stromal cells in the human postnatal thymus. Stem Cells Dev. 2008;17(6):1165–74.
86. Janjanin S, Djouad F, Shanti RM, et al. Human palatine tonsil: a new potential tissue source of multipotent mesenchymal progenitor cells. Arthritis Res Ther. 2008;10(4):R83.
87. Shih YR, Kuo TK, Yang AH, Lee OK, Lee CH. Isolation and characterization of stem cells from the human parathyroid gland. Cell Prolif. 2009;42(4):461–70.
88. Jazedje T, Perin PM, Czeresnia CE, et al. Human fallopian tube: a new source of multipotent adult mesenchymal stem cells discarded in surgical procedures. J Trans Med. 2009;7:46.
89. Digirolamo CM, Stokes D, Colter D, Phinney DG, Class R, Prockop DJ. Propagation and senescence of human marrow stromal cells in culture: a simple colony-forming assay identifies samples with the greatest potential to propagate and differentiate. Br J Haematol. 1999;107(2):275–81.
90. Ylöstalo J, Bazhanov N, Prockop DJ. Reversible commitment to differentiation by human multipotent stromal cells in single-cell-derived colonies. Exp Hematol. 2008;36(10):1390–402.
91. Sengers BG, Dawson JI, Oreffo RO. Characterisation of human bone marrow stromal cell heterogeneity for skeletal regeneration strategies using a two-stage colony assay and computational modelling. Bone. 2010;46(2):496–503.
92. Pevsner-Fischer M, Levin S, Zipori D. The origins of mesenchymal stromal cell heterogeneity. Stem Cell Rev. 2011;7(3):560–8.

Secretome of Mesenchymal Stem Cells

Yuan Xiao, Xin Li, Hong Hao, Yuqi Cui, Minjie Chen, Lingjun Liu, and Zhenguo Liu

Abstract Mesenchymal stem cells (MSCs) are a group of heterogeneous non-hematopoietic cells with self-renewal and multi-lineage differentiation potential, and have been widely used for cell-based therapies. While the mechanisms for the beneficial effects of MSCs on tissue repair and regeneration are complex and not fully understood, paracrine signaling is believed to be at least partially responsible for their therapeutic benefits. MSCs express and secret a large number of paracrine factors with a wide spectrum of biological functions including cell proliferation, differentiation, migration, anti-apoptosis, metabolism, immunomodulation, anti-inflammation, angiogenesis, and tissue remodeling. The regulation on the expression and production of the paracrine factors and related signaling molecules in MSCs are complex, and involves a variety of signaling pathways including Akt, STAT-3, p38 MAPK, and TNF receptors. The paracrine function of MSCs is closely associated with the species, age, and gender of the sources, and environmental factors like hypoxia, as well as the presence of stimuli such as tumor necrosis factor. Some disease conditions especially diabetes mellitus have significant impact on paracrine signaling of MSCs. Significant challenges remain on understanding how paracrine mechanisms work on the target tissues of MSCs, and how to design a therapeutic regimen with different paracrine factors to achieve an optimal outcome for tissue protection and regeneration.

Keywords Mesenchymal stem cell • MSC • Pfactor • Growth factor • Cell signaling

---

Y. Xiao • X. Li • H. Hao • Y. Cui • M. Chen • L. Liu • Z. Liu, M.D., Ph.D. (✉)
Division of Cardiovascular Medicine, Davis Heart and Lung Research Institute,
The Ohio State University Medical Center, Room 200 DHLRI,
473 West 12th Ave, Columbus, OH, USA
e-mail: zhenguo.liu@osumc.edu

R.C. Zhao (ed.), *Essentials of Mesenchymal Stem Cell Biology and Its Clinical Translation*, DOI 10.1007/978-94-007-6716-4_3,

Introduction

Cell therapy with stem cells remains a viable and attractive option for tissue repair and regeneration after injuries including myocardial infarction, stroke, and wound healing [21, 33, 52]. Mesenchymal stem cells (MSCs) are a group of heterogeneous non-hematopoietic cells that were first identified and isolated from the bone marrow in 1960s by Friedenstein and colleagues, and exhibit proliferative and self-renewal potential, and are able to differentiate into multilineage cell types of endodermal, ectodermal, and mesodermal origins including (but not limited to) osteocytes, chondrocytes, endothelial cells, adipocytes, myocytes, cardiomyocytes, neuron, and hepatocytes [36]. Over the past two decades, a number of cell populations with similar characteristics and multilineage differentiation potential have been successfully identified and characterized in many other adult and fetal tissues in addition to bone marrow, including (but not limited to) skin, dental pulp, adipose tissue, synovium, muscle, tonsil, brain, lung tissue, kidney, umbilical cord blood, peripheral blood, and placenta [38, 52]. MSC has been used to describe almost all the progenitor cells with multipotent differentiation potentials from these parenchymal nonhematopoietic tissues. Although there are currently no unique markers to exclusively identify and characterize MSCs or their subpopulation phenotypically, these cells are expected to express at least the stromal markers CD73 and CD105 without expression of the hematopoietic markers CD14, CD34, and CD45 based on the minimal criteria established by the International Society of Cellular Therapy on the nomenclature and definition of the adult tissue-derived undifferentiated progenitor cells with extended proliferative capability and multilineage differentiation potential [5, 12, 36].

Bone marrow-derived mesenchymal stem cells (BM-MSCs) are an attractive and ideal source for cell-based therapy due to the fact that these cells can be easily obtained without ethical concerns, and conveniently expanded ex vivo to clinical scales in a relatively short period of time with minimal loss of potency, and have little (if any) inherent immunogenicity for any adverse immune reactions (even in the setting of xenogeneic transplantation of MSCs) because of their immunosuppressive and/or immunomodulatory properties [4, 7, 36]. Therefore, in this focused review our efforts will be mainly directed to discuss the beneficial effects of cell therapy with BM-MSCs and the mechanism(s) especially the role of paracrine signaling.

Transplantation of MSCs Provides Promising Therapeutic Potential for Tissue Repair and Regeneration

BM-MSCs are considered to be the most utilized and extensively explored stem cell population for both pre-clinical studies and clinical trials on cell-based therapies due to their unique properties as highlighted above (easily isolated and amplified from the bone marrow, immunologically well tolerated, and their multilineage potential). Pre-clinical animal investigations have shown that transplantation of MSCs provides significant beneficial effects in the treatment of a variety of clinical conditions with significant restoration of tissue structures and improvement in organ function including (but not limited to) liver and kidney dysfunction, myocardial infarction, central nervous system disorders, osteoarthritis, autoimmune and inflammatory/degenerative disorders, and cutaneous wound repair [1, 19, 45, 55]. Clinical studies have demonstrated that administration of MSCs (both locally and systematically) in human subjects appeared to be safe, and exhibited promising therapeutic effects for a wide range of disease states like (but not limited to) myocardial infarction, ischemic and non-ischemic cardiomyopathy, ischemic stroke, spinal cord injury, liver diseases, ischemic intestinal diseases, and autoimmune and inflammatory disorders [1, 8, 21, 36, 55]. Obviously, it is beyond the scope of this review to detail all the clinical studies on every individual medical condition using MSCs.

Mechanisms for the Therapeutic Effects of MSCs

It is clear that application of MSCs contributes to the repair and regeneration of damaged tissues with enhanced function and provides significant therapeutic benefits on a variety of disease conditions. However, it is much less clear on the primary mode(s) of action of these cells on achieving their beneficial effects on tissue repair. Initially, it is believed that transplanted MSCs home to the damaged areas, differentiate into the cells specific to the tissue and organ system, thus contributing to tissue repair and regeneration. Indeed, it is observed that the transplanted MSCs integrated into the damaged sites in a variety of tissues where they transform into the cells with the cell markers specific to the cell populations in a individual tissue and organ like hepatocytes in the liver, epithelial cells of the esophagus and small intestine, keratinocytes, and endothelial cells [16, 19, 42], supporting the idea that these cells are capable of homing to and integrating into the damaged tissues and directly contributing to their reconstruction and function recovery. There is no question that a direct engraftment and differentiation into the tissue specific cells and their supporting cells with transplanted MSCs represents an important mechanism for tissue repair and regeneration for some tissues like liver and lung.

MSCs have been shown to display a broad range of important immunomodulatory properties and attenuate tissue damages due to excessive inflammation in the early phase of injuries. These include suppression of T cell and B cell proliferation and terminal differentiation, modulating dendritic cell maturation and activities, and functional modulation of other cells critical to immune responses like natural killer cells and macrophages [40, 55]. The immuno-privileged and immuno-regulatory capabilities as well as anti-inflammatory properties of MSCs certainly contribute (at least partially) to their therapeutic benefits on repair and regeneration not only in autologous but also allogeneic recipients through modification of the local environment of damaged sites and argumentation of the survival and functional recovery of local resident cells with enhanced proliferation, migration, and differentiation, as well as decreased adverse inflammatory and immune reactions and cell apoptosis. This may be the primary mode of action of MSCs on the treatment of immune diseases such as graft-versus-host disease, rheumatoid arthritis, experimental autoimmune encephalomyelitis, sepsis, acute pancreatitis and multiple sclerosis. Although the exact mechanisms for the unusual immunomodulatory and anti-inflammatory effects of MSCs are far from fully understood, they are considered to be mediated through direct cell-cell interactions and/or secreting various immune-related soluble factors or cytokines such as interleukin 6 (IL-6), IL-10, IL-1 β, transforming growth factor-β (TGF-β), interferon-γ (INF- γ), and granulocyte-macrophage colony-stimulating factor (GM-CSF) [24, 36, 40, 55].

Paracrine Mechanisms as a Major Mode of Action for the Therapeutic Effects of MSCs

Recent studies have showed that less than 1% of systemically administered MSCs are still present for longer than a week in any organ system including lung, heart, kidney, liver, spleen, and gut following injection [6, 22, 23, 36, 39]. However, clinically, the beneficial effects are observed much longer than a week in patients who have ischemic heart diseases and receive transplantation of MSCs [46, 53]. It is observed that the differentiation of transplanted MSCs into cardiomyocytes is very inefficient. In some studies, when injected into the myocardium after infarction, MSCs are able to reduce the scar formation, improve angiogenesis, and preserve myocardial function without direct involvement of MSC engraftment into the cardiac compartment (either cardiomyocytes or supporting cells) [20, 37, 46, 53]. When administered to treat animals with acute renal failure, MSCs can prevent apoptosis and promote the proliferation of renal-tubule epithelial cells in a differentiation-independent manner [43, 44]. Cell-free products from human MSCs are reported to effectively enhance wound healing [32]. These observations suggest that paracrine factors and related signaling are a major mechanism responsible (at least partially) for the beneficial effects of MSCs on tissue repair and regeneration and alteration of disease pathophysiology.

Secretion of Paracrine Factors in MSCs

It is well known that MSCs express and produce a wide spectrum of biologically active growth factors and cytokines including, but not limited to, fibroblast growth factor (FGF), IL-1 and 6, TGF-β, and VEGF, are expressed, produced, and released from MSCs [31, 34]. As early as 1996, it was observed that MSCs isolated from human bone marrow constitutively expressed and released G-CSF, SCF, LIF, M-CSF, IL-6, and IL-11 into the in vitro culture medium. These cytokines were

Table 1 Summary of the major paracrine factors from MSCs and their actions on cell protection, tissue repair and regeneration

Paracrine factors	Actions
VEGF, HGF, STC-1, SFRP-2, SDF-1, TGF-β IGF-1, bFGF, TB-4	Cell survival
VEGF, bFGF, IL-1, TNF-α, PDGF-BB, Ang-1, Ang-2 FGF-2, TGF-β, SDF-1, IGF-1, PIGF, MCP, HGF	Angiogenesis
VEGF, FGF-2, HGF, IGF-I, TB4	Anti-apoptosis
VEGF, HGF, IGF-I, TNF-α, TGF-β, G-CSF, SCF, LIF M-CSF, IL-6, IL-11, Activin A	Cell differentiation
BDNF, NGF, neuregulin-1, BNP, IL-6, FGF-2, GDNF VEGF, HGF, FGF-20	Neuroprotection and regeneration
IL-1, IL-10, TB-4, MMP-2, MMP-9, MCP-1, TSP-1 TGF-β, TIMP-1, TIMP-2, TIMP-9, HGF, NGF, ErbB-2	Tissue remodeling
VEGF, bFGF, FGF-2, HGF, TB-4, IGFBP-7	Cell contractility
IL-6, IL-10, IL-1β, TGF-β, INF- γ, GM-CSF, PGE-2, IDO HGF, TNF-α, activin A	Immunomodulatory effects
VEGF, endothelin, Smad-4, Smad-5, glypican-3, FGF-16	Cell proliferation and migration

reported to be involved in the regulation of the differentiation of cells derived from the bone marrow stroma through receptors that were linked to gp130-associated signal transduction pathways [10]. Since then, a long list of biologically active substances (and yet the list is still growing rapidly) such as (but not limited to) VEGF, FGF, MCP-1, HGF, IGF-I, SDF-1, TGF-β, nerve growth factor (NGF), and thrombopoietin have been identified to be expressed and secreted from MSCs derived from bone marrow and a variety of other sources. It has been observed that the conditioned medium from hypoxic MSCs overexpressing Akt gene (Akt-MSCs) markedly inhibits hypoxia-induced apoptosis and triggers vigorous spontaneous contraction of adult rat cardiomyocytes in vitro. Intramyocardial injection of the hypoxic conditioned medium from Akt-MSCs has been shown to significantly reduce the infarct size and improves ventricular function to the same extent as the Akt-MSCs in an acute myocardial infarction rodent model [9], confirming that paracrine actions exerted by the cells through the release of soluble factors are indeed important mechanisms for tissue repair and functional improvement after transplantation of the Akt-MSCs. Accumulating data have demonstrated that the growth factors and cytokines from MSCs exert their beneficial effects on the target cells in their vicinity to facilitate tissue repair and regeneration, including (not limited to) immune response modification, anti-apoptosis, cell survival, metabolism, proliferation, and differentiation, hematopoiesis, myogenesis, angiogenesis, collateral development, remodeling, neuroprotection, renal protection, hair growth, and wound healing [8, 19, 20, 24, 31]. Table 1 summarizes some of the major paracrine factors from MSCs and their actions on cell protection, tissue repair and regeneration. The actions of many other paracrine factors released from MSCs remain to be identified and characterized.

Possible Signaling Pathways Involved in the Secretion of Paracrine Factors in MSCs

Role of Akt Signaling in the Expression of Paracrine Factors

Since MSCs are a mixture of heterogeneous cell populations, and produce a large number of paracrine factors, it may be difficult to investigate the mechanism(s) responsible for the production of individual factors. However, a few signaling pathways have been shown to be critically involved in the expression and production of paracrine factors from MSCs. It has been reported that the expression of VEGF, FGF-2, HGF, IGF-I, and TB4 that are potential mediators of the effects exerted by the Akt-MSC conditioned medium, are significantly up-regulated in the Akt-MSCs especially in response to hypoxia, demonstrating that Akt signaling is important to the regulation on the expression of these factors in MSCs [9].

STAT-3 Signaling Is Important in the Expression of VEGF and TGF-β1 in MSCs

MSCs produce a significant amount of VEGF and TGF-β1 both at basal level and in response to stimuli. The regulation of VEGF expression or production is complex and involves many factors such as hypoxia. It is observed that mouse MSCs release VEGF under normoxia in association with constitutive STAT-3 activity. STAT3 deficiency in STAT-3 knockout mice resulted in a significantly decreased production of VEGF from MSCs. In response to hypoxia or TNF, MSCs produced significantly more VEGF in association with activated p38 MAPK and STAT-3. In addition, STAT-3 ablation neutralized hypoxia-induced release of VEGF from MSCs. Inhibition of p38 MAPK signaling alone has no effect on the release of VEGF from MSCs in normal mice [50].

Multipotent adult progenitor cells (MAPCs) are purified homogeneous MSCs from bone marrow, and are potent source of VEGF and TGF-β1. When the JAK2/STAT-3 signaling pathway in rat MAPCs is blocked with the selective JAK2 phosphorylation inhibitor AG490, VEGF gene expression and protein production are significantly suppressed in the cells [25]. These observations strongly suggest that VEGF expression in MSCs is mediated via JAK2/STAT3 signaling pathway. However, some studies suggest that TGF-α stimulated production of VEGF in human MSCs is mediated via MEK- and PI3-K- but not ERK- or JNK-dependent mechanisms [48]. Very likely, there are different pathways involved in the production of VEGF from MSCs in different species (murine vs human). STAT3 signaling is also critically involved in the regulation of TGF-β1 expression in rat MAPCs. A detectable level of TGF-β1 is expressed in rat MAPCs in culture system. Treatment of the cells with the specific STAT3 phosphorylation inhibitor AG490 significantly blocked STAT3 (Tyr705) phosphorylation, and increased TGF-β1 expression without change in ERK1/2 phosphorylation [26].

Activation of p38 MAPK Signaling Is Involved in the Expression of Paracrine Factors in MSCs

Studies using human MSCs and human adipose progenitor cells demonstrate the secretion of VEGF, HGF, and IGF-I in these cells is significantly increased by stimulation with TNF and is associated with increased activation of p38 mitogen-activated protein kinase (MAPK). Inhibition of p38 MAPK signaling with selective p38 MAPK inhibitor significantly decreased the TNF-stimulated production of VEGF, HGF, and IGF-I in these cells. However, p38 MAPK inhibitor alone had no effect on production of these growth factors without TNF stimulation. These data suggest that TNF enhances the production of paracrine factors in MSCs through a p38 MAPK-dependent mechanism [47]. Inactivation of p38 MAPK signaling is also reported to reduce the expression and production of IL-6, IL-8 and CXCL-1 in MSCs, and decrease wound healing [54], indicating that MSCs promote wound healing through releasing paracrine factors via activation of p38 MAPK signaling.

Role of TNF Receptor-Mediated Mechanism in the Expression of Paracrine Factors in MSCs

Using TNF receptor type 1 (TNFR1) or type 2 (TNFR2) ablation model, it is observed that when MSCs are stimulated with TNF, LPS, or hypoxia for 24 hours, the production of TNF and IGF-1 is decreased in the cells from both knockouts (KOs) as compared with the cells from normal animals. On the other hand, IL-6 secretion is increased in the MSCs from both knockouts over the wild-type cells following TNF stimulation, but is significantly decreased with exposure to LPS. Hypoxia enhances the level of IL-6 in the cells from TNFR1 KO animals, but not in TNFR2 KO cells. TNF stimulation leads to a decreased production of VEGF in TNFR2KO cells, whereas no change in VEGF secretion is observed in TNFR1 KO cells. However, TNFR1 ablation resulted in a decrease in VEGF production in the cells following LPS stimulation with no change in TNFR2 KO cells. With hypoxia, VEGF expression is increased in the TNFR1 KO cells over the normal cells, whereas no difference is present in TNFR2KO cells [29]. These data suggest that TNF receptors and related signaling cascades play a complex role in the regulation on the expression and production of paracrine factors in MSCs in response to different stimuli.

It is also reported that TGF-α promotes the expression of HGF in human MSCs. TGF-α-stimulated production of HGF is effectively prevented by inhibition of MEK, p38, PI-3K signaling, or targeting TNF receptor 2 (TNFR2) using small interfering RNA (siRNA), but not by targeting TNF receptor 1 (TNFR1). However, inhibition of TNFR1 significantly increases basal level of HGF in MSCs. Further investigation indicates that there is a complex interactions between TNF receptors and TGF-α/EGF receptor in the regulation of HGF production in human MSCs via activation of MEK, p38, and PI-3K signaling [49].

Important Factors Associated with the Paracrine Function in MSCs

Paracrine mechanisms for the therapeutic effects of MSCs are very complex, and involve a large number of growth factors and cytokines and related receptors and signaling molecules with a broad range of biological functions. It is important to identify the factors that are critically involved in the regulation on the expression and production of these paracrine factors in MSCs to achieve an optimal therapeutic outcome. In the present focused review, the role of species, age, sex, and environmental factors like hypoxia in the expression of paracrine factors is briefly discussed.

Different Paracrine Factors Are Produced in MSCs from Different Species

Recently, the profiles of paracrine factors from swine and human bone marrow MSCs are characterized and compared in culture system under normoxic or hypoxic conditions [34]. It is shown that the cell markers of swine MSCs are comparable to those of human MSCs with minor differences phenotypically. The majority of paracrine factors including VEGF and Endothelin in the conditioned medium released from swine MSCs are similar to those from human MSCs under normoxic conditions. However, substantial differences in the levels for a number of growth factors and signaling molecules in the conditioned media exist between the two MSCs. Noticeably, a significant amount of FGF-16, frizzled-6, Galectin-3, IL-1 alpha, IL-17E, and Smad-5 are present in the conditioned medium of swine MSCs, while not much in the one of human MSCs. On the other hand, high level of TIMP-1 is detected in the conditioned medium of human MSCs, but not in the one of swine MSCs. When the cells are cultured under hypoxic conditions, only small changes in the paracrine factor profile is observed in the conditioned medium of swine MSCs compared with that under normoxic conditions, while significant changes occur in the paracrine factor profile in the medium of human MSCs. These data suggest that MSCs from different species express and produce different type and / or levels of paracrine factors, and respond differently to the environmental stimulation.

Age Plays an Important Role in the Expression of Paracrine Factors in MSCs

It is well known that the populations of bone marrow MSCs (BMSCs) are closely related to the age of the animal with higher level of MSCs in the younger ones. Although neonatal and adult BMSCs have similar pluripotent potentials and cell surface markers, neonatal bone marrow MSCs (nBMSCs) proliferate faster, and therefore, could be expanded more rapidly than adult bone marrow MSCs (aBMSCs) [30]. It is also reported that MSCs from older hosts are associated with telomere shortening and dysfunction, and a reduced capacity to maintain functional hematopoietic stem cells [15]. MSCs from animals with different ago have been shown to produce different levels of paracrine factors. When compared with aBMSCs, nBMSCs produce lower levels of IL-6 and VEGF, but higher levels of IGF-1 under basal conditions, and after stimulation with TNF. However, there are no differences in LPS-induced production of IL-6, VEGF, or IGF-1 between nBMSCs and adult cells. The difference in cytokine and growth factor production in nBMSCs is considered to be related to p38 and ERK signaling [30]. In a separate study, it is observed that inhibition of NFkB and IKK leads to a significant decrease in VEGF secretion in aBMSCs, but not in nBMSCs [35]. Higher expression of angiogenic growth factors (including HIF-1α, HO-1, VEGF, and FGF-2) is observed in the MSCs from young rats (8–12 weeks old) under anoxia as compared to the cells from old rats (24–26 months) [14]. Clearly, there is a significant difference in the expression and production of paracrine factors between neonatal and adult BMSCs.

Gender Is an Important Determinant in the Production of Paracrine Factors in MSCs

A recent study showed that treatment with MSCs from female donors is associated with greater cardiac protection against acute endotoxemic injury in rats compared with treatment with the cells from male animals [28]. Animal data have shown that MSCs from normal male mice produce significantly greater levels of TNF and IL-6 and significantly less amount of VEGF in response to LPS stimulation and hypoxia than the cells from female animals. A substantial change in the release of TNF, IL-6 and VEGF is observed in MSCs from male TNFR1 knockout mice compared with the cells from the male WT animals, but is not different from female WT MSCs. On the other hand, there is no significant difference in the production of TNF and IL-6 between female WT MSCs and female TNFR1KO MSCs [2, 3]. Apparently, gender differences exist in the therapeutic effects of MSCs and their paracrine function.

Role of Environmental Factors and Disease States in the Expression of Paracrine Factors in MSCs

Although bone marrow is a very hypoxic environment, bone marrow MSCs are very resistant to hypoxic culture condition, survive and function well in hypoxic environment with or without serum [27, 41]. As discussed in this chapter, MSCs produce and secrete a great variety of cytokines and growth factors with a wide spectrum of biological functions that are considered to be largely responsible for the beneficial actions when MSCs are used for tissue repair and regeneration. It is well documented that hypoxia significantly changes the profile of paracrine factors expressed and produced from MSCs. The changes in paracrine factors in response to hypoxia are highly variable and complex with some factors up-regulated, some ones down-regulated, and yet, some factors unchanged [34, 41]. For example, VEGF and TIMP-2 expression is usually up-regulated in bone marrow MSCs from rat, dog, and human under hypoxic condition, while LRP-6 and activin A expression is down-regulated, and no changes in the expression of a number of paracrine factors like (but not limited to) endothelin, frizzled-5, IL-7, IL-27, MMP-16, NCAM-1, and Smad5 in MSCs from both dog and human [34]. There are significant differences in the changes on the expression of some paracrine factors in MSCs from different species in response to hypoxia. It is reported that hypoxia leads to an increased expression of osteoprotegerin, TIMP-2, and IGFBP-7 in human MSCs, while a decrease in their expression in swine MSCs [34].

Some important disease states like hypercholesterolemia and diabetes mellitus (DM) may have significant impact on the paracrine function of MSCs. Indeed, bone marrow-derived progenitor cells from patients with hypercholesterolemia and/or DM exhibit a substantially reduced capacity for neovascularization and decreased paracrine secretion of proangiogenic factors [11, 18]. Recently, it is demonstrated that high glucose culture substantially suppresses VEGF expression in rat MAPCs through inhibition of JAK2/STAT3 signaling [25], and increases TGF-β1 expression in these cells via ERK1/2-induced inhibition of STAT3 signaling [26]. It has also been reported that the levels of IGF-1, FGF-2 and Akt pro-survival factors are significantly decreased in MSCs from type 1 diabetic mice [17]. The protein levels of HIF-1α, VEGF-A, and PDGF-B are shown to be significantly reduced in the conditioned media of mouse MSCs in the presence of high glucose in a dose-dependent manner in association with increased production of intracellular superoxide levels [13]. However, it is shown that high glucose concentration has no effect on the production of VEGF, HGF, or FGF2 from human MSCs at baseline and when treated with TNF-alpha, LPS, or hypoxia [51]. These apparent different results clearly implicate that the effect of high glucose on VEGF expression and production is dependent on specific cell type and/or species.

Conclusion and Future Directions

There is no question that paracrine factors and related signaling molecules represent an important mechanism for the beneficial effects of MSCs on tissue repair and regeneration. It is known that a great number of growth factors and cytokines (the numbers are still rapidly growing) with a broad spectrum of biological functions are expressed and produced in MSCs under different conditions and with different stimuli. A detailed knowledge on how these paracrine molecules mediate the complex protective and regenerative effects of MSCs is of great value to improve their function, viability, homing, retention and integration in the target tissues. Understanding the regulatory mechanisms for the expression and secretion of the paracrine function in MSCs is equally important, allowing engineering the cells for their optimal survival and function. New and more sensitive methods and technology are needed to detect, identify, and characterize new molecules from MSCs that are in small quantity and / or unstable in nature, and yet with powerful therapeutic potential on tissue repair and regeneration. Significant challenges remain on how to design an ideal therapeutic regimen with different paracrine factors to achieve an optimal outcome for tissue protection and regeneration.

Acknowledgements This work was supported by a grant from NIH R01 HL094650 (ZGL).

Disclosures The authors have no conflicts of interest to disclose.

References

1. Bernardo ME, Fibbe WE. Safety and efficacy of mesenchymal stromal cell therapy in autoimmune disorders. Ann N Y Acad Sci. 2012;1266:107–17.
2. Crisostomo PR, Wang M, Herring CM, Markel TA, Meldrum KK, Lillemoe KD, Meldrum DR. Gender differences in injury induced mesenchymal stem cell apoptosis and VEGF, TNF, IL-6 expression: role of the 55 kDa TNF receptor (TNFR1). J Mol Cell Cardiol. 2007;42(1):142–9.
3. Crisostomo PR, Wang M, Herring CM, Morrell ED, Seshadri P, Meldrum KK, Meldrum DR. Sex dimorphisms in activated mesenchymal stem cell function. Shock. 2006;26(6):571–4.
4. Djouad F, Bouffi C, Ghannam S, Noël D, Jorgensen C. Mesenchymal stem cells: innovative therapeutic tools for rheumatic diseases. Nat Rev Rheumatol. 2009;5(7):392–9.
5. Dominici M, Le Blanc K, Mueller I, Slaper-Cortenbach I, Marini F, Krause D, Deans R, Keating A, Prockop DJ, Horwitz E. Minimal criteria for defining multipotent mesenchymal stromal cells. The International Society for Cellular Therapy position statement. Cytotherapy. 2006;8(4):315–7.
6. Gao J, Dennis J, Muzic R, Lundberg M, Caplan A. The dynamic in vivo distribution of bone marrow-derived mesenchymal stem cells after infusion. Cells Tissues Organs. 2001;169:12–20.
7. García-Castro J, Trigueros C, Madrenas J, Pérez-Simón JA, Rodriguez R, Menendez P. Mesenchymal stem cells and their use as cell replacement therapy and disease modelling tool. J Cell Mol Med. 2008;12(6B):2552–65.
8. Gnecchi M, Danieli P, Cervio E. Mesenchymal stem cell therapy for heart disease. Vascul Pharmacol. 2012;57(1):48–55.
9. Gnecchi M, He H, Noiseux N, Liang OD, Zhang L, Morello F, Mu H, Melo LG, Pratt RE, Ingwall JS, Dzau VJ. Evidence supporting paracrine hypothesis for Akt-modified mesenchymal stem cell-mediated cardiac protection and functional improvement. FASEB J. 2006;20(6):661–9.
10. Haynesworth SE, Baber MA, Caplan AI. Cytokine expression by human marrow-derived mesenchymal progenitor cells in vitro: effects of dexamethasone and IL-1 alpha. J Cell Physiol. 1996;166(3):585–92.
11. Heeschen C, Lehmann R, Honold J, Assmus B, Aicher A, Walter DH, Martin H, Zeiher AM, Dimmeler S. Profoundly reduced neovascularization capacity of bone marrow mononuclear cells derived from patients with chronic ischemic heart disease. Circulation. 2004;109:1615–22.
12. Horwitz EM, Le Blanc K, Dominici M, Mueller I, Slaper-Cortenbach I, Marini FC, Deans RJ, Krause DS, Keating A. International Society for Cellular Therapy. Clarification of the nomenclature for MSC: The International Society for Cellular Therapy position statement. Cytotherapy. 2005;7(5):393–5.

13. Ishizuka T, Hinata T, Watanabe Y. Superoxide induced by a high-glucose concentration attenuates production of angiogenic growth factors in hypoxic mouse mesenchymal stem cells. J Endocrinol. 2011;208(2):147–59.
14. Jiang S, Kh Haider H, Ahmed RP, Idris NM, Salim A, Ashraf M. Transcriptional profiling of young and old mesenchymal stem cells in response to oxygen deprivation and reparability of the infarcted myocardium. J Mol Cell Cardiol. 2008;44(3):582–96.
15. Ju Z, Jiang H, Jaworski M, Rathinam C, Gompf A, Klein C, Trumpp A, Rudolph KL. Telomere dysfunction induces environmental alterations limiting hematopoietic stem cell function and engraftment. Nat Med. 2007;13:742–7.
16. Kallis YN, Alison MR, Forbes SJ. Bone marrow stem cells and liver disease. Gut. 2007; 56(5):716–24.
17. Khan M, Akhtar S, Mohsin S, Khan S N, Riazuddin S. Growth factor preconditioning increases the function of diabetes-impaired mesenchymal stem cells. Stem Cells Dev. 2011; 20(1):67–75.
18. Kränkel N, Adams V, Linke A, Gielen S, Erbs S, Lenk K, Schuler G, Hambrecht R. Hyperglycemia reduces survival and impairs function of circulating blood-derived progenitor cells. Arterioscler Thromb Vasc Biol. 2005;25:698–703.
19. Kuroda Y, Kitada M, Wakao S, Dezawa M. Bone marrow mesenchymal cells: how do they contribute to tissue repair and are they really stem cells? Arch Immunol Ther Exp (Warsz). 2011;59(5):369–78.
20. Lai RC, Chen TS, Lim SK. Mesenchymal stem cell exosome: a novel stem cell-based therapy for cardiovascular disease. Regen Med. 2011;6(4):481–92.
21. Lalu MM, McIntyre L, Pugliese C, Fergusson D, Winston BW, Marshall JC, Granton J, Stewart DJ. Canadian Critical Care Trials Group. Safety of cell therapy with mesenchymal stromal cells (SafeCell): a systematic review and meta-analysis of clinical trials. PLoS One. 2012;7(10):e47559.
22. Lee JW, Gupta N, Serikov V, Matthay MA. Potential application of mesenchymal stem cells in acute lung injury. Expert Opin Biol Ther. 2009;9(10):1259–70.
23. Lee RH, Pulin AA, Seo MJ, Kota DJ, Ylostalo J, Larson BL, Semprun-Prieto L, Delafontaine P, Prockop DJ. Intravenous hMSCs improve myocardial infarction in mice because cells embolized in lung are activated to secrete the anti-inflammatory protein TSG-6. Cell Stem Cell. 2009;5(1):54–63.
24. Li H, Fu X. Mechanisms of action of mesenchymal stem cells in cutaneous wound repair and regeneration. Cell Tissue Res. 2012;348(3):371–7.
25. Liu Z, Lei M, Jiang Y, Hao H, Chu L, Xu J, Luo M, Verfaillie CM, Zweier JL, Liu Z. High glucose attenuates VEGF expression in rat multipotent adult progenitor cells in association with inhibition of JAK2/STAT3 signaling. J Cell Mol Med. 2008;13:3427–36.
26. Luo M, Liu Z, Chen G, Hao H, Lu T, Cui Y, Lei M, Verfaillie CM, Liu Z. High glucose enhances TGF-β1 expression in rat bone marrow stem cells via ERK1/2-mediated inhibition of STAT3 signaling. Life Sci. 2012;90(13–14):509–18.
27. Mangi AA, Noiseux N, Kong D, He H, Rezvani M, Ingwall JS, Dzau VJ. Mesenchymal stem cells modified with Akt prevent remodeling and restore performance of infarcted hearts. Nat Med. 2003;9(9):1195–201.
28. Manukyan MC, Weil BR, Wang Y, Abarbanell AM, Herrmann JL, Poynter JA, Brewster BD, Meldrum DR. Female stem cells are superior to males in preserving myocardial function following endotoxemia. Am J Physiol Regul Integr Comp Physiol. 2011;300(6):R1506–14.
29. Markel TA, Crisostomo PR, Wang M, Herring CM, Meldrum DR. Activation of individual tumor necrosis factor receptors differentially affects stem cell growth factor and cytokine production. Am J Physiol Gastrointest Liver Physiol. 2007;293(4):G657–62.
30. Markel TA, Wang M, Crisostomo PR, Manukyan MC, Poynter JA, Meldrum DR. Neonatal stem cells exhibit specific characteristics in function, proliferation, and cellular signaling that distinguish them from their adult counterparts. Am J Physiol Regul Integr Comp Physiol. 2008;294(5):R1491–7.
31. Mirotsou M, Jayawardena TM, Schmeckpeper J, Gnecchi M, Dzau VJ. Paracrine mechanisms of stem cell reparative and regenerative actions in the heart. J Mol Cell Cardiol. 2011;50:280–9.

32. Mishra PJ, Mishra PJ, Banerjee D. Cell-free derivatives from mesenchymal stem cells are effective in wound therapy. World J Stem Cells. 2012;4(5):35–43.
33. Mundra V, Gerling IC, Mahato RI. Mesenchymal stem cell-based therapy. Mol Pharm. 2013;10(1):77–89.
34. Nguyen BK, Maltais S, Perrault LP, Tanguay JF, Tardif JC, Stevens LM, Borie M, Harel F, Mansour S, Noiseux N. Improved function and myocardial repair of infarcted heart by intracoronary injection of mesenchymal stem cell-derived growth factors. J Cardiovasc Transl Res. 2010;3(5):547–58.
35. Novotny NM, Markel TA, Crisostomo PR, Meldrum DR. Differential IL-6 and VEGF secretion in adult and neonatal mesenchymal stem cells: role of NFkB. Cytokine. 2008;43(2):215–9.
36. Parekkadan B, Milwid JM. Mesenchymal stem cells as therapeutics. Annu Rev Biomed Eng. 2010;12:87–117.
37. Pelacho B, Nakamura Y, Zhang J, Ross J, Heremans Y, Nelson-Holte M, Lemke B, Hagenbrock J, Jiang Y, Prosper F, Luttun A, Verfaillie CM. Multipotent adult progenitor cell transplantation increases vascularity and improves left ventricular function after myocardial infarction. J Tissue Eng Regen Med. 2007;1:51–9.
38. Rastegar F, Shenaq D, Huang J, Zhang W, Zhang BQ, He BC, Chen L, Zuo GW, Luo Q, Shi Q, Wagner ER, Huang E, Gao Y, Gao JL, Kim SH, Zhou JZ, Bi Y, Su Y, Zhu G, Luo J, Luo X, Qin J, Reid RR, Luu HH, Haydon RC, Deng ZL, He TC. Mesenchymal stem cells: Molecular characteristics and clinical applications. World J Stem Cells. 2010;2(4):67–80.
39. Schrepfer S, Deuse T, Reichenspurner H, Fischbein M, Robbins R, Pelletier M. Stem cell transplantation: the lung barrier. Transplant Proc. 2007;39:573–6.
40. Shi M, Liu ZW, Wang FS. Immunomodulatory properties and therapeutic application of mesenchymal stem cells. Clin Exp Immunol. 2011;164(1):1–8.
41. Takahashi M, Li TS, Suzuki R, Kobayashi T, Ito H, Ikeda Y, Matsuzaki M, Hamano K. Cytokines produced by bone marrow cells can contribute to functional improvement of the infarcted heart by protecting cardiomyocytes from ischemic injury. Am J Physiol Heart Circ Physiol. 2006;291(2):H886–93.
42. Terai S, Sakaida I, Yamamoto N, Omori K, Watanabe T, Ohata S, Katada T, Miyamoto K, Shinoda K, Nishina H, Okita K. An in vivo model for monitoring trans-differentiation of bone marrow cells into functional hepatocytes. J Biochem. 2003;134(4):551–8.
43. Togel F, Hu Z, Weiss K, Isaac J, Lange C, Westenfelder C. Administered mesenchymal stem cells protect against ischemic acute renal failure through differentiation-independent mechanisms. Am J Physiol Renal Physiol. 2005;289:F31–42.
44. Togel F, Weiss K, Yang Y, Hu Z, Zhang P, Westenfelder C. Vasculotropic, paracrine actions of infused mesenchymal stem cells are important to the recovery from acute kidney injury. Am J Physiol Renal Physiol. 2007;292:F1626–35.
45. Tolar J, Le Blanc K, Keating A, Blazar BR. Concise review: hitting the right spot with mesenchymal stromal cells. Stem Cells. 2010;28(8):1446–55.
46. Tongers J, Losordo DW, Landmesser U. Stem and progenitor cell-based therapy in ischaemic heart disease: promise, uncertainties, and challenges. Eur Heart J. 2011;32(10):1197–206.
47. Wang M, Crisostomo PR, Herring C, Meldrum KK, Meldrum DR. Human progenitor cells from bone marrow or adipose tissue produce VEGF, HGF, and IGF-I in response to TNF by a p38 MAPK-dependent mechanism. Am J Physiol Regul Integr Comp Physiol. 2006; 291(4):R880–4.
48. Wang Y, Crisostomo PR, Wang M, Markel TA, Novotny NM, Meldrum DR. TGF-alpha increases human mesenchymal stem cell-secreted VEGF by MEK- and PI3-K- but not JNK- or ERK-dependent mechanisms. Am J Physiol Regul Integr Comp Physiol. 2008;295(4):R1115–23.
49. Wang Y, Weil BR, Herrmann JL, Abarbanell AM, Tan J, Markel TA, Kelly ML, Meldrum DR. MEK, p38, and PI-3K mediate cross talk between EGFR and TNFR in enhancing hepatocyte growth factor production from human mesenchymal stem cells. Am J Physiol Cell Physiol. 2009;297(5):C1284–93.
50. Wang M, Zhang W, Crisostomo P, Markel T, Meldrum KK, Fu XY, Meldrum DR. STAT3 mediates bone marrow mesenchymal stem cell VEGF production. J Mol Cell Cardiol. 2007;42:1009–15.

51. Weil BR, Abarbanell AM, Herrmann JL, Wang Y, Meldrum DR. High glucose concentration in cell culture medium does not acutely affect human mesenchymal stem cell growth factor production or proliferation. Am J Physiol Regul Integr Comp Physiol. 2009;296:R1735–43.
52. Williams AR, Hare JM. Mesenchymal stem cells: biology, pathophysiology, translational findings, and therapeutic implications for cardiac disease. Circ Res. 2011;109:923–40.
53. Wollert KC, Drexler H. Cell therapy for the treatment of coronary heart disease: a critical appraisal. Nat Rev Cardiol. 2010;7(4):204–15.
54. Yew TL, Hung YT, Li HY, Chen HW, Chen LL, Tsai KS, Chiou SH, Chao KC, Huang TF, Chen HL, Hung SC. Enhancement of wound healing by human multipotent stromal cell conditioned medium: the paracrine factors and p38 MAPK activation. Cell Transplant. 2011;20(5):693–706.
55. Yi T, Song SU. Immunomodulatory properties of mesenchymal stem cells and their therapeutic applications. Arch Pharm Res. 2012;35(2):213–21.

Immunomodulatory Properties of Mesenchymal Stem Cells and Related Applications

Lianming Liao and Robert Chunhua Zhao

Abstract Mesenchymal stem cells (MSCs) are an important cell population that resides in a bone marrow microenvironment. In the past decade, MSCs have been discovered to have profound immunomodulatory functions both *in vitro* and *in vivo*. As MSCs can be expanded rapidly to numbers that are required for clinical application, clinical studies have been performed in immune diseases, bone marrow transplantation and kidney transplantation. In this chapter, the mechanisms underlying MSCs' immunomodulating property and the potential clinical use of MSCs as a modulator of immune responses are reviewed.

Keywords MSC • Immunomodulatory functions • Interactions • T cells • Antigen-presenting cells

Introduction

The use of immunosuppressive agents has led to greatly improved organ graft survival rates and alleviation of autoimmune diseases. Nevertheless, side effects of immunosuppressive agents and patient morbidity due to life-long immunosuppression remain substantial, especially high cancer incidence among the recipients. Active induction of tolerance allowing drug-free allograft acceptance with preserved immunocompetence has long been a goal for both immunologists and

L. Liao (✉)
Academy of Integrative Medicine, Fujian University of Traditional Chinese Medicine, Huatuo Road, No. 1, 350112 Fuzhou, Fujian, P. R. China
e-mail: llm@fjtcm.edu.cn

R.C. Zhao
Center of Excellence in Tissue Engineering, Institute of Basic Medical Sciences and School of Basic Medicine, Chinese Academy of Medical Sciences and Peking, Union Medical College, 5# Dongdansantiao, 100005 Beijing, China, People's Republic
e-mail: zhaoch16@hotmail.com

R.C. Zhao (ed.), *Essentials of Mesenchymal Stem Cell Biology and Its Clinical Translation*, DOI 10.1007/978-94-007-6716-4_4,

clinicians. This method is theoretically the only way to resolve rejection reaction of allogeneic transplantation and simultaneously keep the patients void of the side effects of immunosuppressive medication.

Stable chimerism is linked with permanent tolerance of donor organ or tissue transplants [1]. Induction of mixed hematopoietic chimerism by bone marrow stem cell transplantation, which leads to stable donor-specific tolerance in allogeneic graft, has been reported [2–4]. However, the use of bone marrow transplantation is still largely restricted to patients with malignancies or severe immune diseases. The toxicity of the required host conditioning, the risk of engraftment failure, and the problem of graft-versus-host-disease (GVHD), especially when major histocompatibility complex (MHC) barriers are transgressed, hinder its clinical application. Embryonic-like stem cells transplantations have also been proved to be capable of establishing chimerism and inducing tolerance without supplementary host conditioning in animal model [5]. However, there exists the well-known ethical obstacle for embryonic stem cells and tumorigenesis of embryonic-like stem cells transplantation in human need cautious evaluation. Therefore, development of a new cell population and procedure for donor-specific allograft tolerance induction is still a difficult task in organ transplantation.

In the past decade, bone marrow-derived mesenchymal stem cells (MSCs) are garnering more and more attention for their immunoregulatory activity and regenerative ability. For example, administration of MSCs could prolong donor skin graft survival in mice [6] and nonhuman primates [7]. The ability of MSCs to regulate immune responses could also be harnessed to reduce GVHD at the time of hematopoietic stem cell transplantation [8, 9].

Characteristics of MSC

MSCs are multipotent cells that reside within the bone marrow and can be induced to differentiate into various components of the marrow microenvironment, such as bone, adipose and stromal tissues under proper conditions [10–12]. MSCs support the growth of hematopoietic progenitors by secreting a number of hematopoietic cytokines such as macrophage colony stimulating factor, interleukin (IL)-6, IL-7, IL-8, IL-11, IL-12, IL-14, IL-15 and leukemia inhibitory factor [12, 13]. MSCs have been isolated in different species and are present in the bone marrow at low frequency (1 out of 104–105 mononuclear cells). Although MSCs are originally isolated from bone marrow, they can also be isolated from muscle [14], pancreas [15], dermis [16], adipose tissues [17], lung [18], liver [19], and cord blood [20]. The exact phenotype of MSCs in the tissue (i.e. the cell phenotype prior to culture) is still debated. Simmoms et al. described the first antibody (Stro-1) that targeted MSCs in the bone marrow [21]. Some typical markers of MSCs include CD105 (SH2 or endoglin), CD73 (SH3 or SH4), CD90, CD166, CD44, and CD29.

MSCs Exert Their Immunomodulation Function by Different Mechanisms

The Interaction Between MSCs and T Cells

The key orchestrators of the immune response in transplantation are T cells, which can react to alloantigen both directly, by recognizing intact foreign MHC molecules on donor antigen-presenting cells (APCs), and indirectly, as a result of interactions with processed donor antigens on self APCs [22]. In the thymus, double positive (CD4+CD8+) cells undergo positive and negative selections before leaving the thymus. Positive selection results in survival of T cells with antigen receptors that corecognize self-MHC molecules plus foreign peptides. T cells whose receptors do not detect self-MHC molecules die, presumably by failure to receive critical differentiating signals. Negative selection involves the removal of potentially autoreactive T cells that interact too well with self-MHC molecules plus self-peptides. The majority of cells with specificities for self-antigens are eliminated during development in the thymus [23]. Otherwise, they will mature and migrate to the peripheral lymphoid organs. Cortical epithelial cells are essential for the process of positive selection because they display the self peptide–MHC complexes that are recognized by $CD4^{+}CD8^{+}$ thymocytes and also provide essential differentiation factors [24].

T cells can be physiologically silenced by a number of mechanisms, including deletion in the peripheral immune system; anergy, where they cannot adequately respond following restimulation with antigen; and suppression, which may be mediated by interactions with other cells or with soluble factors [25, 26]. The TH1 cell cytokines IL-2 and interferon gamma (IFN-γ) are considered detrimental to allografts, and the TH2 cell cytokine IL-4 is considered tolerogenic [27]. The same mechanisms act in acquired transplantation tolerance [28] and can be harnessed to achieve donor-specific tolerance by blunting the effects of alloreactive T cells.

Presently, although the relationship between MSCs and T cells is not well defined, several lines of evidence indicate that MSCs may modulate T cells by various mechanisms (Fig. 1). When added into a mixed lymphocyte reaction, either on day 0 or on day 3, baboon MSCs could inhibit an ongoing allogeneic response, leading to a greater than 50 % reduction in proliferative activity [7]. This effect could be maximized by escalating the dose of MSC and could be reduced with the addition of exogenous IL-2. The suppression of proliferative responses by MSCs did not appear to be dependent on the source of MSCs. MSCs were able to inhibit proliferation of T cells independent of whether they were of the same source of the responder, stimulator, or third party. Baboon MSCs could also inhibit proliferative response to potent T-cell mitogen ConA. Similarly, addition of MSCs to T cells stimulated by polyclonal activators (PHA or IL-2) resulted in suppression of proliferation [29]. MSC-inhibited T lymphocytes were not apoptotic and efficiently proliferated on restimulation. MSCs significantly suppressed both $CD4^{+}$ and $CD8^{+}$ T cells [30]. In a study on the immunogenicity and antigen presenting ability of MSCs,

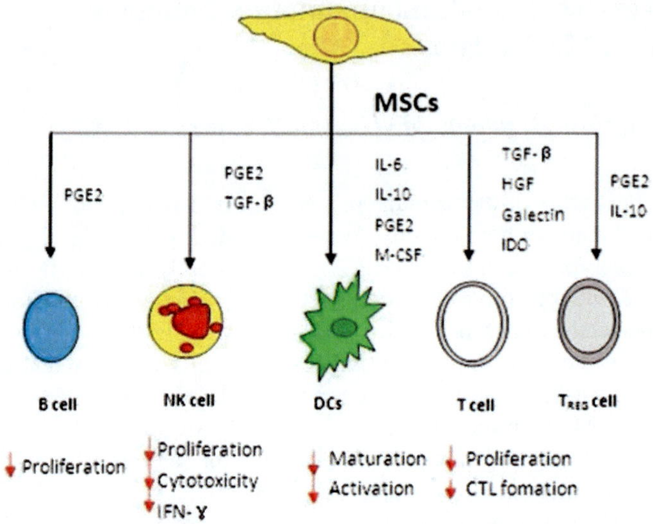

Fig. 1 MSCs inhibit the proliferation of NK cells, B cells and T cells. This effect is mediated through the secretion of various factors, such as prostaglandin E2, human leukocyte antigen-G and hepatocyte growth factor transforming growth factor-β. MSC also inhibit generation and maturation of DCs from monocytes. Abbreviations: *CTL* cytotoxic T cell, *DCs* dendritic cells, *HGF* hepatocyte growth factor, *IDO* indoleamine 2,3-dioxygenase, *MSCs* mesenchymal stem cells, *NK* natural killer, *PGE2* prostaglandin E2, T_{REG} cells, regulatory T cells, *TGF-β* transforming growth factor β

Tse et al. demonstrated human MSCs failed to stimulate allogeneic peripheral blood mononuclear cells or T-cell proliferation in mixed cell cultures [31]. MSCs actively suppressed proliferation of responder peripheral blood mononuclear cells stimulated by third-party allogeneic peripheral blood mononuclear cells as well as T cells stimulated by anti-CD3 and anti-CD28 antibodies. Similarly, marked inhibitory effects of allogeneic and autologous MSCs were also reported both in mixed lymphocyte reaction and after mitogenic lymphocyte stimulation by phytohaemagglutinin, Concanavalin A and protein A [32]. However, little, if any, inhibition occurred after stimulation with pokeweed mitogen. The inhibitory effect was also related to MSCs dose, as a minimum of MSCs was needed. Surprisingly, when the dosage was small, stimulatory effect was noted in some experiments. This phenomenon was also observed when MSCs were cocultured for 3 days with T cells isolated from cord blood and stimulated with phytohaemagglutinin for 60 h, in which high concentration of MSCs most often resulted in inhibition, while low concentration resulted in stimulation of T cell proliferation [33]. In another study, effect of MSCs on response of naive and memory T cells to their cognate antigenic epitopes was evaluated [34]. For this purpose, murine male transplantation antigens, HY, was selected to trigger immune response. C57BL/6 female mice immunized with male cells were the source of memory T cells, whereas C6 mice transgenic for HY-specific T-cell receptor provided naive T cells. Responder cells were stimulated in vitro with male spleen cells or HY peptides in the presence or absence of MSCs. MSCs inhibited HY-specific naive and memory T cells in a dose-dependent fashion and affected cell proliferation, cytotoxicity, and the number of interferon-gamma–producing HY peptide-specific T cells. However, MSC did not selectively target antigen-reactive T cells. The expression of MHC molecules and the presence in culture of antigen-presenting cells or of CD4$^+$/CD25$^+$ regulatory T cells were not required for MSCs to inhibit. This data demonstrate that autologous or allogeneic MSCs strongly suppress lymphocyte proliferation that is triggered by both cellular as well as nonspecific mitogenic stimuli in different in vitro models; this phenomenon has no immunologic restriction.

Suppression of lymphocyte proliferation by MSCs is likely due to the production of soluble factors. By using the transwell system, i.e., when MSCs were separated from T cells physically by a permeable membrane, T-cell proliferation was also significantly inhibited. After simultaneous addition of anti–transforming growth factor-β1 and anti-hepatocyte growth factor antibodies to bone marrow stromal cells-containing mixed lymphocyte reactions, T-cell proliferation was restored at values that were comparable to those detected in mixed lymphocyte reactions without bone marrow stromal cells, indicating transforming growth factor-β1 and hepatocyte growth factor were the mediators of marrow stromal cells' effects [29]. Chen et al. showed secretion of transforming growth factor-β1 by MSCs reached to 1 ng/ml in 72 h [30]. Tse et al. demonstrated IL-10 secreted by MSCs also accounted for the suppressive activity by MSCs [31]. However, the inhibitory activity was abrogated when MSCs were replaced by MSC culture supernatant [34]. This may suggest that pretreatment of MSCs with lymphocytes is necessary for MSCs to secret inhibitory factors.

In addition, the tryptophan catabolizing enzyme indoleamine 2,3-dioxygenase (IDO) has been suggested to play a role in the suppression of T cell proliferation by MSCs [35] Upon stimulation with IFN-γ, MSCs express IDO activity that degrades essential tryptophan and results in kynurenine breakdown products, resulting in reduced lymphocyte proliferation.

The family of human galectins are key players in the regulation of the innate and adaptive immune response [36]. One family member, galectin-1, is a negative regulator of immune responses. Galectin-1 may inhibit proliferation and induce apoptosis of activated T cells [37–39]. Galectin-1 protein was detected intracellularly and on the cell surface of MSCs. It was reported that galectin-1 released into the cell culture supernatant by MSCs modulated the release of cytokines involved in GVHD and autoimmunity (e.g., tumor necrosis factor-α [TNF-α], IFN-γ, interleukin-2 [IL-2], and IL-10). Thus, galectin-1 may mediate the immunomodulatory effect of MSCs on allogeneic T cells [40].

MSC-induced T-cell anergy has been proposed as another potential mechanism of immune suppression. MSCs lack surface expression of costimulatory molecules, such as CD80 (B7-1) and CD86 (B7-2), and it is believed that MSCs can render T cells anergic [41]. Another level at which MSCs may modulate immune responses is through the induction of regulatory T cells (Treg). MSCs might induce formation of CD4$^+$CD25$^+$ regulatory T cells that were responsible for inhibition of allogeneic lymphocyte proliferation [42–44]. In the experimental autoimmune uveitis in mice, intraperitoneal injection of MSCs was able to significantly attenuate uveitis and that a significantly higher percentage of adaptive Treg was present in MSC-treated mice than in MSC-untreated animals. Furthermore, induction of antigen-specific Treg by MSCs was due at least in part to the secretion og TGF-β [43].

The Interactions Between MSCs and Antigen-Presenting Cells

MSCs may also regulate the immune response through their interaction with dendritic cells (DCs) which play a key role in the induction of immunity (Fig. 1). MSCs may induce tolerance by inhibiting DC maturation and function, inhibit *in vitro* differentiation of DCs from monocytes and $CD34^+$ progenitors, and reduce secretion of pro-inflammatory cytokines (IL-12, IFN-γ, and TNF-α) and increase IL-10 secretion [45]. The production of IL-6, PGE2, IL-10 and M-CSF by MSCs may contribute to the inhibitory effect of MSCs on DC differentiation, although blocking studies indicate that these factors are not solely responsible for the inhibitory effect. DCs generated in the presence of MSCs were impaired in their response to maturation signals and exhibited no expression of CD83 or up-regulation of HLA-DR and costimulatory molecules [46–48]. Immature DCs generated in the presence of MSCs were strongly hampered in their ability to induce activation of T cells. These results suggest that MSCs suppress the differentiation of DCs, resulting in the formation of immature DCs that exhibit a suppressor or inhibitory phenotype.

The Interaction Between MSCs and B-Cells

MSCs could inhibit the proliferation of B cells stimulated with anti-CD40L and IL-4, or with pokeweed mitogen [49]. B-cells were arrested in the G0/G1 phase of the cell cycle. Similar to T-cell suppression, the major mechanism of B-cell suppression is MSC production of soluble factors, as indicated by transwell experiments. In addition, B-cell differentiation was inhibited because IgM, IgG, and IgA production was significantly impaired. What is more, CXCR4, CXCR5, and CCR7 B-cell expression, as well as chemotaxis to CXCL12, the CXCR4 ligand, and CXCL13, the CXCR5 ligand, were significantly down-regulated by MSCs, suggesting that these cells affect chemotactic properties of B cells [50]. Allogeneic MSCs have been shown to inhibit the proliferation, activation and IgG secretion of B cells from BXSB mice that are used as an experimental model for human systemic lupus erythematosus (Fig. 1) [51].

Interaction Between MSCs and Natural Killer Cells

Natural killer (NK) cells exhibit spontaneous cytolytic activity that mainly targets cells that lack expression of HLA class I molecules. MSCs suppressed IL-2 or IL-15 driven NK-cell proliferation and IFN-γ production, and NK cells cultured for 4–5 days with IL-2 in the presence of MSCs have a reduced cytotoxic potential against K562 target cells [52]. Experiments with transwell culture systems indicated that MSCs suppressed the proliferation and cytokine production of IL-15 stimulated NK cells via soluble factors. At low NK-to-MSC ratios, MSCs alter the phenotype of NK cells and suppress proliferation, cytokine secretion, and cyto-toxicity against HLA-class I-expressing targets. Some of these effects required cell-to-cell contact, whereas others were mediated by soluble factors, including PGE2 and transforming growth factor-β [53]. Indoleamine 2,3-dioxygenase also mediates MSC-induced inhibition of NK cells (Fig. 1) [54].

MSCs secrete the soluble isoform HLA-G5. Blocking experiments using neutralizing anti-HLA-G antibody demonstrate that HLA-G5 contributed first to the suppression of allogeneic T-cell proliferation and then to the expansion of regulatory T cells. MSCs inhibited both NK cell-mediated cytolysis and interferon-gamma secretion through HLA-G5 [55].

Taken together, numerous studies convincingly demonstrate that MSCs are able to modulate the function of different immune cells in vitro, particularly involving the suppression of T-cell proliferation. However, the mechanisms underlying the immunosuppressive effects of MSCS are still unclear.

MSCs Survive and Induce Immune Tolerance in the Host

It has been suggested that MSCs escape the immune system after they are infused to allogeneic recipient because they possess a cell surface phenotype that reflects poor recognition by T cells. For example, injection of genetically modified MSCs in baboon was not followed by their rejection because of the lack of immunogenicity of MSCs [7, 56]. Indeed, the distinct immunophenotype profile of MSCs, i.e., no expression of costimulatory molecules B7-1, B7-2, CD40 and CD40L associated with the absence of MHC class I and II expression, suggests that they may not be recognized by allogenic T-cells and can escape host immune system' rejection. MSCs can easily migrate and reside in various tissues, which may result from their expression of a variety of adhesion molecule. In our study, when allogeneic murine MSCs were transplanted into lethally irradiated recipient mice 150 days before allogeneic skin transplantation, allogeneic donor skins were successful transplanted and have survived for more than 100 days without any rejection reaction [6]. Immunohistochemistry staining showed donor MSCs could established long term residency in gastrointestinal tissues, kidney, lung, liver, thymus, and skin [57]. In a baboon model, following systemic infusion of GFP-marked MSC into an immunocompetent host, MSCs could be detected in a wide non-hematopoietic tissue distribution between 9 and 21 months later, including gastrointestinal tissues (colon, duodenum, jejunum, and ileum), kidney, skin, lung, thymus, and liver. Importantly, the results suggested that tissue distribution of MSC following systemic infusion was not affected by histocompatibility or prior conditioning. In the non-conditioned recipient, engraftment of MSC in these tissues was also achieved, although less abundant. When reinfused in nonhuman models, ex vivo–expanded human MSCs migrated to and became incorporated into several tissues of the recipient animals where MSCs were capable to elicit tissue-specific differentiation programs, indicating that MSCs have multiorgan homing capacity and an intrinsic degree of plasticity [58–61]. Studies involving direct injection of MSCs into the rodent brain reported migration of cells within the brain and differentiation into glial populations. This approach has used xenogeneic transplant of human cells into the rat brain, as well as homologous mouse/mouse tracking studies [58]. When purified human MSCs from adult bone marrow were injected into the left ventricle of CB17 SCID/beige adult mice, a limited number of cells survived and over time morphologically resembled the surrounding host cardiomyocytes. Immunohistochemistry revealed de novo expression of desmin, beta-myosin heavy chain, alpha-actinin, cardiac troponin T, and phospholamban at levels comparable to those of the host cardiomyocytes [59]. In another xenograft model, bone marrow stromal cells were isolated from C57B1/6 mice and injected into immunocompetent adult Lewis rats. One week later, the recipient animals underwent coronary artery ligation and were sacrificed at various time points ranging from 1 day to 12 weeks after ligation. Labeled mice cells engrafted into the bone marrow cavities of the recipient rats for at least 13 weeks after transplantation without any immunosuppression. In the heart, some of these cells subsequently showed positive staining for cardiomyocyte specific proteins, while other labeled cells participated in angiogenesis in the infarcted area. These findings indicate marrow stromal cells are adult stem cells with unique immunologic tolerance allowing their engraftment into a xenogeneic environment, while preserving their ability to be recruited to an injured myocardium to form a stable cardiac chimera [60]. Similarly, human MSCs engrafted into fetal lamb could persist in multiple tissues for as long as 13 months after transplantation. Transplanted human cells underwent site-specific differentiation into chondrocytes, adipocytes, myocytes, cardiomyocytes, bone marrow stromal cells and thymic stroma. Unexpectedly, there was long-term engraftment even when cells were transplanted after the expected development of immunocompetence [61]. This is in contrast to the fact that fetal lambs develop the capacity to reject allogeneic skin grafts [62] and demonstrate allogeneic or xenogeneic hematopoietic engraftment failure [63] after 75 days of gestation. Thus, MSCs maintain their multipotential capacity after transplantation, and seem to have unique immunologic characteristics that allow persistence in a xenogeneic environment. It is tempting to hypothesize that such wide distribution of infused MSCs in the host may induce peripheral tolerance.

Another finding that may account for MSCs' immunomodulation effect is that bone marrow stromal cells could migrate to the thymus after transplantation and thus may exert their immunomodulation effect there. This is first demonstrated by Li et al. who found donor-derived bone marrow stromal cells could migrate into the thymus and participate in the positive selection of T lymphocytes after bone marrow transplantation plus bone grafts [64]. It therefore seems that bone marrow stromal cells may provide a scaffold for the adhesion of early T cells and, at least in culture, supply the appropriate stimuli for thymus precursor cell proliferation [65]. MSCs could secrete transforming growth factor-β1 in vitro22 and transforming growth factor-β1 is a potent inhibitor of T-cell proliferation, both in IL-2 and IL-4 derived response [66–68]. Transforming growth factor-β1 also significantly inhibited triple-negative ($CD3^-CD4^-CD8^-$) thymocytes in vitro [69]. Thus, MSCs migrating to thymus may inhibit proliferation of T-cell by secretion of transforming growth factor-β1.

Preclinical animal studies demonstrated that MSCs can prolong allograft survival and alleviate autoimmune disease. When donor MSCs were intravenously administrated to MHC-mismatched recipient baboons prior to placement of autologous, donor, and third-party skin grafts, MSCs led to prolonged skin graft survival when compared to control animals (11.3 vs 7 days) [7]. In a murine allograft system, we showed that allogeneic donor skins were successful transplanted and have survived for more than 100 days without any rejection reaction with pre-infusion of donor MSCs [6].

Allogeneic bone marrow transplantation associated to bone grafts was found to be efficient in the treatment of autoimmune disorders, such as in the MLR/lpr mouse model of lupus [70]. In these experiments, stromal cells have been assumed to play a critical role as compared to hematopoietic stem cells. In order to determine the real impact of MSCs in these experiments, the adherent cells were removed from the total bone marrow samples before transplantation. In this case, 75 % of the treated animals died within 90 days. In contrast, complementation of adherent cell-depleted bone marrow with stromal cells permitted the mice to survive 48 weeks and cured the autoimmune disease, suggesting that MSCs play a critical role in the complex immunoregulation of T- and B-cells.

Applications of MSCs

GVHD

A potential application of MSCs in bone marrow transplantation is the prevention and treatment of steroid-resistant GVHD. Severe GVHD is a life-threatening complication after HSC transplantation. Unfortunately steroids, the first-line treatment for GVHD, have a response rate of 30–50 %. In patients with severe steroid-resistant acute GVHD, the overall survival is low [71]. In a case report, a 20-year-old woman with high-risk acute myelogenous leukemia was transplanted with granulocyte colony stimulating factor-mobilized peripheral blood CD34+ hematopoietic stem cells and bone-marrow-derived MSCs from her HLA haplotype-mismatched father after myeloablative conditioning therapy. The patient engrafted rapidly and had no acute or chronic GVHD. Since transplantation, the patient has shown an enduring trilineage haematological complete response without any evidence of leukemia relapse at 31 months [45]. Several pilot studies have shown the efficiency of MSCs in treating steroid-resistant acute GVHD [72–74]. These findings were confirmed by a phase II study [75]. Recently, Weng et al. reported that MSCs derived from HLA-identical sibling donors or HLA-disparate third-party donors were also effective as a salvage therapy for refractory chronic GVHD [76].

Organ Transplantation

MSCs may also offer therapeutic opportunities in organ transplantation by inhibiting T-cell proliferation, cytotoxic T-cell activity, B cell activation and differentiation and DC maturation and thereby blunting the effector arm of the alloresponse. In a baboon skin transplant model, a single intravenous administration of donor type MSCs into MHC-mismatched recipients resulted in significant prolonged graft survival [7]. In a mouse transplant model, intraportal administration of MSCs extended heart allograft survival from 10 days in untreated controls to a median survival time of 40 days, with 33 % of MSC-treated recipients showing long-term tolerance [77]. Ding et al. showed that MSCs protected islet allografts from rejection [78]. In the life-sustaining mouse islet allograft model the allogeneic islets were rejected within 30 days. Surprisingly, administration of MSCs prevented rejection and led to long-term normoglycemia. In a recent report, 1 and 2×106 MSC/kg recipient body weight were infused at the time of renal transplantation and at 2 weeks post transplant respectively. Preliminary results indicate that induction therapy with MSC appears to be equally effective as Basiliximab in the prevention of acute rejection and is associated with better clinical outcomes as far as early renal graft function and rate of infections [79]. In another pilot study, donor-derived bone marrow MSCs combined with a sparing dose of tacrolimus (50 % of standard dose) were shown to have a comparable effects with standard dose of tacrolimus in terms of acute rejection, graft function, and patient and graft survival within 12 months after kidney transplantation [80].

Chronic Inflammatory Autoimmune Diseases

MSCs have shown promise in exerting an anti-inflammatory immunomodulatory role in some autoimmune disease with little evidence of toxicity. They are effective for the treatment of autoimmune disease in various animal models, such as systemic lupus erythematosus (SLE), autoimmune enteropathy, autoimmune encephalomyelitis, autoimmune type 1 diabetes, and autoimmune rheumatic diseases [81–84]. Clinical studies for refractory SLE patients using allogeneic MSCs demonstrated improvement in serological markers and renal function [84, 85]. Clinical trials with MSCs for diabetes and lupus nephritis are underway (http://clinicaltrials.gov/). Autologous BM-derived MSCs have been shown to be potently antiproliferative to stimulated T-cells from both healthy donors and autoimmune patients (RA, systemic sclerosis, Sjoegren's, SLE) [86]. Interestingly, it was reported that functional abnormalities existed in BM-derived MSCs from both patients with SLE and MRL/lpr mice [87], which suggests that abnormal MSCs may contribute to the development of SLE and allogeneic MSCs from healthy donors may be superior to autologous ones in treating SLE. Recently investigators also tried to treat Crohn's Disease with MSCs. Crohn's disease is chronic inflammatory disorder of the gastrointestinal tract. Refractory patients do not respond to steroids, immunosuppressive agents (e.g., azathioprine, 6-mercaptopurine and methotrexate) or anti-TNF therapy and suffer from a poor quality of life. Duijvestein et al. reported that autologous BM-derived MSC therapy in patients with refractory Crohn's disease was promising. MSCs were infused intravenously at a target dose of 1–2 × 106 cells/kg body weight. In eight patients treated, Crohn's disease activity index scores improved in five patients, clinical response was seen in three patients at week 6 [88]. In another pilot study, ten patients with fistulising Crohn's disease were treated with autologous BM-derived MSCs [89]. MSCs were injected both into the lumen and the wall of the fistula tracks. Twelve months afterwards the sustained complete closure (seven cases) or incomplete closure (three cases) of fistula tracks with a parallel reduction of Crohn's disease and perianal disease activity indexes ($p < 0.01$ for both), and rectal mucosal healing were achieved.

Conclusions and Future Directions

Ex vivo–generated MSCs might be useful in clinical situations in which engraftment failure is high, such as human leukocyte antigen-mismatched sibling, matched unrelated donor marrow, and umbilical cord blood transplantation, and may decrease GVHD and facilitate the engraftment and proliferation of hemotopoietic progenitors. Reinfusion of MSCs aimed at exploiting immunoregulatory role might eventually be of relevance also in the setting of allografting with reduced conditioning regimens. The mechanism of its ability in immune treatment and its direct immunomodulatory therapeutic effect are not well understood and await further research. For example, although MSCs do not express MHC antigens at the time of in vitro culture, they are certain to express these antigens after they differentiate into committed cells in vivo. Why they still can stay in the host cannot be explained by their lacking of immunogenicity. As MSCs may be expanded as many as 40 generations and result in an increase of more than 104-fold in number while still maintain their multipotent mesenchymal lineages capability and phenotype, they are feasible for ex vivo implantation in clinical settings. Moreover, MSCs do not present alloantigen and do not require MHC expression to exert their inhibitory effect, suggesting that they can be derived from a donor irrespective of their MHC haplotype and be prepared as an "off-the-shelf" reagent for any patient.

References

1. Wekerle T, Sykes M. Mixed chimerism and transplantation tolerance. Annu Rev Med. 2001;52:353–70.
2. Sayegh MH, Fine NA, Smith JL, Rennke HG, Milford EL, Tilney NL. Immunologic tolerance to renal allografts after bone marrow transplants from the same donors. Ann Intern Med. 1991;114:954–5.

3. Helg C, Chapuis B, Bolle JF, Morel P, Salomon D, Roux E, Antonioli V, Jeannet M, Leski M. Renal transplantation without immunosuppression in a host with tolerance induced by allogeneic bone marrow transplantation. Transplantation. 1994;58:1420–2.
4. Sorof JM, Koerper MA, Portale AA, Potter D, DeSantes K, Cowan M. Renal transplantation without chronic immunosuppression after T cell-depleted, HLA-mismatched bone marrow transplantation. Transplantation. 1995;59:1633–5.
5. Fandrich F, Lin X, Chai GX, Schulze M, Ganten D, Bader M, Holle J, Huang DS, Parwaresch R, Zavazava N, Binas B. Preimplantation-stage stem cells induce long-term allogeneic graft acceptance without supplementary host conditioning. Nat Med. 2002;8:171–8.
6. Han Q, Deng W, Li C, Zhao RC. Allogeneic adult stem cells establish long-term residence in recipient tissues and facilitate skin transplantation. Exp Hematol. 2003;31:158.
7. Bartholomew A, Sturgeon C, Siatskas M, Ferrer K, McIntosh K, Patil S, Hardy W, Devine S, Ucker D, Deans R, Moseley A, Hoffman R. Mesenchymal stem cells suppress lymphocyte proliferation in vitro and prolong skin graft survival in vivo. Exp Hematol. 2002;30:42–8.
8. Frassoni FLM, Bacigalupo A, Gluckman E. Expanded mesenchymal stem cells (MSC), coinfused with HLA identical hemopoietic stem cell transplants, reduce acute and chronic graft versus host disease: a matched pair analysis. Bone Marrow Transplant. 2002;29:75.
9. Lazarus H, Curtin P, Devine S. Role of mesenchymal stem cells in allogeneic transplantation: early phase 1 clinical results. Blood. 2000;96:392.
10. Fridenstein A. Stromal bone marrow cells and the hematopoietic microenvironment. Arkh Patol. 1982;44:3–11.
11. Pittenger MF, Mackay AM, Beck SC, Jaiswal RK, Douglas R, Mosca JD, Moorman MA, Simonetti DW, Craig S, Marshak DR. Multi-lineage potential of adult human mesenchymal stem cells. Science. 1999;284:143–7.
12. Deans RJ, Moseley AB. Mesenchymal stem cells: biology and potential clinical uses. Exp Hematol. 2000;28:875–84.
13. Deans RJ. Mesenchymal stem cells: cell and gene therapy applications. Eur Cytokine Netw. 2000;11:323–4.
14. Asakura A, Komaki M, Rudnicki M. Muscle satellite cells are multipotential stem cells that exhibit myogenic, osteogenic, and adipogenic differentiation. Differentiation. 2001;68: 245–53.
15. Hu Y, Liao L, Wang Q, Ma L, Ma G, Jiang X, Zhao RC. Isolation and identification of mesenchymal stem cells from human fetal pancreas. J Lab Clin Med. 2003;141:342–9.
16. Young HE, Steele TA, Bray RA, Hudson J, Floyd JA, Hawkins K, Thomas K, Austin T, Edwards C, Cuzzourt J, Duenzl M, Lucas PA, Black Jr AC. Human reserve pluripotent mesenchymal stem cells are present in the connective tissues of skeletal muscle and dermis derived from fetal, adult, and geriatric donors. Anat Rec. 2001;264:51–62.
17. Zuk PA, Zhu M, Mizuno H, Huang J, Futrell JW, Katz AJ, Benhaim P, Lorenz HP, Hedrick MH. Multilineage cells from human adipose tissue: implications for cell-based therapies. Tissue Eng. 2001;7:211–28.
18. Noort WA, Kruisselbrink AB, in't Anker PS, Kruger M, van Bezooijen RL, de Paus RA, Heemskerk MH, Löwik CW, Falkenburg JH, Willemze R, Fibbe WE. Mesenchymal stem cells promote engraftment of human umbilical cord blood-derived CD34+ cells in NOD/SCID mice. Exp Hematol. 2002;30:870–8.
19. Hu Y, Zhang LY, Ma GJ, Jiang XY, Zhao CH. Phenotypical and biological characteristics of human fetal marrow and liver mesenchymal stem cells. Zhongguo Shi Yan Xue Ye Xue Za Zhi. 2001;9:289–93.
20. Erices A, Conget P, Minguell JJ. Mesenchymal progenitor cells in human umbilical cord blood. Br J Haematol. 2000;109:235–42.
21. Simons PG, Torok-Storb B. Identification of stromal cell precursors in human bone marrow by a novel monoclonal antibody STRO-1. Blood. 1991;78:55–62.
22. Sayegh MH, Turka LA. The role of T-cell costimulatory activation pathways in transplant rejection. N Engl J Med. 1998;338:1813–21.

23. von Boehmer H, Kisielow P. Self-nonself discrimination by T cells. Science. 1990;248:1369–73.
24. Anderson G, Jenkinson EJ, Moore NC, Owen JJ. MHC class II-positive epithelium and mesenchyme cells are both required for T-cell development in the thymus. Nature. 1993;362:70–3.
25. Van Parijs L, Abbas AK. Homeostasis and self-tolerance in the immune system: turning lymphocytes off. Science. 1998;280:243–8.
26. Charlton B, Auchincloss Jr H, Fathman CG. Mechanisms of transplantation tolerance. Annu Rev Immunol. 1994;12:707–34.
27. Nickerson P, Steurer W, Steiger J, Zheng X, Steele AW, Strom TB. Cytokines and the Th1/Th2 paradigm in transplantation. Curr Opin Immunol. 1994;6:757–64.
28. Li XC, Strom TB, Turka LA, Wells AD. T cell death and transplantation tolerance. Immunity. 2001;14:407–16.
29. Di Nicola M, Carlo-Stella C, Magni M, Milanesi M, Longoni PD, Matteucci P. Human bone marrow stromal cells suppress T-lymphocyte proliferation induced by cellular or nonspecific mitogenic stimuli. Blood. 2002;99:3838–43.
30. Chen JL, Guo ZK, Xu C, Li YH, Hou CM, Mao N, Chen H. Mesenchymal stem cells suppress allogeneic T cell responses by secretion of TGF-beta1. Zhongguo Shi Yan Xue Ye Xue Za Zhi. 2002;10:285–8.
31. Tse WT, Pendleton JD, Beyer WM, Egalka MC, Guinan EC. Suppression of allogeneic T-cell proliferation by human marrow stromal cells: implications in transplantation. Transplantation. 2003;75:389–97.
32. Le Blanc K, Tammik L, Sundberg B, Haynesworth SE, Ringden O. Mesenchymal stem cells inhibit and stimulate mixed lymphocyte cultures and mitogenic responses independently of the major histocompatibility complex. Scand J Immunol. 2003;57:11–20.
33. He J, Zhang Y, Jiang XX, Liu G, Liu YL, Li HL, Mao N. Effect of human bone marrow mesenchymal stem cell on cord blood T lymphocyte transformation. Zhongguo Shi Yan Xue Ye Xue Za Zhi. 2003;11:11–4.
34. Krampera M, Glennie S, Dyson J, Scott D, Laylor R, Simpson E, Dazzi F. Bone marrow mesenchymal stem cells inhibit the response of naive and memory antigen-specific T cells to their cognate peptide. Blood. 2003;101:3722–9.
35. Meisel R, Zibert A, Laryea M, Göbel U, Däubener W, Dilloo D. Human bone marrow stromal cells inhibit allogeneic T-cell responses by indoleamine 2,3-dioxygenase-mediated tryptophan degradation. Blood. 2004;103:4619–21.
36. Yang RY, Rabinovich GA, Liu FT. Galectins: structure, function and therapeutic potential. Expert Rev Mol Med. 2008;10:e17.
37. Blaser C, Kaufmann M, Müller C, Wells V, Mallucci L, Pircher H. Beta galactoside-binding protein secreted by activated T cells inhibits antigen-induced proliferation of T cells. Eur J Immunol. 1998;28:2311–9.
38. Rabinovich GA, Ramhorst RE, Rubinstein N, Corigliano A, Daroqui MC, Kier-Joffé EB, Fainboim L. Induction of allogenic T-cell hyporesponsiveness by galectin-1-mediated apoptotic and non-apoptotic mechanisms. Cell Death Differ. 2002;9:661–70.
39. Perillo NL, Pace KE, Seilhamer JJ, Baum LG. Apoptosis of T cells mediated by galectin-1. Nature. 1995;378:736–9.
40. Gieseke F, Böhringer J, Bussolari R, Dominici M, Handgretinger R, Müller I. Human multipotent mesenchymal stromal cells use galectin-1 to inhibit immune effector cells. Blood. 2010;116:3770–9.
41. Glennie S, Soeiro I, Dyson PJ, Lam EW, Dazzi F. Bone marrow mesenchymal stem cells induce division arrest anergy of activated T cells. Blood. 2005;105:2821–7.
42. Maccario R, Podesta M, Moretta A, Cometa A, Comoli P, Montagna D, Daudt L, Ibatici A, Piaggio G, Pozzi S, Frassoni F, Locatelli F. Interaction of human mesenchymal stem cells with cells involved in alloantigen-specific immune response favors the differentiation of CD4+ T-cell subsets expressing a regulatory/suppressive phenotype. Haematologica. 2005;90:516–25.

43. Tasso R, Ilengo C, Quarto R, Cancedda R, Caspi RR, Pennesi G. Mesenchymal stem cells induce functionally active T-regulatory lymphocytes in a paracrine fashion and ameliorate experimental autoimmune uveitis. Invest Ophthalmol Vis Sci. 2012;53:786–93.
44. Carrion F, Nova E, Ruiz C, Diaz F, Inostroza C, Rojo D, Mönckeberg G, Figueroa FE. Autologous mesenchymal stem cell treatment increased T regulatory cells with no effect on disease activity in two systemic lupus erythematosus patients. Lupus. 2010;19:317–22.
45. Beyth S, Borovsky Z, Mevorach D, Liebergall M, Gazit Z, Aslan H, Galun E, Rachmilewitz J. Human mesenchymal stem cells alter antigen-presenting cell maturation and induce T-cell unresponsiveness. Blood. 2005;105:2214–9.
46. Jiang XX, Zhang Y, Liu B, Zhang SX, Wu Y, Yu XD, Mao N. Human mesenchymal stem cells inhibit differentiation and function of monocyte-derived dendritic cells. Blood. 2005;105:4120–6.
47. Nauta AJ, Kruisselbrink AB, Lurvink E, Willemze R, Fibbe WE. Mesenchymal stem cells inhibit generation and function of both CD34+ –derived and monocyte-derived dendritic cells. J Immunol. 2006;177:2080–7.
48. Zhang W, Ge W, Li C, You S, Liao L, Han Q, Deng W, Zhao RC. Effects of mesenchymal stem cells on differentiation, maturation, and function of human monocyte-derived dendritic cells. Stem Cells Dev. 2004;13:263–71.
49. Augello A, Tasso R, Negrini SM, Amateis A, Indiveri F, Cancedda R, Pennesi G. Bone marrow mesenchymal progenitor cells inhibit lymphocyte proliferation by activation of the programmed death 1 pathway. Eur J Immunol. 2005;35:1482–90.
50. Corcione A, Benvenuto F, Ferretti E, Giunti D, Cappiello V, Cazzanti F, Risso M, Gualandi F, Mancardi GL, Pistoia V, Uccelli A. Human mesenchymal stem cells modulate B-cell functions. Blood. 2006;107:367–72.
51. Deng W, Han Q, Liao L, You S, Deng H, Zhao RC. Effects of allogeneic bone marrow-derived mesenchymal stem cells on T and B lymphocytes from BXSB mice. DNA Cell Biol. 2005;24:458–63.
52. Krampera M, Cosmi L, Angeli R, Pasini A, Liotta F, Andreini A, Santarlasci V, Mazzinghi B, Pizzolo G, Vinante F, Romagnani P, Maggi E, Romagnani S, Annunziato F. Role for interferon-gamma in the immunomodulatory activity of human bone marrow mesenchymal stem cells. Stem Cells. 2006;24:386–98.
53. Sotiropoulou PA, Perez SA, Gritzapis AD, Baxevanis CN, Papamichail M. Interactions between human mesenchymal stem cells and natural killer cells. Stem Cells. 2006;24:74–85.
54. Spaggiari GM, Capobianco A, Abdelrazik H, Becchetti F, Mingari MC, Moretta L. Mesenchymal stem cells inhibit natural killer-cell proliferation, cytotoxicity, and cytokine production: role of indoleamine 2,3-dioxygenase and prostaglandin E2. Blood. 2008;111:1327–33.
55. Selmani Z, Naji A, Zidi I, Favier B, Gaiffe E, Obert L, Borg C, Saas P, Tiberghien P, Rouas-Freiss N, Carosella ED, Deschaseaux F. Human leukocyte antigen-G5 secretion by human mesenchymal stem cells is required to suppress T lymphocyte and natural killer function and to induce CD4+ CD25highFOXP3+ regulatory T cells. Stem Cells. 2008;26:212–22.
56. Devine SM, Cobbs C, Jennings M, Bartholomew A, Hoffman R. Mesenchymal stem cells distribute to a wide range of tissues following systemic infusion into nonhuman primates. Blood. 2003;101:2999–3001.
57. Deng W, Han Q, Liao L. Long-term distribution of adult stem cells in different tissues after transplantation. Stem Cell Cell Therapy. 2003;1:62–5.
58. Azizi SA, Stokes D, Augelli BJ, DiGirolamo C, Prockop DJ. Engraftment and migration of human bone marrow stromal cells implanted in the brains of albino rats – similarities to astrocyte grafts. Proc Natl Acad Sci USA. 1998;95:3908–14.
59. Toma C, Pittenger MF, Cahill KS, Byrne BJ, Kessler PD. Human mesenchymal stem cells differentiate to a cardiomyocyte phenotype in the adult murine heart. Circulation. 2002;105:93–8.
60. Saito T, Kuang JQ, Bittira B, Al-Khaldi A, Chiu RC. Xenotransplant cardiac chimera: immune tolerance of adult stem cells. Ann Thorac Surg. 2002;74:19–24.

61. Liechty KW, MacKenzie TC, Shaaban AF, Radu A, Moseley AM, Deans R, Marshak DR, Flake AW. Human mesenchymal stem cells engraft and demonstrate site-specific differentiation after in utero transplantation in sheep. Nat Med. 2000;6:1282–6.
62. Silverstein AM, Prendergast RA, Kraner KL. Fetal response to antigenic stimulus IV. Rejection of skin homografts by the fetal lamb. J Exp Med. 1964;119:955–64.
63. Zanjani E, Almeida-Porada G, Ascensao J, Mackintosh F, Flake A. Transplantation of hematopoietic stem cells in utero. Stem Cells. 1997;15:79–93.
64. Li Y, Hisha H, Inaba M, Lian Z, Yu C, Kawamura M, Yamamoto Y, Nishio N, Toki J, Fan H, Ikehara S. Evidence for migration of donor bone marrow stromal cells into recipient thymus after bone marrow transplantation plus bone grafts: a role of stromal cells in positive selection. Exp Hematol. 2000;28:950–60.
65. BardaSaad M, Rozenszajn LA, Ashush H, Shav-Tal Y, Nun AB, Zipori D. Adhesion molecules involved in the interactions between early T cells and mesenchymal bone marrow stromal cells. Exp Hematol. 1999;27:834–44.
66. Ruegemer JJ, Ho SN, Augustine JA, Schlager JW, Bell MP, McKean DJ, Abraham RT. Regulatory effects of transforming growth factor-β on IL-2- and IL-4-dependent T cell cycle progression. J Immunol. 1990;144:1769–76.
67. Ahuja SS, Paliogianni F, Yarnada H, Balow JE, Boumpas DT. Effect of transforming growth factor-β on early and late activation events in human T cells. J Immunol. 1993;150:3109.
68. Fox FE, Ford HC, Douglas R, Cherian S, Nowell PC. Evidence that TGF-β can inhibit T-lymphocyte proliferation through paracrine and autocrine mechanisms. Cell Immunol. 1993;150:45–58.
69. Mossalayi MD, Mentz F, Ouaaz F, Dalloul AH, Blanc C, Debre P, Ruscetti FW. Early human thymocyte proliferation is regulated by an externally controlled autocrine transforming growth factor-β1 mechanism. Blood. 1995;85:3594–601.
70. Ishida T, Inaba M, Hisha H, Sugiura K, Adachi Y, Nagata N, Ogawa R, Good RA, Ikehara S. Requirement of donor-derived stromal cells in the bone marrow for successful allogeneic bone marrow transplantation. Complete prevention of recurrence of autoimmune diseases in MRL/MP-Ipr/Ipr mice by transplantation of bone marrow plus bones (stromal cells) from the same donor. J Immunol. 1994;152:3119–27.
71. Deeg HJ. How I treat refractory acute GVHD. Blood. 2007;109:4119–26.
72. Ringdén O, Uzunel M, Rasmusson I, Remberger M, Sundberg B, Lönnies H, Marschall HU, Dlugosz A, Szakos A, Hassan Z, Omazic B, Aschan J, Barkholt L, Le Blanc K. Mesenchymal stem cells for treatment of therapy-resistant graft-versus-host disease. Transplantation. 2006;81:1390–7.
73. Fang B, Song Y, Liao L, Zhang Y, Zhao RC. Favorable response to human adipose tissue-derived mesenchymal stem cells in steroid-refractory acute graft-versus-host disease. Transplant Proc. 2007;39:3358–62.
74. Fang B, Song YP, Liao LM, Han Q, Zhao RC. Treatment of severe therapy-resistant acute graft-versus-host disease with human adipose tissue-derived mesenchymal stem cells. Bone Marrow Transplant. 2006;38:389–90.
75. Le Blanc K, Frassoni F, Ball L, Locatelli F, Roelofs H, Lewis I, Lanino E, Sundberg B, Bernardo ME, Remberger M, Dini G, Egeler RM, Bacigalupo A, Fibbe W, Ringdén O. Developmental Committee of the European Group for Blood and Marrow Transplantation. Mesenchymal stem cells for treatment of steroid-resistant, severe, acute graft-versus-host disease: a phase II study. Lancet. 2008;371:1579–86.
76. Weng JY, Du X, Geng SX, Peng YW, Wang Z, Lu ZS, Wu SJ, Luo CW, Guo R, Ling W, Deng CX, Liao PJ, Xiang AP. Mesenchymal stem cell as salvage treatment for refractory chronic GVHD. Bone Marrow Transplant. 2010;45:1732–40.
77. Casiraghi F, Azzollini N, Cassis P, Imberti B, Morigi M, Cugini D, Cavinato RA, Todeschini M, Solini S, Sonzogni A, Perico N, Remuzzi G, Noris M. Pretransplant infusion of mesenchymal stem cells prolongs the survival of a semiallogeneic heart transplant through the generation of regulatory T cells. J Immunol. 2008;181:3933–46.

78. Ding Y, Xu D, Feng G, Bushell A, Muschel RJ, Wood KJ. Mesenchymal stem cells prevent the rejection of fully allogenic islet grafts by the immunosuppressive activity of matrix metalloproteinase-2 and -9. Diabetes. 2009;58:1797–806.
79. Tan J, Wu W, Xu X, Liao L, Zheng F, Messinger S, Sun X, Chen J, Yang S, Cai J, Gao X, Pileggi A, Ricordi C. Induction therapy with autologous mesenchymal stem cells in living-related kidney transplants: a randomized controlled trial. JAMA. 2012;307:1169–77.
80. Peng Y, Ke M, Xu L, Liu L, Chen X, Xia W, Li X, Chen Z, Ma J, Liao D, Li G, Fang J, Pan G, Xiang AP. Donor-derived mesenchymal stem cells combined with low-dose tacrolimus prevent acute rejection after renal transplantation: a clinical pilot study. Transplantation. 2013;95:161–8.
81. Parekkadan B, Tilles AW, Yarmush ML. Bone marrow-derived mesenchymal stem cells ameliorate autoimmune enteropathy independently of regulatory T cells. Stem Cells. 2008;26:1913–9.
82. Zappia E, Casazza S, Pedemonte E, Benvenuto F, Bonanni I, Gerdoni E, Giunti D, Ceravolo A, Cazzanti F, Frassoni F, Mancardi G, Uccelli A. Mesenchymal stem cells ameliorate experimental autoimmune encephalomyelitis inducing T-cell anergy. Blood. 2005;106:1755–61.
83. Fiorina P, Jurewicz M, Augello A, Vergani A, Dada S, La Rosa S, Selig M, Godwin J, Law K, Placidi C, Smith RN, Capella C, Rodig S, Adra CN, Atkinson M, Sayegh MH, Abdi R. Immunomodulatory function of bone marrow-derived mesenchymal stem cells in experimental autoimmune type 1 diabetes. J Immunol. 2009;183:993–1004.
84. Sun L, Akiyama K, Zhang H, Yamaza T, Hou Y, Zhao S, Xu T, Le A, Shi S. Mesenchymal stem cell transplantation reverses multiorgan dysfunction in systemic lupus erythematosus mice and humans. Stem Cells. 2009;27:1421–32.
85. Liang J, Zhang H, Hua B, Wang H, Lu L, Shi S, Hou Y, Zeng X, Gilkeson GS, Sun L. Allogenic mesenchymal stem cells transplantation in refractory systemic lupus erythematosus: a pilot clinical study. Ann Rheum Dis. 2010;69:1423–9.
86. Bocelli-Tyndall C, Bracci L, Spagnoli G, Braccini A, Bouchenaki M, Ceredig R, Pistoia V, Martin I, Tyndall A. Bone marrow mesenchymal stromal cells (BM-MSCs) from healthy donors and auto-immune disease patients reduce the proliferation of autologous and allogeneic-stimulated lymphocytes in vitro. Rheumatology. 2007;46:403–8.
87. Sun LY, Zhang HY, Feng XB, Hou YY, Lu LW, Fan LM. Abnormality of bone marrow-derived mesenchymal stem cells in patients with systemic lupus erythematosus. Lupus. 2007;16:121–8.
88. Duijvestein M, Vos AC, Roelofs H, Wildenberg ME, Wendrich BB, Verspaget HW, Kooy-Winkelaar EM, Koning F, Zwaginga JJ, Fidder HH, Verhaar AP, Fibbe WE, van den Brink GR, Hommes DW. Autologous bone marrow-derived mesenchymal stromal cell treatment for refractory luminal Crohn's disease: results of a phase I study. Gut. 2010;59:1662–9.
89. Ciccocioppo R, Bernardo ME, Sgarella A, Maccario R, Avanzini MA, Ubezio C, Minelli A, Alvisi C, Vanoli A, Calliada F, Dionigi P, Perotti C, Locatelli F, Corazza GR. Autologous bone marrow-derived mesenchymal stromal cells in the treatment of fistulising Crohn's disease. Gut. 2011;60:788–98.

Mesenchymal Stem Cell Homing to Injured Tissues

Yaojiong Wu and Robert Chunhua Zhao

Abstract A growing body of preclinical evidence suggests that mesenchymal stem cells (MSCs) are effective for the structural and functional recovery of many damaged organs. Accordingly, a large number of clinical trials have been underway to determine the benefit of MSC-based therapies. While systemic infusion is a minimal invasive administration route of MSCs and has been used extensive in clinical studies, culture expanded MSCs appear to have significantly impaired homing capacity, resulting in low levels of engraftment to injured tissues. Meanwhile, the therapeutic effect of MSCs in tissue repair and regeneration is likely to correlate to the number of MSCs that have engrafted into the tissue. Considerable progresses have been made in the past in understanding the molecular mechanisms of the trafficking, migration and engraftment of MSCs. In consideration of the profound therapeutic potential in tissue repair/regeneration that MSCs have displayed after direct intra-organ delivery, improving the homing ability of cultured expanded MSCs will certainly enhance their therapeutic efficacy after systemic infusion.

Keywords Mesenchymal stem cells • Homing • Engraftment • Chemokines • Tissue repair and regeneration

Y. Wu, M.D., Ph.D. (✉)
Life Science Division, Tsinghua University Graduate School at Shenzhen, L406A, Tsinghua Campus, The University Town, Shenzhen, China
e-mail: wu.yaojiong@sz.tsinghua.edu.cn

R.C. Zhao
Center of Excellence in Tissue Engineering, Institute of Basic Medical Sciences and School of Basic Medicine, Chinese Academy of Medical Sciences and Peking Union Medical College, 5# Dongdansantiao, 100005 Beijing, China, People's Republic
e-mail: zhaoch16@hotmail.com

R.C. Zhao (ed.), *Essentials of Mesenchymal Stem Cell Biology and Its Clinical Translation*, DOI 10.1007/978-94-007-6716-4_5,

Introduction

Stem cell homing is a phenomenon that was initially related to hematopoietic stem cells (HSCs). Accordingly, HSCs are considered to have the capacity to migrate through the bloodstream to different organs and return to their niches in the BM under the guidance of chemical signaling navigation [1]. Bone marrow-derived mesenchymal stem cells (BM-MSCs), which represent ~0.001–0.01 % of all nucleated BM cells and are ~10-fold less abundant than HSCs, have been assumed to exhibit a similar homing property. The homing capacity of bone marrow stem cells facilitates them to migrate and engraft into injured tissues. Correspondingly, several clinical trials are underway that administer BM-MSCs via an intravenous delivery route, thereby eliminating the potential damages that could otherwise cause from the use of intra-organ delivery routes [2, 3]. However, recent studies have indicated that expansion of BM-MSCs *ex vivo* can severely impair their homing capacity [4], resulting in low levels engraftment into target tissues [5–7]. Accordingly, there are concerns over the influence of the poor engraftment on the outcomes of BM-MSC-based therapies.

In this chapter, we provide an overview of the biology of MSC homing, particularly the mediators for MSC trafficking, migration and engraftment. We will also discuss influences of culture expansion on their homing capacity to target tissues and potentially the therapeutic effects.

Homing Capacity of MSCs

Several lines of evidence indicate the homing capacity of MSCs. For example, 24 h after transplantation of uncultured enhanced green fluorescence protein (EGFP)-transgenic BM-MSCs into sublethally irradiated mice, as many as 55–65 % of the EGFP cells were recovered from the BM, and 4–7 % were recovered from the spleen [4]. This high seeding efficiency compares favorably with that of HSCs, where ~25 % of HSCs homed to the BM [8]. These results suggest that primitive BM-MSCs exhibit profound capacity for homing. When BM-MSCs are systemically administered after stroke, the cells home and migrate toward the brain and are shown to acquire a neuronal phenotype with expression of nerve cell specific markers NeuN and MAP-2 and GFAP, and improve functional outcome of the rats [9]. In mice receiving bleomycin, an agent causing lung inflammation and subsequent fibrosis, BM-MSCs are found to home to the lungs and reduce inflammation and fibrosis [10]. Systemic infusion of ex vivo expanded BM-MSCs in baboons after lethal total body irradiation results in long term engraftment of the cells in multiple organs including the skin, intestines, liver, lungs, thymus and kidney [11]. In an ischemic stroke model, human MSCs were found in the injured brain following both ipsilateral and contralateral injections [12], implying profound migration capacity of the cells. These results indicate that BM-MSC homing and engraftment to injured tissues are independent of causes of injuries and tissue types.

The homing capacity of BM-MSCs is severely impaired after culture expansion [13]. When BM-MSCs were subjected to 24 h of *ex vivo* culturing, the seeding fraction of BM-MSCs detected in the BM decreased dramatically to 10 % compared to non-cultured BM-MSCs. Correspondingly, EGFP-BM-MSCs cultured for 48 h were not detected in lymphohematopoietic organs following systemic infusion [4]. In consistence, we found that ~1–2 % of systemically infused allogeneic BM-MSCs that underwent five rounds of passaging *ex vivo* were detected in the infarcted heart 72 h after injection in mice. This percentage is approximately fivefold less than the number of expanded but non-passaged CD34(+) endothelial progenitor cells in parallel experiments (unpublished data). These results are consistent with several studies which also detected low rates of engraftment into the ischemic myocardium by cultured BM-MSCs [14].

It appears that impaired trafficking by BM-MSCs alters their fates after infusion. For example, a recent study demonstrated that up to 70 % of allogeneic BM-MSCs expanded *ex vivo* were trapped in the lungs following intravenous injection in rats, and only 6.3 % of the cells were detected in the ischemic heart, which was slightly more than the cells in the non-ischemic heart (4.6 %) [5]. This finding is consistent with previous studies that also detected a considerable entrapment of infused BM-MSCs in the lungs [6, 7]. Delivery by left ventricular cavity infusion thus bypassing the first pass through lungs resulted in drastically lower lung uptake and specifically higher uptake in infarcted myocardium compared [6]. However, several studies suggest that intra-artery delivered MSCs entrap at the precapillary level probably because of their large size and form microemboli [15, 16]. Moreover, a study showed that intracoronary injection of BM-MSCs caused myocardial infarction in a dog model, as indicated by ECG changes, increased troponin I levels and histological data [17]. Although preliminary clinical studies suggest that intracoronary infusion of MSCs is safe and could be of benefit to patients [18], microemboli formed by MSCs as shown in the above studies deserve more vigorous investigations to verify its significance. In addition to enlargement in size, BM-MSCs undergo dynamic changes in their molecular phenotype, though they may maintain stable expression of representative cell surface markers such as CD73, CD90 and CD105 during *ex vivo* expansion. Indeed, a previous study showed that human BM-MSCs in late passages lost their surface expression of chemokine receptors CCR1, CCR7, CCR9, CXCR4, CXCR5, and CXCR6 and chemotactic responses to corresponding ligands [19]. Therefore, it is necessary to optimize culture conditions for BM-MSCs to sustain their expression of receptors for homing.

Homing Ability of MSCs and Their Therapeutic Effects

Numerous animal studies have shown that direct delivery of MSCs to injured tissues can significantly promote their structural and functional recovery [20–22]. Catheter delivery of allogeneic BM-MSCs to infarcted myocardium resulted in profound reduction in scar formation and improvement in cardiac function in pigs

[20]. Endocardial injection of BM-MSCs in pigs also showed cardioprotective effects in chronic ischemic myocardium with significantly decreased fibrosis and improved left ventricular function [23]. Moreover, direct application of a monolayer sheet of MSCs onto the scarred myocardium resulted in reversed wall thinning in the scar area and improved cardiac function in rats [24]. However, comparative studies to evaluate the impact of cell delivery routes on cell engraftment rates and therapeutic effects are limited [3]. In a few recent studies, three routes (intramyocardial, intracoronary and intravenous) for BM-MSC administration to treat myocardial infarction have been compared. A direct intramyocardial injection of BM-MSCs resulted in the highest rates of BM-MSC engraftment and functional improvement of the left ventricle on animal MI models [16, 25]. However, intracoronary infusion of BM-MSCs caused a reduction in coronary blood flow and microinfarction. Thirdly, intravenous infusion resulted in a much lower engraftment rate for BM-MSCs to the ischemic myocardium than either of the two previously described approaches [14]. Similarly, in a recent study, intra-carotid arterial administration of human MSCs lead to significantly higher engraftment to the injured brain in rats compared to intravenous transplantation [26, 27]. Whether this administration route will result in an enhanced therapeutic effect of MSCs on brain injuries awaits further investigations. Collectively, whether low engraftment of systemically infused MSCs to target organ(s) ultimately compromise their therapeutic benefit needs to be clarified.

Mechanisms of Leukocyte Trafficking into Sites of Inflammation

Homing and engraftment is a prerequisite for therapeutic cells to play their roles in the target tissue particularly when cells are infused systemically [28]. The molecular mechanisms of homing of BM-MSCs have not been fully understood, but a preliminary study suggests that BM-MSCs exhibit certain activities such as rolling similar to leukocyte homing to inflammatory sites [29]. Therefore, knowledge that has been learnt in leukocyte trafficking, adhesion and migration may help study MSC homing. During inflammation, the recruitment of leukocytes to the inflammatory site requires a coordinated sequence of molecular and cellular events. The initial activity of leukocytes homing involves the rolling of the cells on the endothelium mediated by selectins, which are expressed in inflamed venules [30–33]. This process allows leukocytes to become activated by cytokines and express high affinity intergrins on the surface including CD18 and CD49d which lead to the subsequent step of leukocyte arrest/firm adhesion on endothelial cells [34]. The transendothelial migration of leukocytes involves signals derived from clustering of apically disposed intercellular adhesion molecule-1 and vascular cell adhesion molecule-1, disassociation of endothelial junctions, and targeted recycling of platelet/endothelial cell adhesion molecule from the lateral border recycling compartment of the endothelial cells [35].

Finally, the migration/invasion of leucocytes in the ECM are mediated by integrin-ECM molecules interactions and matrix degradation [36].

Potential Ligands/Receptors for MSC Homing

Molecular mediators involved with BM-MSC migration and engraftment into injured tissues have not been fully understood. Based on the hypothesis that specific chemoattractant molecules and adhesion molecules are present in the tissue following injuries, and are up-regulated to induce the migration and engraftment of BM-MSCs, we developed a functional genomics strategy to identify mediators of BM-MSCs in the ischemic myocardium in mice [37, 38]. Genes that were significantly up-regulated following MI included CC ligands 2, 6, 7 and 9, CXC ligands 1, 2, and 12, cytokines including interleukin (IL)-1 and -6, transforming growth factor (TGF) β1 and β2, tumor necrosis factor receptor II (TNFRII), and cell adhesion molecules including fibronectin-1, laminin-1, intercellular adhesion molecule (ICAM)-1, vascular cell adhesion molecule (VCAM)-1, E-selectin, thrombospondin 1, and tenascin C [37, 38]. Chemokines up-regulated by ischemia has been carefully reviewed by Frangogiannis [39]. The expression of nine corresponding receptors including CCR2, CXCR4, IL6RA, E-selectin ligand, CD29, CD49d, CD49f, integrin α8, and integrin α9 were detected in early passage murine BM-MSCs [37, 38]. Recently, several additional ligand/receptor pairs have been suggested to be potentially involved in the homing of BM-MSCs to injured tissues, which include hyaluronic acid/CD44 [40], N-formylated peptides/N-formyl peptide receptor (FPR) and the formyl peptide receptor-like-1 (FPRL1) [41], platelet-derived growth factor-AB (PDGF-AB)/PDGF-receptor alpha and beta, and insulin-like growth factor 1 (IGF-1)/IGF receptor [42]. Further studies will determine their functional roles in BM-MSC homing.

It remains unclear whether MSCs utilize the same molecules for transendothelial migration as leukocytes do. Several molecules such as CD18 and CD49d which are critical for the trafficking and migration of leukocytes are either not expressed, or are expressed at very low levels on the surface of MSCs following *ex vivo* culturing [38, 43]. For leukocyte rolling along endothelial cells, the interactions of selectins with their corresponding ligands are considered essential [33]. Correspondingly, MSCs derived from BM and placenta were observed to roll and adhere to postcapillary venules of a mouse model in a P-selectin-dependent manner [29]. However, it is controversial over the capacity for MSCs to express functional selectin ligands particularly after culture [44]. Nevertheless, culture expanded BM-MSCs have recently been shown to exhibit increased rolling activity on activated endothelia and homing capacity to inflamed tissue after surface modification with a nanometer-scale polymer construct containing sialyl Lewisx (sLex) [45]. Similarly, overexpression of CD49d in MSCs was shown to increase their homing to the bone marrow [46]. These studies suggest that adhesion molecules affect the homing capacity of MSCs.

Chemokines for MSC Homing

Currently, over 50 chemokines and 20 chemokine receptors have been discovered [47]. While most chemokines are secreted and are 67–127 amino acids in length, some chemokines are membrane-bound (i.e., CXCL16 and CX3CL1). Chemokines are divided into four groups depending on the spacing of their first two cysteine residues near the NH2-terminus: CXC chemokines, CC chemokines, C chemokines and CX3C chemokine [48].

Chemokine induction is one of the prominent features in ischemic myocardium. Diverse chemokine/chemokine receptor axes participate in the regulation of chemotactic activities for various cell types besides leukocytes. Following acute myocardial ischemia, several chemokines are up-regulated immediately [37–39]. However, the up-regulation is transient for many cytokines, and expression levels usually decline markedly within the first 72 h following acute MI [37, 38]. This chemokine expression pattern is consistent with the timing course of BM-MSC homing, and the freshly infarcted myocardium appears to be more chemoattractive to BM-MSCs [49]. The expression of chemokine receptors in BM-MSCs and their corresponding ligands are have been well documented previously [50].

CCR1 and CCR2 in MSC Homing

A functional role for chemokines in the recruitment of BM-MSCs into the infarcted myocardium tissue has been implicated in several recent studies. CCL7, also known as monocyte-specific chemokine 3 (MCP3), has previously been shown to specifically attract monocytes and to regulate macrophage functions [51]. Recently the chemokine has been found to be crucially involved in BM-MSC homing to the ischemic myocardium and in their intramyocardial migration and survival. Over expression of CCL7 at the site of old myocardial infarct recruited systemically administered BM-MSCs and improved cardiac structure and function in rats [49]. Consistently, increased expression of CCR1 (a receptor of CCL7) by murine BM-MSCs has been shown to dramatically increase BM-MSC chemotactic migration and to increase BM-MSCs survival following intramyocardial delivery in mice [52]. As a result, the engraftment of BM-MSCs into the ischemic myocardium increased which was associated with an improved therapeutic effect to the infarcted heart.

In addition to CCR1, CCR2 is also a significant factor in the homing of BM-MSCs. In a recent study, MCP-1 (also known as CCL2, a ligand of CCR2) in the myocardium recruited systemically administered murine BM-derived multipotent adult mesenchymal stem cells (MASC) [53]. MASCs used in the study represented a CD34-/Sca-1^{high} subpopulation of BM cells and showed a high potential for differentiation and proliferation [53, 54]. Following myocardial ischemia, the expression of MCP-1 is up-regulated immediately [37–39]. Correspondingly,

MASCs home to regions of the myocardium damaged by ischemia/reperfusion in wild type mice [53]. *Ex vivo* treatment of wild type cells with MCP-1 induces a clustering of CCR2 receptors on the cell surface and affects cell polarization. Furthermore, in these assays, the intracellular adaptor molecule, FROUNT, which interacts with CCR2, appeared to be required [53]. In combination, these results indicate that myocardial overexpression of MCP-1, but not CXCL12 alone, is associated with the recruitment of BM-MSCs to the myocardium in the absence of ischemic injuries [55], suggesting a central role for MCPs in BM-MSC homing to the ischemic myocardium.

CXCR4/SDF-1 Axis in MSC Homing

CXCL12 (SDF-1) has been shown to enhance the migration of HSCs to ischemic myocardium [55, 56]. For this, the CXCR4/SDF-1 axis is assumed to play an important role in BM-MSC homing. However, unlike HSCs, BM-MSCs express low levels of CXCR4 in minor subpopulations in early passages, and the expression levels decline progressively with successive culture expansion [57, 58]. Notably, the expression of intracellular CXCR4 appear to sustain in BM-MSCs for passages in culture [59], and short-term exposure of human Flk1(+) BM-MSCs to a cocktail of cytokines consisting of Flt-3 ligand, stem cell factor (SCF), IL-6, hepatocyte growth factor (HGF) and IL-3 significant up-regulate both cell surface and intracellular levels of CXCR4 [59]. Functionally, this phenotype of BM-MSCs is associated with an increase in migration capacity *ex vivo* in response to SDF-1, and an increase in homing efficiency to the BM of irradiated mice [59]. Based on these results, it is hypothesized that the expression levels and distribution patterns of CXCR4 are regulated by cytokines. Correspondingly, pre-conditioning BM-MSCs in hypoxia increased their expression of CXCR4 and migration to ischemic kidneys in mice [60]. Overexpression of CXCR4 in cultured BM-MSCs resulted in an increase in their recruitment to acutely infarcted myocardium in rats [61]. However, controversial studies have been reported. In an acute kidney injury mouse model, separate and simultaneous overexpression of CXCR4 and CXCR7 (another receptor of SDF-1) in BM-MSCs did not increase their presence in the injured kidney [62]. Moreover, blockade of CXCR4 did not affect the intramyocardial migration of murine BM-MSCs to ischemic areas in mice [37]. These results suggest that the SDF-1/CXCR4 axis may be largely involved in the extravasation of BM-MSCs, and the even requires the presence of other mediators. Nevertheless, preliminary data suggest that there is no correlation between serum SDF-1 levels and the number of circulating putative MSCs in patients with ischemic heart diseases. For example, in patients with MI, an increase in blood SDF-1 levels was found to coincide with elevated levels of CD34(+) cells [63], but not MSCs [64]. Therefore, although the role of the CXCR4/SDF-1 axis in HSCs has been well recognized, the role of this signaling axis in the homing of BM-MSCs requires further investigation.

Other Chemokines

Recently, several other chemokines, including fractalkine (CX3CL1) and CCL25 (TECK), have been shown to influence MSC chemotaxis. Human BM-MSCs migrated in response to CCL25 in a dose-dependent manner ex vivo [65]. CX3CR1 has been found to mediate both leukocyte migration and adhesion. Fractalkine expression is up-regulated in the brain following ischemia in rats [66]. When BM-MSCs were exposed to low levels of oxygen, their expression levels of CX3CR1 and their migration to fractalkine increased in a *ex vivo* study [67]. Accordingly, knockdown of CX3CR1 expression in human BM-MSCs significantly reduced their recruitment to ischemic brain tissue after intravenous injection in rats [66]. These results suggest that fractalkine/CX3CR1 interactions are likely to play a role in the homing of BM-MSCs to tissue injuries.

Summary and Prospective

MSCs hold great promise for improving repair/regeneration of injured or diseased organs. However, the efficacy of this treatment may largely rely on sufficient recruitment of MSCs to the target tissue. Therefore, the identification of homing mediators and the development of improved culturing systems to maintain the primitive homing capacity of MSCs are crucial and hold profound promise to improve the therapeutic effect of systemically administered MSCs. Chemokine receptors appear to be key mediators for BM-MSCs homing to ischemic tissues. But other mediators are likely involved in the complex homing process. Moreover, MSCs with efficient homing capacity may be used as a vehicle to deliver bioactive agents to remote target organs.

Disclosures The authors indicate no potential conflicts of interest.

References

1. Lapidot T, Dar A, Kollet O. How do stem cells find their way home? Blood. 2005;106:1901–10.
2. Hare JM, Traverse JH, Henry TD, Dib N, Strumpf RK, Schulman SP, Gerstenblith G, DeMaria AN, Denktas AE, Gammon RS, Hermiller Jr JB, Reisman MA, Schaer GL, Sherman W. A randomized, double-blind, placebo-controlled, dose-escalation study of intravenous adult human mesenchymal stem cells (prochymal) after acute myocardial infarction. J Am Coll Cardiol. 2009;54:2277–86.
3. Salem HK, Thiemermann C. Mesenchymal stromal cells: current understanding and clinical status. Stem Cells. 2010;28:585–96.
4. Rombouts WJ, Ploemacher RE. Primary murine MSC show highly efficient homing to the bone marrow but lose homing ability following culture. Leukemia. 2003;17:160–70.

5. Assis AC, Carvalho JL, Jacoby BA, Ferreira RL, Castanheira P, Diniz SO, Cardoso VN, Goes AM, Ferreira AJ. Time-dependent migration of systemically delivered bone marrow mesenchymal stem cells to the infarcted heart. Cell Transplant. 2010;19:219–30.
6. Barbash IM, Chouraqui P, Baron J, Feinberg MS, Etzion S, Tessone A, Miller L, Guetta E, Zipori D, Kedes LH, Kloner RA, Leor J. Systemic delivery of bone marrow-derived mesenchymal stem cells to the infarcted myocardium: feasibility, cell migration, and body distribution. Circulation. 2003;108:863–8.
7. Kraitchman DL, Tatsumi M, Gilson WD, Ishimori T, Kedziorek D, Walczak P, Segars WP, Chen HH, Fritzges D, Izbudak I, Young RG, Marcelino M, Pittenger MF, Solaiyappan M, Boston RC, Tsui BMW, Wahl RL, Bulte JWM. Dynamic imaging of allogeneic mesenchymal stem cells trafficking to myocardial infarction. Circulation. 2005;112:1451–61.
8. Vos O, Luiten F, Ploemacher RE. Lodging of CFU(S) under various circumstances in bone marrow, spleen and liver. Exp Hematol. 1980;8:860–6.
9. Chen J, Li Y, Wang L, Zhang Z, Lu D, Lu M, Chopp M. Therapeutic benefit of intravenous administration of bone marrow stromal cells after cerebral ischemia in rats. Stroke. 2001;32:1005–11.
10. Gussoni E, Soneoka Y, Strickland CD, Buzney EA, Khan MK, Flint AF, Kunkel LM, Mulligan RC. Dystrophin expression in the mdx mouse restored by stem cell transplantation. Nature. 1999;401:390–4.
11. Devine SM, Cobbs C, Jennings M, Bartholomew A, Hoffman R. Mesenchymal stem cells distribute to a wide range of tissues following systemic infusion into nonhuman primates. Blood. 2003;101:2999–3001.
12. Kim D, Chun BG, Kim YK, Lee YH, Park CS, Jeon I, Cheong C, Hwang TS, Chung H, Gwag BJ, Hong KS, Song J. In vivo tracking of human mesenchymal stem cells in experimental stroke. Cell Transplant. 2008;16:1007–12.
13. Karp JM, Leng Teo GS. Mesenchymal stem cell homing: the devil is in the details. Cell Stem Cell. 2009;4:206–16.
14. Freyman T, Polin G, Osman H, Crary J, Lu M, Cheng L, Palasis M, Wilensky RL. A quantitative, randomized study evaluating three methods of mesenchymal stem cell delivery following myocardial infarction. Eur Heart J. 2006;27:1114–22.
15. Toma C, Wagner WR, Bowry S, Schwartz A, Villanueva F. Fate of culture-expanded mesenchymal stem cells in the microvasculature: in vivo observations of cell kinetics. Circ Res. 2009;104:398–402.
16. Perin EC, Silva GV, Assad JA, Vela D, Buja LM, Sousa AL, Litovsky S, Lin J, Vaughn WK, Coulter S, Fernandes MR, Willerson JT. Comparison of intracoronary and transendocardial delivery of allogeneic mesenchymal cells in a canine model of acute myocardial infarction. J Mol Cell Cardiol. 2008;44:486–95.
17. Vulliet PR, Greeley M, Halloran SM, MacDonald KA, Kittleson MD. Intra-coronary arterial injection of mesenchymal stromal cells and microinfarction in dogs. Lancet. 2004; 363:783–4.
18. Chen SL, Fang WW, Ye F, Liu YH, Qian J, Shan SJ, Zhang JJ, Chunhua RZ, Liao LM, Lin S, Sun JP. Effect on left ventricular function of intracoronary transplantation of autologous bone marrow mesenchymal stem cell in patients with acute myocardial infarction. Am J Cardiol. 2004;94:92–5.
19. Honczarenko M, Le Y, Swierkowski M, Ghiran I, Glodek AM, Silberstein LE. Human bone marrow stromal cells express a distinct set of biologically functional chemokine receptors. Stem Cells. 2006;24:1030–41.
20. Amado LC, Saliaris AP, Schuleri KH, St. John M, Xie J-S, Cattaneo S, Durand DJ, Fitton T, Kuang JQ, Stewart G, Lehrke S, Baumgartner WW, Martin BJ, Heldman AW, Hare JM. Cardiac repair with intramyocardial injection of allogeneic mesenchymal stem cells after myocardial infarction. Proc Natl Acad Sci USA. 2005;102:11474–9.
21. Mangi AA, Noiseux N, Kong D, He H, Rezvani M, Ingwall JS, Dzau VJ. Mesenchymal stem cells modified with Akt prevent remodeling and restore performance of infarcted hearts. Nat Med. 2003;9:1195–201.

22. Charwat S, Gyongyosi M, Lang I, Graf S, Beran G, Hemetsberger R, Nyolczas N, Sochor H, Glogar D. Role of adult bone marrow stem cells in the repair of ischemic myocardium: current state of the art. Exp Hematol. 2008;36:672–80.
23. Schneider C, Krause K, Jaquet K, Geidel S, Malisius R, Boczor S, Rau T, Zienkiewicz T, Hennig D, Kuck KH. Intramyocardial transplantation of bone marrow-derived stem cells: ultrasonic strain rate imaging in a model of hibernating myocardium. J Card Fail. 2008;14:861–72.
24. Miyahara Y, Nagaya N, Kataoka M, Yanagawa B, Tanaka K, Hao H, Ishino K, Ishida H, Shimizu T, Kangawa K, Sano S, Okano T, Kitamura S, Mori H. Monolayered mesenchymal stem cells repair scarred myocardium after myocardial infarction. Nat Med. 2006;12:459–65.
25. Heldman AW, Hare JM. Cell therapy for myocardial infarction: special delivery. J Mol Cell Cardiol. 2008;44:473–6.
26. Lundberg J, Sodersten E, Sundstrom E, Le Blanc K, Andersson T, Hermanson O, Holmin S. Targeted intra-arterial transplantation of stem cells to the injured CNS is more effective than intravenous administration: engraftment is dependent on cell type and adhesion molecule expression. Cell Transplant. 2012;21:333–43.
27. Walczak P, Zhang J, Gilad AA, Kedziorek DA, Ruiz-Cabello J, Young RG, Pittenger MF, van Zijl PC, Huang J, Bulte JW. Dual-modality monitoring of targeted intraarterial delivery of mesenchymal stem cells after transient ischemia. Stroke. 2008;39:1569–74.
28. Chavakis E, Urbich C, Dimmeler S. Homing and engraftment of progenitor cells: a prerequisite for cell therapy. J Mol Cell Cardiol. 2008;45:514–22.
29. Ruster B, Gottig S, Ludwig RJ, Bistrian R, Muller S, Seifried E, Gille J, Henschler R. Mesenchymal stem cells display coordinated rolling and adhesion behavior on endothelial cells. Blood. 2006;108:3938–44.
30. McEver RP. Rolling back neutrophil adhesion. Nat Immunol. 2010;11:282–4.
31. Imhof BA, Aurrand-Lions M. Adhesion mechanisms regulating the migration of monocytes. Nat Rev Immunol. 2004;4:432–44.
32. Kamei M, Carman CV. New observations on the trafficking and diapedesis of monocytes. Curr Opin Hematol. 2010;17:43–52.
33. Katayama Y, Hidalgo A, Furie BC, Vestweber D, Furie B, Frenette PS. PSGL-1 participates in E-selectin-mediated progenitor homing to bone marrow: evidence for cooperation between E-selectin ligands and {alpha}4 integrin. Blood. 2003;102:2060–7.
34. Hynes RO. Integrins: bidirectional, allosteric signaling machines. Cell. 2002;110:673–87.
35. Muller WA. Mechanisms of transendothelial migration of leukocytes. Circ Res. 2009;105:223–30.
36. Sorokin L. The impact of the extracellular matrix on inflammation. Nat Rev Immunol. 2010;10:712–23.
37. Ip JE, Wu Y, Huang J, Zhang L, Pratt RE, Dzau VJ. Mesenchymal stem cells use integrin beta1 not CXC chemokine receptor 4 for myocardial migration and engraftment. Mol Biol Cell. 2007;18:2873–82.
38. Wu Y, Ip JE, Huang J, Zhang L, Matsushita K, Liew CC, Pratt RE, Dzau VJ. Essential role of ICAM-1/CD18 in mediating EPC recruitment, angiogenesis, and repair to the infarcted myocardium. Circ Res. 2006;99:315–22.
39. Frangogiannis NG. Chemokines in the ischemic myocardium: from inflammation to fibrosis. Inflamm Res. 2004;53:585–95.
40. Herrera MB, Bussolati B, Bruno S, Morando L, Mauriello-Romanazzi G, Sanavio F, Stamenkovic I, Biancone L, Camussi G. Exogenous mesenchymal stem cells localize to the kidney by means of CD44 following acute tubular injury. Kidney Int. 2007;72:430–41.
41. Viswanathan A, Painter RG, Lanson Jr NA, Wang G. Functional expression of N-formyl peptide receptors in human bone marrow-derived mesenchymal stem cells. Stem Cells. 2007;25:1263–9.
42. Ponte AL, Marais E, Gallay N, Langonne A, Delorme B, Herault O, Charbord P, Domenech J. The in vitro migration capacity of human bone marrow mesenchymal stem cells: comparison of chemokine and growth factor chemotactic activities. Stem Cells. 2007;25:1737–45.

43. Pittenger MF, Mackay AM, Beck SC, Jaiswal RK, Douglas R, Mosca JD, Moorman MA, Simonetti DW, Craig S, Marshak DR. Multilineage potential of adult human mesenchymal stem cells. Science. 1999;284:143–7.
44. Brooke G, Tong H, Levesque JP, Atkinson K. Molecular trafficking mechanisms of multipotent mesenchymal stem cells derived from human bone marrow and placenta. Stem Cells Dev. 2008;17:929–40.
45. Sarkar D, Spencer JA, Phillips JA, Zhao W, Schafer S, Spelke DP, Mortensen LJ, Ruiz JP, Vemula PK, Sridharan R, Kumar S, Karnik R, Lin CP, Karp JM. Engineered cell homing. Blood. 2011;118:e184–91.
46. Kumar S, Ponnazhagan S. Bone homing of mesenchymal stem cells by ectopic alpha 4 integrin expression. FASEB J. 2007;21:3917–27.
47. Bromley SK, Mempel TR, Luster AD. Orchestrating the orchestrators: chemokines in control of T cell traffic. Nat Immunol. 2008;9:970–80.
48. Lazennec G, Richmond A. Chemokines and chemokine receptors: new insights into cancer-related inflammation. Trends Mol Med. 2010;16:133–44.
49. Schenk S, Mal N, Finan A, Zhang M, Kiedrowski M, Popovic Z, McCarthy PM, Penn MS. Monocyte chemotactic protein-3 is a myocardial mesenchymal stem cell homing factor. Stem Cells. 2007;25:245–51.
50. Wu Y, Zhao RC. The role of chemokines in mesenchymal stem cell homing to myocardium. Stem Cell Rev. 2012;8:243–50.
51. McQuibban GA, Gong JH, Tam EM, McCulloch CA, Clark-Lewis I, Overall CM. Inflammation dampened by gelatinase A cleavage of monocyte chemoattractant protein-3. Science. 2000;289:1202–6.
52. Huang J, Zhang Z, Guo J, Ni A, Deb A, Zhang L, Mirotsou M, Pratt RE, Dzau VJ. Genetic modification of mesenchymal stem cells overexpressing CCR1 increases cell viability, migration, engraftment, and capillary density in the injured myocardium. Circ Res. 2010; 106:1753–62.
53. Belema-Bedada F, Uchida S, Martire A, Kostin S, Braun T. Efficient homing of multipotent adult mesenchymal stem cells depends on FROUNT-mediated clustering of CCR2. Cell Stem Cell. 2008;2:566–75.
54. Belema Bedada F, Technau A, Ebelt H, Schulze M, Braun T. Activation of myogenic differentiation pathways in adult bone marrow-derived stem cells. Mol Cell Biol. 2005;25:9509–19.
55. Abbott JD, Huang Y, Liu D, Hickey R, Krause DS, Giordano FJ. Stromal cell-derived factor-1alpha plays a critical role in stem cell recruitment to the heart after myocardial infarction but is not sufficient to induce homing in the absence of injury. Circulation. 2004;110:3300–5.
56. Askari AT, Unzek S, Popovic ZB, Goldman CK, Forudi F, Kiedrowski M, Rovner A, Ellis SG, Thomas JD, DiCorleto PE, Topol EJ, Penn MS. Effect of stromal-cell-derived factor 1 on stem-cell homing and tissue regeneration in ischaemic cardiomyopathy. Lancet. 2003;362:697–703.
57. Wynn RF, Hart CA, Corradi-Perini C, O'Neill L, Evans CA, Wraith JE, Fairbairn LJ, Bellantuono I. A small proportion of mesenchymal stem cells strongly expresses functionally active CXCR4 receptor capable of promoting migration to bone marrow. Blood. 2004;104:2643–5.
58. Sordi V, Malosio ML, Marchesi F, Mercalli A, Melzi R, Giordano T, Belmonte N, Ferrari G, Leone BE, Bertuzzi F, Zerbini G, Allavena P, Bonifacio E, Piemonti L. Bone marrow mesenchymal stem cells express a restricted set of functionally active chemokine receptors capable of promoting migration to pancreatic islets. Blood. 2005;106:419–27.
59. Shi M, Li J, Liao L, Chen B, Li B, Chen L, Jia H, Zhao RC. Regulation of CXCR4 expression in human mesenchymal stem cells by cytokine treatment: role in homing efficiency in NOD/SCID mice. Haematologica. 2007;92:897–904.
60. Liu H, Liu S, Li Y, Wang X, Xue W, Ge G, Luo X. The role of SDF-1-CXCR4/CXCR7 axis in the therapeutic effects of hypoxia-preconditioned mesenchymal stem cells for renal ischemia/reperfusion injury. PLoS One. 2012;7:e34608.
61. Cheng Z, Ou L, Zhou X, Li F, Jia X, Zhang Y, Liu X, Li Y, Ward CA, Melo LG, Kong D. Targeted migration of mesenchymal stem cells modified with CXCR4 gene to infarcted myocardium improves cardiac performance. Mol Ther. 2008;16:571–9.

62. Gheisari Y, Azadmanesh K, Ahmadbeigi N, Nassiri SM, Golestaneh AF, Naderi M, Vasei M, Arefian E, Mirab-Samiee S, Shafiee A, Soleimani M, Zeinali S. Genetic modification of mesenchymal stem cells to overexpress CXCR4 and CXCR7 does not improve the homing and therapeutic potentials of these cells in experimental acute kidney injury. Stem Cells Dev. 2012;21:2969–80.
63. Wojakowski W, Tendera M, Michalowska A, Majka M, Kucia M, Maslankiewicz K, Wyderka R, Ochala A, Ratajczak MZ. Mobilization of CD34/CXCR4+, CD34/CD117+, c-met+ stem cells, and mononuclear cells expressing early cardiac, muscle, and endothelial markers into peripheral blood in patients with acute myocardial infarction. Circulation. 2004;110:3213–20.
64. Wang Y, Johnsen HE, Mortensen S, Bindslev L, Ripa RS, Haack-Sorensen M, Jorgensen E, Fang W, Kastrup J. Changes in circulating mesenchymal stem cells, stem cell homing factor, and vascular growth factors in patients with acute ST elevation myocardial infarction treated with primary percutaneous coronary intervention. Heart. 2006;92:768–74.
65. Binger T, Stich S, Andreas K, Kaps C, Sezer O, Notter M, Sittinger M, Ringe J. Migration potential and gene expression profile of human mesenchymal stem cells induced by CCL25. Exp Cell Res. 2009;315:1468–79.
66. Zhu J, Zhou Z, Liu Y, Zheng J. Fractalkine and CX3CR1 are involved in the migration of intravenously grafted human bone marrow stromal cells toward ischemic brain lesion in rats. Brain Res. 2009;1287:173–83.
67. Hung SC, Pochampally RR, Hsu SC, Sanchez C, Chen SC, Spees J, Prockop DJ. Short-term exposure of multipotent stromal cells to low oxygen increases their expression of CX3CR1 and CXCR4 and their engraftment in vivo. PLoS One. 2007;2:e416.

Major Signaling Pathways Regulating the Proliferation and Differentiation of Mesenchymal Stem Cells

Joseph D. Lamplot, Sahitya Denduluri, Xing Liu, Jinhua Wang, Liangjun Yin, Ruidong Li, Wei Shui, Hongyu Zhang, Ning Wang, Guoxin Nan, Jovito Angeles, Lewis L. Shi, Rex C. Haydon, Hue H. Luu, Sherwin Ho, and Tong-Chuan He

Abstract Various highly regulated, complex signaling pathways govern the lineage-specific differentiation of mesenchymal stem cells (MSCs). The unique properties of MSCs have allowed us to characterize the differentiation pathways and proliferative stimuli of these lineages. While the specific signaling cascades controlling differentiation and proliferation are unique among each lineage, some pathways are critical in the differentiation of multiple lineages. Furthermore, a considerable amount of crosstalk exists between the major signaling pathways. The TGF-β

J.D. Lamplot • S. Denduluri • J. Angeles • L.L. Shi • R.C. Haydon • H.H. Luu • S. Ho
Molecular Oncology Laboratory, Department of Orthopaedic Surgery,
The University of Chicago Medical Center, 5841 South Maryland
Avenue, MC 3079, Chicago, IL 60637, USA

X. Liu • G. Nan
Molecular Oncology Laboratory, Department of Orthopaedic Surgery,
The University of Chicago Medical Center, 5841 South Maryland
Avenue, MC 3079, Chicago, IL 60637, USA

Stem Cell Biology and Therapy Laboratory of the Key Laboratory for
Pediatrics Designated by Chinese Ministry of Education, The Children's
Hospital of Chongqing Medical University, Chongqing 400014, China

J. Wang • L. Yin • R. Li • W. Shui • H. Zhang
Molecular Oncology Laboratory, Department of Orthopaedic Surgery,
The University of Chicago Medical Center, 5841 South Maryland
Avenue, MC 3079, Chicago, IL 60637, USA

The Affiliated Hospitals, Chongqing Medical University, Chongqing 400016, China

N. Wang
Molecular Oncology Laboratory, Department of Orthopaedic Surgery,
The University of Chicago Medical Center, 5841 South Maryland
Avenue, MC 3079, Chicago, IL 60637, USA

Affiliated Southwest Hospital of the Third Military Medical University,
Chongqing 400038, China

R.C. Zhao (ed.), *Essentials of Mesenchymal Stem Cell Biology and Its Clinical Translation*, DOI 10.1007/978-94-007-6716-4_6,

superfamily has significant effects on proliferation and differentiation, including key roles in osteogenic and chondrogenic differentiation. Bone morphogenetic proteins (BMPs) are members of the TGF-β superfamily with well-described effects on osteogenesis and chondrogenesis. Demonstrating crosstalk with BMPs, other signaling pathways implicated in osteogenic and chondrogenic differentiation include the Wnt signaling pathway and the FGF family. Hedgehog and notch signaling also crosstalk with Wnts, with diverse effects on osteogenic, myogenic and adipogenic differentiation. PPARγ is the master regulator of adipogenesis and is essential for the induction of normal adipogenesis, and C/EBPs modulate the expression and function of PPARγ. Crosstalk exists between the major signaling pathways governing adipogenesis and osteogenesis, as differential expression of BMPs and PPARγ significantly alters adipogenic and osteogenic differentiation. While the major signaling pathways mediating lineage-specific differentiation are well-studied, the complex crosstalk between these pathways and lineage-specific cascades makes elucidation of specific mechanisms quite difficult.

Keywords BMPs • Cell signaling • FGF • MSC differentiation • MSC proliferation • PPARγ • Wnt

Introduction

As multipotent cells with the ability to differentiate into conventional mesodermal lineages and other tissues, including muscle, fat, skin and cardiac tissues [1, 2], the unique properties of MSCs have allowed us to characterize the differentiation pathways and proliferative stimuli of various lineages. The process of osteoblastic differentiation is complex, tightly regulated and influenced by a variety of factors and signaling pathways including BMPs, PPARγ and Wnt/β-catenin [3–5]. Various exogenous cytokines and growth factors promote chondrogenesis with considerable overlap between osteogenesis and chondrogenesis. Both TGF-β cytokines, including BMPs, and the fibroblast Growth Factor (FGF) family of cytokines stimulate chondrogenic differentiation [6–9]. Adipogenesis is regulated by a complex signaling cascade involving changes in the expression and/or activity of transcription factors including PPARγ, regulating expression of many adipocyte-specific genes [2, 10–12].

T.-C. He, M.D., Ph.D. (✉)
Molecular Oncology Laboratory, Department of Orthopaedic Surgery, The University of Chicago Medical Center, 5841 South Maryland Avenue, MC 3079, Chicago, IL 60637, USA

Stem Cell Biology and Therapy Laboratory of the Key Laboratory for Pediatrics Designated by Chinese Ministry of Education, The Children's Hospital of Chongqing Medical University, Chongqing 400014, China

The Affiliated Hospitals, Chongqing Medical University, Chongqing 400016, China
e-mail: tche@uchicago.edu

PPARγ also interacts with members of the CCAAT/enhancer binding proteins (C/EBP) family to regulate adipogenesis [2, 13]. During the process of myogenesis, the transition of uncommitted mesoderm to myoblasts is regulated by numerous genes and cell signaling molecules including the MyoD family of basic helix-loop-helix transcription factors [14]. Our advancing knowledge of the signaling pathways governing lineage-specific proliferation and differentiation will facilitate the development of novel therapies with widespread clinical applications including regenerative medicine.

Mesenchymal Stem Cell Lineages

Various signaling pathways govern the lineage-specific differentiation and proliferative capacity of MSCs. While the complex signaling cascades controlling the processes of differentiation and proliferation are unique among each lineage, there is a considerable amount of crosstalk existing between the pathways. A general description of lineage-specific differentiation will be followed by a discussion of the major signaling pathways governing these processes.

Osteogenic Lineage

Bone is a highly vascularized tissue undergoing constant remodeling via a dynamic process of breakdown and rebuilding by hematopoietic-derived osteoclasts and mesenchymal-derived osteoblasts [15–17]. A local balance of osteoclastic resorption and osteoblastic rebuilding is mediated by various signaling molecules including morphogens, hormones, growth factors, cytokines, matrix proteins and transcription factors [18–23]. Residing near the surface of bone, committed pre-osteoblasts secrete alkaline phosphatase, an early marker of osteogenic differentiation [19, 20, 22, 23]. Upon differentiation of the preosteoblast into an active mature osteoblast, the cell phenotype changes to a larger nucleus, enlarged Golgi and more extensive endoplasmic reticulum; these changes support secretion of bone matrix proteins [16]. Soon after, the cell becomes terminally differentiated into an osteocyte, providing mechanical support and regulation of mineral deposition. Stages of osteoblast differentiation including proliferation, matrix maturation and mineralization correlate with expression of osteoblast phenotypic markers reflecting the various stages of differentiation [16]. The proliferative stage features upregulation of cell cycle-associated genes, while the maturation phase features expression of the early osteogenic marker alkaline phosphatase. Mineralization features expression of late markers of differentiation including osteocalcin and osteopontin [16]. The process of osteoblastic differentiation is complex, tightly regulated and influenced by a variety of factors and signaling pathways including BMPs, PPARγ and Wnt/β-catenin [3–5].

Chondrogenic Lineage

Similar to adipogenesis, formation of cartilage occurs as MSCs differentiate into precursor cells called chondroblasts [6–9, 24]. Chondrogenesis is a multi-step event requiring the commitment of MSCs followed by aggregation and differentiation into chondrocytes [2, 25]. Aggregation of chondroprogenitor MSCs into precartilage condensations is one of the earliest events in chondrogenesis and depends on signals mediated by cell-cell and cell-matrix interactions [24]. Various cytokines and growth factors promote chondrogenesis, with considerable overlap between osteogenesis and chondrogenesis. TGF-β is among the earliest signals in chondrogenic condensation [24]. The two main effectors of TGF-β signaling in chondrogenic differentiation are the Smad [26] and MAPK [27] pathways [6–9]. Both TGF-β cytokines and the fibroblast growth Factor (FGF) family of cytokines stimulate chondrogenic differentiation through MAPK signaling, activating the master regulator of chondrogenesis Sox9 [6–9]. Meanwhile, parathyroid hormone-related peptide and Indian Hedgehog seem to play a role in chondrogenesis via crosstalk with the FGF signaling cascade [28, 29]. Insulin-like growth factor-1 (IGF-I) also stimulates chondrogenesis [30] via the Inositol 3-Kinase pathway [31], synergizing with TGF- β1 [32] or FGF-2 [33, 34]. Furthermore, overlap exists between the highly regulated processes of chondrogenesis and the conversion of perichondrium to perisoteum. FGF and the bone morphogenetic proteins (BMPs) regulate chondrocyte proliferation, maturation and hypertrophy [29, 35, 36] but also regulate the conversion of perichondrium to perisoteum [37]. Wnt signaling also plays a critical role in the induction of FGFs [38].

Adipogenic Lineage

Adipocytes form from MSCs by a sequence of events characterized by two phases [39, 40]. The first phase is called determination and is defined by the commitment of a multipotent MSC to a preadipocyte or adipoblast. Following determination, the preadipocyte loses its ability to differentiate along any other lineage yet is phenotypically the same as its undetermined precursor. The second phase is called differentiation, in which the preadipocyte or adipoblast matures into an adipocyte, thereby gaining the capacity to synthesize lipids and triglycerides, perform lipolysis and secrete adipocyte-specific factors. During adipogenesis, there is a shift in gene expression patterns of uncommitted MSCs promoting the phenotypic properties of mature adipocytes [11, 12]. Adipogenesis is regulated by complex signaling cascades involving changes in the expression and/or activity of transcription factors including PPARγ, which regulates many adipocyte-specific genes [2, 10–12]. PPARγ also interacts with members of the CCAAT/enhancer binding proteins (C/EBP) family to regulate adipogenesis [2, 13].

Myogenic Lineage

Muscle development occurs with the transition of uncommitted mesoderm from somites, giving rise to a population of committed, dividing muscle precursor cells called myoblasts [41, 42]. During myogenesis, myogenic precursor cell fate is first determined by Pax3/Pax7 [14] followed by a myogenic differentiation program by muscle regulatory factors including the MyoD family of basic helix-loop-helix transcription factors, Hedgehog proteins, the Wnt family proteins and BMPs [32, 43–45]. After transitioning from mesoderm to myoblast, myoblasts cease dividing and express muscle-specific genes, terminally differentiating into myocytes; this process is dependent on the MyoD family and the Mef2 family [46]. The FGF family of transcription factors has been implicated in myoblast proliferation arrest and subsequent expression of muscle-specific genes. Terminally differentiated myocytes then fuse with one another, forming multinucleate myofibers comprising skeletal muscle [47–49].

Major Signaling Pathways Controlling MSC Differentiation

TGFβ/BMP Superfamily

The transforming growth factor beta (TGFβ) superfamily is made up of more than 30 secreted dimeric polypeptides playing crucial roles in the regulation of various cellular functions including proliferation, differentiation and embryonic development [18, 19, 50–57]. There are three isoforms, TGFβ1, TGFβ2 and TGFβ3, which are highly conserved and share a cluster of conserved cysteine residues [53]. Mutations in this pathway are associated with many disorders affecting the skeletal, muscular and cardiovascular system as well as tumor development [21, 58–60]. TGFβ acts synergistically with Runx2, the regulatory gene responsible for early osteogenic differentiation [2, 61], to upregulate expression of interleukin 11 (IL-11); this upregulation reduces adipogenesis while promoting osteoblastic differentiation [62]. TGFβ also plays an important role in chondrogenic differentiation, and these effects may be enhanced when certain BMPs are co-administered. BMPs are members of the TGFβ1 superfamily, with more than 30 identified in mammals and 14 in humans [18, 19, 63–65]. BMPs play an important role in the regulation of cell proliferation and differentiation during development [66, 67] as well as in stem cell biology [68, 69], with genetic disruptions resulting in skeletal and extraskeletal abnormalities [70].

BMP signaling transduction begins with the binding of a heterodimeric complex of two transmembrane serine/threonine kinase receptors, BMPR type 1 and BMPR type 2 [71, 72]; these activated receptor kinases in turn transduce signals by phosphorylating the transcription factors Smad 1, 5 and/or 8 [73]. Phosphorylated Smads then form a heterodimeric complex with Smad 4 which is translocated into the nucleus, activating the transcription of target genes [53, 74, 75]. The Smad family is comprised of eight members including the Receptor-regulated or R-Smads (Smads 1, 2, 3, 5 and 8), the Co-Smads (Smad 4) and the Inhibitory Smads (Smad 6 and 7) [76]. Inhibitory Smads inhibit TGF-β signaling by blocking the phosphorylation of R-Smads by activated Type I TGF-β receptors [50, 74, 77–80].

Several BMPs, namely BMP2, BMP6 and BMP9 (and to a lesser extent BMP7) are potent inducers of osteogenesis both *in vitro* and *in vivo* [18, 19, 81, 82]. Exposure of MSCs to osteogenic BMPs causes increased expression of osteoblast-specific markers including connective tissue growth factor (CTGF), Inhibitor of DNA Binding (Id), Alkaline Phosphatase (ALP) and Cbfa1/Runx2 [19, 83–88]. BMP2, BMP4 and BMP7, in coordination with other signaling molecules and cofactors, also promote preadipocyte differentiation [89]. PPARγ is upregulated by the osteogenic BMPs, while PPARγ knockout mice demonstrate decreased osteogenic differentiation in response to BMP stimulation [86, 89, 90].

Kang et al. conducted a comprehensive analysis of 14 types of BMPs to assess their abilities to induce lineage-specific differentiation of MSCs [89]. While BMP2, -4, -6, -7 and -9 induced both adipogenic and osteogenic differentiation of MSCs *in vitro* and *in vivo*, commitment of MSCs to either lineage was mutually exclusive. Overexpression of the essential osteogenic transcription factor Runx2 synergized with BMP-induced osteogenic differentiation without effects on adipogenesis, and overexpression of PPARγ2 together with BMP-2, -6 or -9 stimulation promoted both osteogenic and adipogenic differentiation. Knockdown or deletion of PPARγ2 expectedly inhibited adipogenic differentiation but furthermore decreased BMP-induced ossification, demonstrating the important role that PPARγ2 may play in osteogenic differentiation.

Recent studies have demonstrated BMP9 to be among the most potent BMPs in inducing osteogenic differentiation of MSCs [81, 82, 86]. Increased expression of the early osteogenic marker ALP was seen in C3H10T1/2 MSCs, preosteoblastic C2C12 cells and osteoblastic TE85 cells [81, 82, 86]. Furthermore, BMP9-sitmulated C2C12 cells also demonstrated increased expression of the late osteoblastic marker osteocalcin and mineralized osteoid nodules as indicated by Alizarin Red immunohistochemical staining. Furthermore, BMP3, a known inhibitor of BMP2 and BMP7-mediated osteogenesis, does not inhibit BMP9-mediated osteogenesis. Non-adenoviral delivery of BMP9 has also resulted in potent osteoinduction of MSCs using methods including sonoporation of rhBMP9, nucleofection of BMP9 and peptide derived from BMP9 [91–93]. Important mediators of BMP9-induced osteogenic signaling include the Inhibitors of Differentiation (Ids) HLH factors [87], connective tissue growth factor (CTGF) [85] and Hey 1 bHLH Factor [89, 94]. Crosstalk between other signaling pathways and BMP9-mediated osteogenesis also exists. Specifically, the Wnt/β-catenin signaling pathway [95], PPARγ [89], the IGF signaling pathways [96] and retinoid signaling pathways [97] appear to be involved in BMP9-induced osteogenesis.

BMPs also play a major role in chondrogenesis and were first identified as molecules inducing ectopic endochondral ossification [24, 98]. BMPs initiate chondroprogenitor cell differentiation as well as late stages of chondrocyte maturation and terminal differentiation to the hypertrophic phenotype [24]. In fact, the balance of signaling between BMPs and FGFs throughout chondrogenesis determines the rate of cell proliferation, adjusting the pace of chondrocyte terminal differentiation to the proliferation rate [36]. BMPs, namely BMP2, -4 and -7, are required for the differentiation of chondrocytic precursors into chondrocytes and for the formation of precartilaginous condensations [99, 100]. BMP2 and -6 are found exclusively in hypertrophic chondrocytes, while BMP7 is expressed in proliferating chondrocytes [24, 36]. The nuclear transcription factor Sox9, called the master regulator of chondrogenesis, is one of the earliest markers expressed in progenitors undergoing condensations and is required for expression of cartilage-specific matrix proteins including type II collagen (Col2a1) [4, 101–103]. Two other members of the Sox family, Sox5 and Sox6 are co-expressed with Sox9 during chondrogenic differentiation [104], forming homo- or heterodimers which bind more efficiently to pairs of HMG box sites than single sites. Expression of SOX proteins depends upon BMP signaling via BMPR1A and BMPR1B [100]. In summary, the TGF-β superfamily, including BMPs, play critical roles in proliferation, differentiation and embryonic development with well-described effects on the processes of osteogenesis and chondrogenesis.

Wnt Signaling

Playing an important role in embryonic development and tissue induction [105–114], aberrations in the Wnt pathway have been associated with a variety of human diseases ranging from cancer to degenerative diseases [115–117]. The Wnt family is made up of many secreted glycoproteins [105, 108, 114], and the canonical Wnt/β-catenin pathway plays a critical role in bone development [118].

The canonical Wnt/β-catenin pathway is activated when the Wnt ligand binds the 7-transmembrane domain-spanning Frizzled (Frz) receptor and the LRP5/6 co-receptors [105, 108, 109, 114]. This interaction leads to phosphorylation of the Disheveled (Dvl) protein, which in turn interacts with Axin, Frat-1 and APC tumor suppressor, thereby preventing GSK3β from phosphorylating β-catenin [105, 108, 109, 114]. Unphosphorylated β-catenin is thus stabilized and avoids degradation by β-TrCP, facilitating translocation of β-catenin into the nucleus [105, 108, 109, 114]. Upon entering the nucleus, β-catenin interacts with the transcription factors LEF/Tcf4 to activate expression of downstream genes including c-Myc, cyclin D1, PPARγ, WISPs, CTGF, Cyr61 and various other targets promoting cell proliferation, tissue expansion and cell fate determination [85, 90, 109, 119–124].

Wnt signaling plays an important role in skeletal development and osteoblastic differentiation, as several Wnt genes are expressed in the developing limb and have been implicated in mesenchymal chondrogenesis [125–132]. Wnt signaling also plays a key role as a negative regulator in adipogenic differentiation [131–134]. Canonical Wnt/β-catenin synergizes with the essential osteogenic regulator Runt-related transcription factor 2 (Runx2) in promoting the differentiation pathway of

MSCs toward osteogenic precursors [135, 136]. Meanwhile, the non-canonical Wnt pathway suppresses the adipogenic regulator PPARγ while enhancing Runx2, inducing osteogenesis [131, 137]. Wnt5a promotes osteoblastogenesis while serving as a co-repressor of PPARγ-mediated adipogenesis [137], and these counter-regulatory properties are essential for terminal osteogenic differentiation [138]. Furthermore, terminal differentiation of osteoblasts requires both Wnt antagonists Dkk1 and Dkk2 [139, 140].

The Wnt co-receptor LRP5 plays a role in bone mass regulation; loss-of-function mutations in human LRP5 are associated with low bone mineral density and skeletal fragility [118], while activation mutations in the LRP5 are associated with high bone mineral density [141, 142]. When loss of LRP6 is coupled to loss of LRP5, low bone mineral density is further exacerbated, demonstrating that the LRP5 and LRP6 co-receptors both participate in the effects of Wnt signaling on bone mass [143]. Furthermore, disruption of LRP inhibitors including Dkk1 [141] and Sclerostin [144] permit unphosphorylated β-catenin to stimulate osteogenesis, while overexpression of the Wnt antagonist Dkk1 is associated with the presence of lytic bone lesions [46]. Wnt3a has been shown to promote proliferation while suppressing osteogenic differentiation of MSCs [145]. Conversely, it has been suggested that Wnt/β-catenin signaling in osteoblasts may coordinate postnatal bone acquisition by regulating the differentiation of both osteoblasts and osteoclasts [146]. Furthermore, Wnt/β-catenin signaling in MSCs has been shown to control osteoblast and chondrocyte differentiation during skeletogenesis [147, 148]. Stabilization of β-catenin in differentiated osteoblasts results in high bone mass, while its deletion from differentiated osteoblasts may lead to osteopenia via osteoprotegerin-mediated bone resorption [133]. These findings suggest that Wnt/β--catenin signaling promotes the ability of differentiated osteoblasts to inhibit osteoclast differentiation. Finally, non-canonical Wnt signaling was shown to promote bone formation via G-protein-linked PKC-γ activation [149].

Wnt proteins also regulate chondrogenic and myogenic development [150, 151]. Canonical Wnt ligands suppress chondrogenesis [135, 136], while inhibition of Wnt signaling induces transdifferentiation of myoblasts to adipocytes [152].

Wnts crosstalk with bone morphogenetic proteins (BMPs) in the regulation of MSC osteogenic differentiation. BMPs 2, 6 and 9 are the major osteogenic BMPs [33] and Wnt3a synergizes with BMP9 to induce osteogenic differentiation. Meanwhile, β-catenin knockdown or Frz antagonist overexpression blocks BMP9-mediated osteogenic differentiation and bone formation [95]. β-catenin alone is able to induce osteogenic differentiation but not ectopic bone formation, and the addition of BMPs to β-catenin is essential for ectopic bone formation [19, 153]. Wnt antagonist Dkk1 or β-catenin null inhibit BMP-mediated bone formation, demonstrating that the Wnt and BMP pathways may crosstalk with Smad-4 and β-catenin interaction [154].

Wnts may also crosstalk with Hedgehog (Hh) and Notch signaling pathways during osteogenic regulation [138]. Wnt signaling regulates Gli2 and Gli3, key mediators in Hh signal transduction [155]. A conserved domain appears to exist between Notch signaling molecules and the transcription factors Tcf-Lef-1, allowing Notch signaling mediators to inhibit the canonical Wnt pathway [156]. Other pathways implicated in osteogenic induction through Wnt signaling include PKC-δ, Src/ERK and PI3K/Akt [157]. To summarize, Wnt signaling has a critical role in skeletal development and osteogenic differentiation and also crosstalks with many other signaling pathways important in osteogenic regulation.

FGF Signaling

Consisting of 23 members expressed in nearly all tissue types during development [158, 159], fibroblast growth factors (FGFs) are important in chondrogenesis and osteogenic differentiation [160–163]. FGF ligands are 20–35 kDa and bind the FGF receptor extracellular ligand binding domain as well as a highly conserved intracellular signaling domain containing intrinsic kinase and tyrosine residues [164]. FGFs dimerize upon ligand binding, causing autophosphorylation of the intrinsic kinase residues and initializing the FGF signaling cascade [158, 164]. The FGF Receptor Substrate 2 (FRS2) protein is then phosphorylated, recruiting the Grb2/SOS complex to the plasma membrane and activating the MAPK pathways including ERK1/2, p38 MAPK, SAPK/JNK, PKC and PI3K [164, 165].

FGF ligands and receptors play a key role in osteogenic differentiation [160, 162, 163, 165, 166]. Specifically, FGF-2 induces ALP activity in rat bone marrow precursor cells, while FGF-2, -4 and -8 induce expression of the essential osteogenic transcription factor Runx2 [166]. FGF-9 induces expression of the late osteogenic marker osteocalcin, while FGF-2, -9 and -18 are critical in matrix mineralization [166]. There is conflicting evidence regarding the effects of FGF ligands on osteoblast proliferation (Deng 238, 244). FGFR1 plays a dominant role in osteogenic differentiation, while FGFR2 plays a role in both osteogenic proliferation and differentiation. FGF3 controls chondrocyte proliferation during endochondral ossification and may also play a crucial role in osteogenesis. Mice lacking FGFR3 are osteopenic [167], while mice with constitutive activation of FGFR3 due to the Gly369Cys homozygous mutation have defects in endochondral ossification with upregulation of the osteogenic markers osteocalcin, osteopontin and osteonectin [168].

FGFs also play a critical role in chondrogenesis [24]. FGF signaling in chondrogenesis depends on the temporal and spatial regulation of both FGF ligands and receptors [162]. FGFR2 is expressed early within condensing mesenchyme, while FGFR3 is expressed in proliferating chondrocytes within the central core of mesenchymal condensations [24]. Within the growth plate, FGFR3 is the master inhibitor of chondrocyte proliferation via phosphorylation of the Stat1 transcription factor, thereby increasing expression of the cell cycle inhibitor p21 [169]. FGF-18 seems to be the preferred ligand of FGFR3, as both FGF18-deficient and FGFR-3 deficient mice have an expanded zone of proliferating chondrocytes [35]. Altogether, FGF signaling plays an essential role in osteogenic and chondrogenic proliferation and differentiation.

Parathyroid Hormone-Related Peptide (PTHrP)

Parathyroid hormone (PTH) and parathyroid hormone-related peptide (PTHrP) are critical in maintaining skeletal homeostasis via a balance of osteoclastic bone resorption and osteoblastic bone formation [7]. PTHrP is secreted during fetal life by perichondral cells and early proliferative chondrocytes, playing a role in endochondral bone formation by keeping proliferating chondrocytes within the proliferative pool [170–172]. Both PTH and PTHrP bind to the same G-protein coupled receptors (GPCR) [173] which are expressed at low levels by proliferating chondrocytes and high levels by prehypertrophic and early hypertrophic chondrocytes. Furthermore, crosstalk between the Indian Hedgehog (Ihh) and PTHrP pathways controls the cell-fate of chondrocytes leaving the proliferative pool via a feedback loop [174, 175]. While PTH and PTHrP lack substantial sequence homology, they bind and activate the same GPCR with nearly the same affinity; this receptor mediates the endocrine functions of PTH and the autocrine/paracrine actions of PTHrP [171, 176]. By well-characterized mechanisms, PTHrP plays a key and apparently overlapping role in both chondrogenesis and osteogenesis.

Notch Signaling

Notch signaling plays a critical role in cell fate decision during development, homeostasis in adults [177–183] and skeletal development [184]. Activated by a membrane-bound ligand, the notch gene encodes a single pass transmembrane receptor. Four notch receptors (Notch 1–4) and five ligands (δ-like1, δ-like3, δ-like4, Jagged 1 and Jagged 2) exist [177, 179–183, 185]. Notch receptor binding with ligand induces proteolytic cleavage of the receptor and release of the Notch Intracellular Domain (NICD) which subsequently translocates into the nucleus and binds transcription factors [186, 187], upregulating the expression of primary target genes including HES (Hairy Enhancer of Split) and HES-related Repressor Protein (HERP) [177, 179–183, 185].

Notch signaling plays a critical role in somite formation during embryogenesis, and mutations in Notch1 cause embryonic defects in somite segmentation [177, 179–183, 188]. Notch signaling regulates adult stem cell differentiation, affecting osteogenesis by poorly defined mechanisms [156, 189, 190]. While Notch/TGF-β crosstalk between Notch1 and BMP2 promotes osteogenic differentiation, Notch1 overexpression inhibits osteogenesis by repressing Wnt/β-catenin but not BMP signaling [191]. Synergy between Notch and BMP4 inhibits myogenic differentiation [192], and inhibition of the Notch Pathway interferes with adipogenesis *in vitro* [193]. Altogether, Notch signaling seems to play a role in osteogenic differentiation of MSCs as well as myogenic and adipogenic differentiation, but the specific mechanisms governing these processes remain poorly defined.

Hedgehog (Hh) Signaling

Hedgehog proteins are secreted factors critical in many developmental processes [194–198]. The three members of this family are Sonic Hedgehog (Shh), Indian Hedgehog (Ihh) and Desert Hedgehog (Dhh), each having a distinct set of functions in the regulation of developmental processes [195, 196]. Sonic Hedgehog regulates central nervous system development and skeletal patterning [195, 196]. Desert Hedgehog plays a crucial role in peripheral nerve development and spermatogenesis [195, 196]. While Indian Hedgehog shares many functions with Sonic Hedgehog, its most important role is in endochondral bone development [199–202]. Secreted active Hedgehog fragments are catalytically cleaved from a 45 kDa precursor to a 19 kDa N-terminal fragment then modified by the attachment of palmitic acid and cholesterol. These active molecules regulate the activities of both neighboring and distant cells by binding to the receptor Patched (Ptc) and the coreceptor Smoothened (Smo), two transmembrane proteins on the cell surface [200, 203–205]. Without Hedgehog ligands, Ptc represses Smo activity and converts the transcription factor Gli to its repressor form, but when Hedgehog binds to Ptc, Smo is released and activates the Gli family transcription factors [194, 204–206]. Accumulation of Gli in the nucleus induces Hedgehog target gene expression including Patched-1 and Glis.

Shh is upregulated during the early phases of bone repair and triggers expression of angiogenic growth factors including VEGF, suggesting that Shh may be one of the key molecules linking osteogenesis and angiogenesis during bone repair [207]. The effects of Shh on osteogenic differentiation are associated with increased expression of bone morphogenetic proteins, including BMP2 [208] and BMP4 [209], as well as PTHrP [203, 210]. Several studies have demonstrated that Shh signaling drives osteogenic differentiation of MSCs [211–213]. Like Shh, Ihh is upregulated in the bone marrow [214] during the early phases of bone repair. Furthermore, Ihh is produced by both prehypertrophic and hypertrophic chondrocytes [215, 216] during the process of endochondral ossification which induces upregulation of PTHrP, leading to hypertrophic cartilage differentiation. To summarize, Hh signaling plays a diverse role in developmental processes and progenitor cell differentiation, with important roles in osteogenic and chondrogenic differentiation of MSCs.

Nuclear Receptor PPARγ

PPARγ, a member of the nuclear hormone receptor gene superfamily of ligand-activated transcription factors, is expressed as two isoforms (PPARγ1 and 2) and is commonly referred to as the master regulator of adipogenesis; no factor has been identified which can induce normal adipogenesis without PPARγ [11]. While PPARγ2 is the predominant form expressed in adipose tissue, both PPARγ1 and PPARγ2 expression are highly induced during adipogenesis and likely play a critical role in adipogenesis [217, 218]. Ectopic expression of PPARγ alone can induce adipocyte differentiation in uncommitted fibroblasts [219], and all critical cell signaling pathways involved in adipogenesis converge on PPARγ [12, 220].

Adipogenesis of MSCs demonstrates a significant increase in PPARγ expression [221], directly inducing expression of the majority of genes characterizing the adipocyte phenotype including fatty acid synthase, Glut4 and acetyl CoA carboxylase [11]. Other target genes of PPARγ include Ap2, lipoprotein lipase (LPL), acyl-Coa synthetase (ACS) and CD36 [5, 222, 223]. PPARγ binds to fatty acids and derivatives including linoleic acid and docosahexaenoic acid (DHA) [5]. Upon binding of their respective ligands, PPARγ and the retinoid X receptor (RXR) are activated and form a heterodimer, which translocates into the nucleus and regulates downstream target genes in concert with nuclear receptor coactivators, including the C/EBP family [5]. Treatment of MSCs with PPARγ agonists, such as thiazolidinediones, induces adipogensis [224, 225]. Binding of these ligands activates PPARγ, stimulating the differentiation of preadipocytes and the initiation of steps required for lipid storage [5]. The presence of a ligand appears to be required for commitment of cells to the adipocyte lineage but not to maintain differentiation of mature adipocytes: Differentiation of nonadipogenic fibroblasts with overexpression of PPARγ required exposure to ligand, while adipogenic differentiation of preadipocytes occurred even in the absence of ligand [226]. Thus, while PPARγ expression is critical in both phases of adipogenesis, its ligands are only required in the determination phase. Although endogenous fatty-acid derived molecules have been demonstrated to bind and activate PPARγ [226, 227], none have been directly implicated in adipogenesis [11].

PPARγ also plays a significant role in osteogenesis [5]. Progenitor cells with homozygous deficiency of PPARγ spontaneously differentiate into osteoblasts, while heterozygous PPARγ deficiency causes increased bone formation *in vivo* [228]. Meanwhile, PPARγ is significantly upregulated by the osteogenic bone morphogenetic proteins (BMPs) [86, 89]. Overexpression of PPARγ2 promotes BMP-induced osteogenesis and adipogenesis, while silencing of PPARγ2 inhibits adipogenic differentiation while stimulating osteogenic differentiation [89]. Although the specific mechanisms relating PPARγ and osteogenesis remain to be clearly defined, nuclear competition between PPARγ and other members of the nuclear receptor superfamily may play a role; regulation of the osteogenic promoter osteocalcin by glucocorticoids, vitamin D and thyroid hormone occurs via the same nuclear pathway as PPARγ [229–231]. Furthermore, PPARγ activation by fatty acid ligands and derivatives may slow osteoblast differentiation, explaining the tendency to shift toward adipogenesis with increased availability of these ligands [5]. In summary, PPARγ plays a critical and well-characterized role in adipogenic differentiation and an important but poorly characterized role in osteogenic differentiation of MSCs.

CCAAT/Enhancer Binding Proteins (C/EBPs)

The CCAAT/enhancer binding proteins (C/EBPs) are members of the basic-leucine zipper class of transcription factors and consist of six isoforms including C/EBPα, C/EBPβ, C/EBPγ, C/EBPδ and transcription factor homologous to CCAAT/enhancer-binding protein (CHOP), each of which are expressed in adipocytes [11]. C/EBPα, C/

EBPβ and C/EBPδ promote adipogenesis while C/EBPγ and CHOP inhibit adipogenesis through the formation of dimers and inactivation of C/EBPβ [11]. Adipogenesis within MSCs demonstrates a rapid induction of C/EBPβ and C/EBPδ preceding induction of the major adipogenic transcription factors C/EBPα and PPARγ [11, 232]. While C/EBPβ and C/EBPδ are not required for C/EBPα and PPARγ expression, they demonstrate a synergistic effect on adipogenesis via induction of these two major adipogenic transcription factors. In human MSCs, adipogenesis demonstrates early and late rounds of C/EBPα and C/EBPβ induction prior to terminal differentiation, suggesting involvement of these factors in both the determination and differentiation phases of MSC adipogenesis [221]. C/EBPα directly activates adipogenic genes including PPARγ which are required for the development of mature adipocytes [11]. Once activated, PPARγ and C/EBPα stimulate the expression of one another, remaining elevated throughout the life of mature adipocytes [219, 233]. Ectopic expression of C/EBPα drives adipogenic differentiation of cell lines otherwise not undergoing adipogenesis, including mouse fibroblasts [234, 235]. Furthermore, fibroblasts without C/EBPα expression have significantly reduced adipogenic potential and PPARγ expression; however both are reversed upon administration of PPARγ. Conversely, administration of C/EBPα in fibroblasts lacking PPARγ does not rescue the decreased adipogenic potential [236]. These findings suggest that C/EBPα promotes adipogenesis in a PPARγ-dependent manner. Altogether, C/EBPs play a key role in adipogenesis largely through modulation of PPARγ expression and function [11].

MyoD

MyoD is a basic helix-loop-helix (bHLH) transcription factor characterized as a muscle regulatory factor (MRF) [14]. Other MRFs include Myf5, MyoD, Myog and Mrf4, all of which demonstrate significantly increased expression following the onset of myogenesis [193]. MyoD, along with other MRFs, are essential for myoblast determination [14]. When discovered by Weintraub, MyoD alone was found to convert 10T1/2 fibroblasts into myoblasts [14, 237]. More recently, double Myf5/MyoD mutants failed to develop skeletal muscle secondary to a lack of precursor myoblast cells [238].

MyoD and other bHLH MRF factors bind the E-box sequence (CANNTG) within promoters of downstream target genes, driving transcription of muscle-related genes in combination with myocyte enhancer factor 2 (Mef2) [14, 239]. bHLH MRFs dimerize with E-proteins (E12, E47 and HEB) to activate downstream gene expression [14, 239]. Inhibitor of DNA binding (Id) proteins (Id1-Id4) serves as negative regulators of myogenesis and prevent MyoD-mediated activation of downstream target genes [14, 240]. Ids heterodimerize with E-proteins, and to a lesser extent MyoD, attenuating the function of MyoD [14]. It has been recently shown that the transcriptional repressor RP58 is expressed in early differentiated muscle and functions to repress Id2/Id3 expression, permitting MyoD to promote muscle differentiation [14]. From these recent studies, MyoD seems to both activate (via the MRF Myog) and repress (via RP58) a distinct set of genes allowing the progression of skeletal myogenesis to late differentiation [14, 241]. In summary, MyoD is a well-established and critical transcription factor in muscle development and myogenic differentiation of MSCs.

Perspectives

While we continue to make advances in our understanding of the signaling pathways governing proliferation and differentiation of MSCs, many questions remain to be answered. We must not only gain a deeper understanding of the specific mechanisms governing lineage-specific differentiation, but since many of the signaling molecules regulating these processes are from different families, we must better elucidate the contribution of each family as well as the crosstalk that exists between these signaling pathways. With further investigation, these pathways and their association with one another will become better understood. With these advances, manipulation of the proliferative capacity and differentiation pathways of MSCs will allow us to develop much-needed novel therapies which will translate to the clinical setting (Fig. 1).

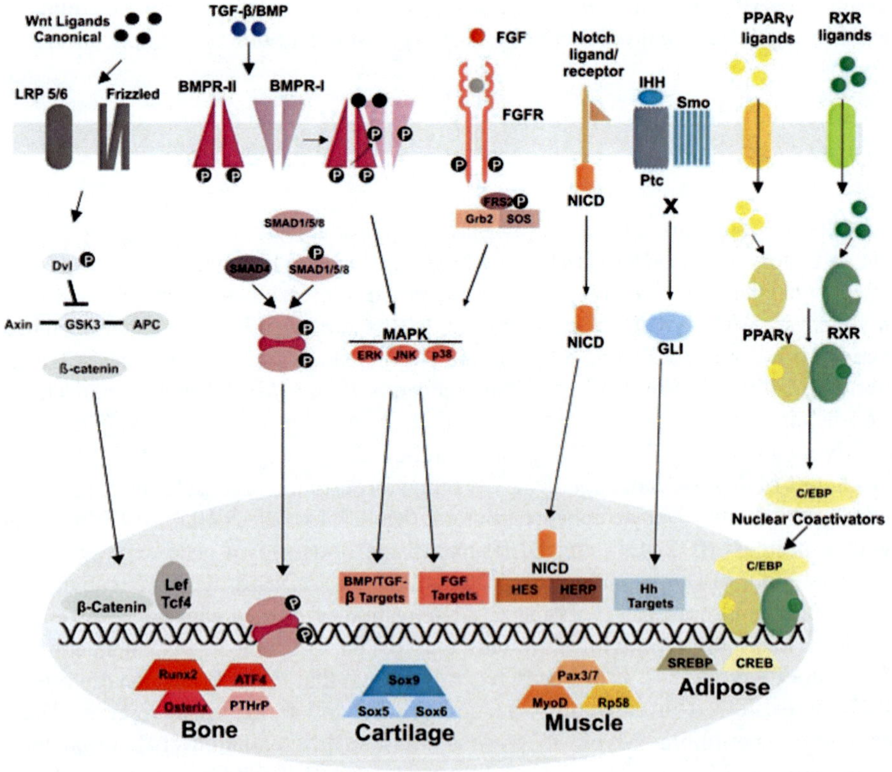

Fig. 1 Schematic representation of the major signaling pathways that regulate lineage commitment and differentiation of MSCs

References

1. Arthur A, Zannettino A, Gronthos S. The therapeutic applications of multipotential mesenchymal/stromal stem cells in skeletal tissue repair. J Cell Physiol. 2009;218:237–45.
2. Rastegar F, Shenaq D, Huang J, Zhang W, Zhang BQ, He BC, Chen L, Zuo GW, Luo Q, Shi Q, Wagner ER, Huang E, Gao Y, Gao JL, Kim SH, Zhou JZ, Bi Y, Su Y, Zhu G, Luo J, Luo X, Qin J, Reid RR, Luu HH, Haydon RC, Deng ZL, He TC. Mesenchymal stem cells: molecular characteristics and clinical applications. World J Stem Cells. 2010;2:67–80.
3. Luther G, Wagner ER, Zhu G, Kang Q, Luo Q, Lamplot J, Bi Y, Luo X, Luo J, Teven C, Shi Q, Kim SH, Gao JL, Huang E, Yang K, Rames R, Liu X, Li M, Hu N, Liu H, Su Y, Chen L, He BC, Zuo GW, Deng ZL, Reid RR, Luu HH, Haydon RC, He TC. BMP-9 induced osteogenic differentiation of mesenchymal stem cells: molecular mechanism and therapeutic potential. Curr Gene Ther. 2011;11:229–40.
4. Tang N, Song WX, Luo J, Haydon RC, He TC. Osteosarcoma development and stem cell differentiation. Clin Orthop Relat Res. 2008;466:2114–30.
5. Wagner ER, He BC, Chen L, Zuo GW, Zhang W, Shi Q, Luo Q, Luo X, Liu B, Luo J, Rastegar F, He CJ, Hu Y, Boody B, Luu HH, He TC, Deng ZL, Haydon RC. Therapeutic implications of PPARgamma in human osteosarcoma. PPAR Res 2010;2010:956427.
6. de Crombrugghe B, Lefebvre V, Behringer RR, Bi W, Murakami S, Huang W. Transcriptional mechanisms of chondrocyte differentiation. Matrix Biol. 2000;19:389–94.
7. de Crombrugghe B, Lefebvre V, Nakashima K. Regulatory mechanisms in the pathways of cartilage and bone formation. Curr Opin Cell Biol. 2001;13:721–7.
8. DeLise AM, Fischer L, Tuan RS. Cellular interactions and signaling in cartilage development. Osteoarthritis Cartilage. 2000;8:309–34.
9. Lefebvre V, Smits P. Transcriptional control of chondrocyte fate and differentiation. Birth Defects Res C Embryo Today. 2005;75:200–12.
10. Rosen ED, Spiegelman BM. Molecular regulation of adipogenesis. Annu Rev Cell Dev Biol. 2000;16:145–71.
11. Muruganandan S, Roman AA, Sinal CJ. Role of chemerin/CMKLR1 signaling in adipogenesis and osteoblastogenesis of bone marrow stem cells. J Bone Miner Res. 2010;25:222–34.
12. Rosen ED, Walkey CJ, Puigserver P, Spiegelman BM. Transcriptional regulation of adipogenesis. Genes Dev. 2000;14:1293–307.
13. Farmer SR. Regulation of PPARgamma activity during adipogenesis. Int J Obes. 2005;29 Suppl 1:S13–6.
14. Yokoyama S, Ito Y, Ueno-Kudoh H, Shimizu H, Uchibe K, Albini S, Mitsuoka K, Miyaki S, Kiso M, Nagai A, Hikata T, Osada T, Fukuda N, Yamashita S, Harada D, Mezzano V, Kasai M, Puri PL, Hayashizaki Y, Okado H, Hashimoto M, Asahara H. A systems approach reveals that the myogenesis genome network is regulated by the transcriptional repressor RP58. Dev Cell. 2009;17:836–48.
15. Harada S, Rodan GA. Control of osteoblast function and regulation of bone mass. Nature. 2003;423:349–55.
16. Olsen BR, Reginato AM, Wang W. Bone development. Annu Rev Cell Dev Biol. 2000; 16:191–220.
17. Ralston SH, de Crombrugghe B. Genetic regulation of bone mass and susceptibility to osteoporosis. Genes Dev. 2006;20:2492–506.
18. Luo J, Sun MH, Kang Q, Peng Y, Jiang W, Luu HH, Luo Q, Park JY, Li Y, Haydon RC, He TC. Gene therapy for bone regeneration. Curr Gene Ther. 2005;5:167–79.
19. Luu HH, Song WX, Luo X, Manning D, Luo J, Deng ZL, Sharff KA, Montag AG, Haydon RC, He TC. Distinct roles of bone morphogenetic proteins in osteogenic differentiation of mesenchymal stem cells. J Orthop Res. 2007;25:665–77.
20. Karsenty G. The genetic transformation of bone biology. Genes Dev. 1999;13:3037–51.
21. Li X, Cao X. BMP signaling and skeletogenesis. Ann N Y Acad Sci. 2006;1068:26–40.
22. Lian JB, Stein GS, Stein JL, van Wijnen AJ. Transcriptional control of osteoblast differentiation. Biochem Soc Trans. 1998;26:14–21.

23. Reddi AH. Bone morphogenetic proteins: an unconventional approach to isolation of first mammalian morphogens. Cytokine Growth Factor Rev. 1997;8:11–20.
24. Goldring MB, Tsuchimochi K, Ijiri K. The control of chondrogenesis. J Cell Biochem. 2006;97:33–44.
25. Akiyama H. Control of chondrogenesis by the transcription factor Sox9. Mod Rheumatol. 2008;18:213–9.
26. Massague J, Wotton D. Transcriptional control by the TGF-beta/Smad signaling system. EMBO J. 2000;19:1745–54.
27. Tuli R, Seghatoleslami MR, Tuli S, Howard MS, Danielson KG, Tuan RS. p38 MAP kinase regulation of AP-2 binding in TGF-beta1-stimulated chondrogenesis of human trabecular bone-derived cells. Ann N Y Acad Sci. 2002;961:172–7.
28. Chen L, Li C, Qiao W, Xu X, Deng C. A Ser(365) → Cys mutation of fibroblast growth factor receptor 3 in mouse downregulates Ihh/PTHrP signals and causes severe achondroplasia. Hum Mol Genet. 2001;10:457–65.
29. Naski MC, Ornitz DM. FGF signaling in skeletal development. Front Biosci. 1998;3:d781–94.
30. Makower AM, Wroblewski J, Pawlowski A. Effects of IGF-I, rGH, FGF, EGF and NCS on DNA-synthesis, cell proliferation and morphology of chondrocytes isolated from rat rib growth cartilage. Cell Biol Int Rep. 1989;13:259–70.
31. Hidaka K, Kanematsu T, Takeuchi H, Nakata M, Kikkawa U, Hirata M. Involvement of the phosphoinositide 3-kinase/protein kinase B signaling pathway in insulin/IGF-I-induced chondrogenesis of the mouse embryonal carcinoma-derived cell line ATDC5. Int J Biochem Cell Biol. 2001;33:1094–103.
32. Fukumoto T, Sperling JW, Sanyal A, Fitzsimmons JS, Reinholz GG, Conover CA, O'Driscoll SW. Combined effects of insulin-like growth factor-1 and transforming growth factor-beta1 on periosteal mesenchymal cells during chondrogenesis in vitro. Osteoarthritis Cartilage. 2003;11:55–64.
33. Frenz DA, Liu W, Williams JD, Hatcher V, Galinovic-Schwartz V, Flanders KC, Van de Water TR. Induction of chondrogenesis: requirement for synergistic interaction of basic fibroblast growth factor and transforming growth factor-beta. Development. 1994;120:415–24.
34. Heng BC, Cao T, Lee EH. Directing stem cell differentiation into the chondrogenic lineage in vitro. Stem Cells. 2004;22:1152–67.
35. Liu Z, Xu J, Colvin JS, Ornitz DM. Coordination of chondrogenesis and osteogenesis by fibroblast growth factor 18. Genes Dev. 2002;16:859–69.
36. Minina E, Kreschel C, Naski MC, Ornitz DM, Vortkamp A. Interaction of FGF, Ihh/Pthlh, and BMP signaling integrates chondrocyte proliferation and hypertrophic differentiation. Dev Cell. 2002;3:439–49.
37. Vortkamp A, Lee K, Lanske B, Segre GV, Kronenberg HM, Tabin CJ. Regulation of rate of cartilage differentiation by Indian hedgehog and PTH-related protein. Science. 1996;273:613–22.
38. Niswander L. Interplay between the molecular signals that control vertebrate limb development. Int J Dev Biol. 2002;46:877–81.
39. Otto TC, Lane MD. Adipose development: from stem cell to adipocyte. Crit Rev Biochem Mol Biol. 2005;40:229–42.
40. Zuk PA, Zhu M, Mizuno H, Huang J, Futrell JW, Katz AJ, Benhaim P, Lorenz HP, Hedrick MH. Multilineage cells from human adipose tissue: implications for cell-based therapies. Tissue Eng. 2001;7:211–28.
41. Bailey DM, Davies B, Milledge JS, Richards M, Williams SR, Jordinson M, Calam J. Elevated plasma cholecystokinin at high altitude: metabolic implications for the anorexia of acute mountain sickness. High Alt Med Biol. 2000;1:9–23.
42. Buckingham M, Bajard L, Chang T, Daubas P, Hadchouel J, Meilhac S, Montarras D, Rocancourt D, Relaix F. The formation of skeletal muscle: from somite to limb. J Anat. 2003;202:59–68.
43. De Ugarte DA, Morizono K, Elbarbary A, Alfonso Z, Zuk PA, Zhu M, Dragoo JL, Ashjian P, Thomas B, Benhaim P, Chen I, Fraser J, Hedrick MH. Comparison of multi-lineage cells from human adipose tissue and bone marrow. Cells Tissues Organs. 2003;174:101–9.

44. Digirolamo CM, Stokes D, Colter D, Phinney DG, Class R, Prockop DJ. Propagation and senescence of human marrow stromal cells in culture: a simple colony-forming assay identifies samples with the greatest potential to propagate and differentiate. Br J Haematol. 1999;107:275–81.
45. Noth U, Osyczka AM, Tuli R, Hickok NJ, Danielson KG, Tuan RS. Multilineage mesenchymal differentiation potential of human trabecular bone-derived cells. J Orthop Res. 2002;20:1060–9.
46. Black BL, Olson EN. Transcriptional control of muscle development by myocyte enhancer factor-2 (MEF2) proteins. Annu Rev Cell Dev Biol. 1998;14:167–96.
47. Chen JC, Goldhamer DJ. Skeletal muscle stem cells. Reprod Biol Endocrinol. 2003;1:101.
48. Hughes SM. Muscle development: reversal of the differentiated state. Curr Biol. 2001;11: R237–9.
49. Wagers AJ, Conboy IM. Cellular and molecular signatures of muscle regeneration: current concepts and controversies in adult myogenesis. Cell. 2005;122:659–67.
50. Attisano L, Wrana JL. Signal transduction by the TGF-beta superfamily. Science. 2002;296:1646–7.
51. Blobe GC, Schiemann WP, Lodish HF. Role of transforming growth factor beta in human disease. N Engl J Med. 2000;342:1350–8.
52. Feng XH, Zhang Y, Wu RY, Derynck R. The tumor suppressor Smad4/DPC4 and transcriptional adaptor CBP/p300 are coactivators for smad3 in TGF-beta-induced transcriptional activation. Genes Dev. 1998;12:2153–63.
53. Massague J. TGF-beta signal transduction. Annu Rev Biochem. 1998;67:753–91.
54. Mishra L, Derynck R, Mishra B. Transforming growth factor-beta signaling in stem cells and cancer. Science. 2005;310:68–71.
55. Shi Y, Massague J. Mechanisms of TGF-beta signaling from cell membrane to the nucleus. Cell. 2003;113:685–700.
56. ten Dijke P, Hill CS. New insights into TGF-beta-Smad signalling. Trends Biochem Sci. 2004;29:265–73.
57. Waite KA, Eng C. From developmental disorder to heritable cancer: it's all in the BMP/TGF-beta family. Nat Rev Genet. 2003;4:763–73.
58. Harradine KA, Akhurst RJ. Mutations of TGFbeta signaling molecules in human disease. Ann Med. 2006;38:403–14.
59. Serra R, Chang C. TGF-beta signaling in human skeletal and patterning disorders. Birth Defects Res C Embryo Today. 2003;69:333–51.
60. Zhao GQ. Consequences of knocking out BMP signaling in the mouse. Genesis. 2003;35:43–56.
61. Fujita T, Azuma Y, Fukuyama R, Hattori Y, Yoshida C, Koida M, Ogita K, Komori T. Runx2 induces osteoblast and chondrocyte differentiation and enhances their migration by coupling with PI3K-Akt signaling. J Cell Biol. 2004;166:85–95.
62. Enomoto H, Furuichi T, Zanma A, Yamana K, Yoshida C, Sumitani S, Yamamoto H, Enomoto-Iwamoto M, Iwamoto M, Komori T. Runx2 deficiency in chondrocytes causes adipogenic changes in vitro. J Cell Sci. 2004;117:417–25.
63. Hoffmann A, Gross G. BMP signaling pathways in cartilage and bone formation. Crit Rev Eukaryot Gene Expr. 2001;11:23–45.
64. Hogan BL. Bone morphogenetic proteins: multifunctional regulators of vertebrate development. Genes Dev. 1996;10:1580–94.
65. Rosen V, Thies RS. The BMP proteins in bone formation and repair. Trends Genet. 1992; 8:97–102.
66. Hall PA, Watt FM. Stem cells: the generation and maintenance of cellular diversity. Development. 1989;106:619–33.
67. Weissman IL, Anderson DJ, Gage F. Stem and progenitor cells: origins, phenotypes, lineage commitments, and transdifferentiations. Annu Rev Cell Dev Biol. 2001;17:387–403.
68. Keller G. Embryonic stem cell differentiation: emergence of a new era in biology and medicine. Genes Dev. 2005;19:1129–55.

69. Vats A, Bielby RC, Tolley NS, Nerem R, Polak JM. Stem cells. Lancet. 2005;366:592–602.
70. Eckfeldt CE, Mendenhall EM, Verfaillie CM. The molecular repertoire of the 'almighty' stem cell. Nat Rev. 2005;6:726–37.
71. Massague J, Weis-Garcia F. Serine/threonine kinase receptors: mediators of transforming growth factor beta family signals. Cancer Surv. 1996;27:41–64.
72. Yamashita H, Ten Dijke P, Heldin CH, Miyazono K. Bone morphogenetic protein receptors. Bone. 1996;19:569–74.
73. Wozney JM, Rosen V, Celeste AJ, Mitsock LM, Whitters MJ, Kriz RW, Hewick RM, Wang EA. Novel regulators of bone formation: molecular clones and activities. Science. 1988;242:1528–34.
74. Heldin CH, Miyazono K, ten Dijke P. TGF-beta signalling from cell membrane to nucleus through SMAD proteins. Nature. 1997;390:465–71.
75. Wrana JL. Regulation of Smad activity. Cell. 2000;100:189–92.
76. Massague J, Seoane J, Wotton D. Smad transcription factors. Genes Dev. 2005;19:2783–810.
77. Derynck R, Zhang Y, Feng XH. Smads: transcriptional activators of TGF-beta responses. Cell. 1998;95:737–40.
78. Itoh S, Itoh F, Goumans MJ, Ten Dijke P. Signaling of transforming growth factor-beta family members through Smad proteins. Eur J Biochem/FEBS. 2000;267:6954–67.
79. Miyazono K, ten Dijke P, Heldin CH. TGF-beta signaling by Smad proteins. Adv Immunol. 2000;75:115–57.
80. Ten Dijke P, Goumans MJ, Itoh F, Itoh S. Regulation of cell proliferation by Smad proteins. J Cell Physiol. 2002;191:1–16.
81. Cheng H, Jiang W, Phillips FM, Haydon RC, Peng Y, Zhou L, Luu HH, An N, Breyer B, Vanichakarn P, Szatkowski JP, Park JY, He TC. Osteogenic activity of the fourteen types of human bone morphogenetic proteins (BMPs). J Bone Joint Surg. 2003;85-A:1544–52.
82. Kang Q, Sun MH, Cheng H, Peng Y, Montag AG, Deyrup AT, Jiang W, Luu HH, Luo J, Szatkowski JP, Vanichakarn P, Park JY, Li Y, Haydon RC, He TC. Characterization of the distinct orthotopic bone-forming activity of 14 BMPs using recombinant adenovirus-mediated gene delivery. Gene Ther. 2004;11:1312–20.
83. He TC. Distinct osteogenic activity of BMPs and their orthopaedic applications. J Musculoskelet Neuronal Interact. 2005;5:363–6.
84. Lian JB, Stein GS, Javed A, van Wijnen AJ, Stein JL, Montecino M, Hassan MQ, Gaur T, Lengner CJ, Young DW. Networks and hubs for the transcriptional control of osteoblastogenesis. Rev Endocr Metab Disord. 2006;7:1–16.
85. Luo Q, Kang Q, Si W, Jiang W, Park JK, Peng Y, Li X, Luu HH, Luo J, Montag AG, Haydon RC, He TC. Connective tissue growth factor (CTGF) is regulated by Wnt and bone morphogenetic proteins signaling in osteoblast differentiation of mesenchymal stem cells. J Biol Chem. 2004;279:55958–68.
86. Peng Y, Kang Q, Cheng H, Li X, Sun MH, Jiang W, Luu HH, Park JY, Haydon RC, He TC. Transcriptional characterization of bone morphogenetic proteins (BMPs)-mediated osteogenic signaling. J Cell Biochem. 2003;90:1149–65.
87. Peng Y, Kang Q, Luo Q, Jiang W, Si W, Liu BA, Luu HH, Park JK, Li X, Luo J, Montag AG, Haydon RC, He TC. Inhibitor of DNA binding/differentiation helix-loop-helix proteins mediate bone morphogenetic protein-induced osteoblast differentiation of mesenchymal stem cells. J Biol Chem. 2004;279:32941–9.
88. Yamaguchi A, Komori T, Suda T. Regulation of osteoblast differentiation mediated by bone morphogenetic proteins, hedgehogs, and Cbfa1. Endocr Rev. 2000;21:393–411.
89. Kang Q, Song WX, Luo Q, Tang N, Luo J, Luo X, Chen J, Bi Y, He BC, Park JK, Jiang W, Tang Y, Huang J, Su Y, Zhu GH, He Y, Yin H, Hu Z, Wang Y, Chen L, Zuo GW, Pan X, Shen J, Vokes T, Reid RR, Haydon RC, Luu HH, He TC. A comprehensive analysis of the dual roles of BMPs in regulating adipogenic and osteogenic differentiation of mesenchymal progenitor cells. Stem Cells Dev. 2009;18:545–59.
90. Si W, Kang Q, Luu HH, Park JK, Luo Q, Song WX, Jiang W, Luo X, Li X, Yin H, Montag AG, Haydon RC, He TC. CCN1/Cyr61 is regulated by the canonical Wnt signal and plays an

important role in Wnt3A-induced osteoblast differentiation of mesenchymal stem cells. Mol Cell Biol. 2006;26:2955–64.
91. Aslan H, Zilberman Y, Arbeli V, Sheyn D, Matan Y, Liebergall M, Li JZ, Helm GA, Gazit D, Gazit Z. Nucleofection-based ex vivo nonviral gene delivery to human stem cells as a platform for tissue regeneration. Tissue Eng. 2006;12:877–89.
92. Santos JL, Pandita D, Rodrigues J, Pego AP, Granja PL, Tomas H. Non-viral gene delivery to mesenchymal stem cells: methods, strategies and application in bone tissue engineering and regeneration. Curr Gene Ther. 2011;11:46–57.
93. Sheyn D, Kimelman-Bleich N, Pelled G, Zilberman Y, Gazit D, Gazit Z. Ultrasound-based nonviral gene delivery induces bone formation in vivo. Gene Ther. 2008;15:257–66.
94. Sharff KA, Song WX, Luo X, Tang N, Luo J, Chen J, Bi Y, He BC, Huang J, Li X, Jiang W, Zhu GH, Su Y, He Y, Shen J, Wang Y, Chen L, Zuo GW, Liu B, Pan X, Reid RR, Luu HH, Haydon RC, He TC. Hey1 basic helix-loop-helix protein plays an important role in mediating BMP9-induced osteogenic differentiation of mesenchymal progenitor cells. J Biol Chem. 2009;284:649–59.
95. Tang N, Song WX, Luo J, Luo X, Chen J, Sharff KA, Bi Y, He BC, Huang JY, Zhu GH, Su YX, Jiang W, Tang M, He Y, Wang Y, Chen L, Zuo GW, Shen J, Pan X, Reid RR, Luu HH, Haydon RC, He TC. BMP-9-induced osteogenic differentiation of mesenchymal progenitors requires functional canonical Wnt/beta-catenin signalling. J Cell Mol Med. 2009;13:2448–64.
96. Chen L, Jiang W, Huang J, He BC, Zuo GW, Zhang W, Luo Q, Shi Q, Zhang BQ, Wagner ER, Luo J, Tang M, Wietholt C, Luo X, Bi Y, Su Y, Liu B, Kim SH, He CJ, Hu Y, Shen J, Rastegar F, Huang E, Gao Y, Gao JL, Zhou JZ, Reid RR, Luu HH, Haydon RC, He TC, Deng ZL. Insulin-like growth factor 2 (IGF-2) potentiates BMP-9-induced osteogenic differentiation and bone formation. J Bone Miner Res. 2010;25:2447–59.
97. Zhang W, Deng ZL, Chen L, Zuo GW, Luo Q, Shi Q, Zhang BQ, Wagner ER, Rastegar F, Kim SH, Jiang W, Shen J, Huang E, Gao Y, Gao JL, Zhou JZ, Luo J, Huang J, Luo X, Bi Y, Su Y, Yang K, Liu H, Luu HH, Haydon RC, He TC, He BC. Retinoic acids potentiate BMP9-induced osteogenic differentiation of mesenchymal progenitor cells. PLoS One. 2010;5:e11917.
98. Urist MR. Bone: formation by autoinduction. Science. 1965;150:893–9.
99. Yoon BS, Lyons KM. Multiple functions of BMPs in chondrogenesis. J Cell Biochem. 2004;93:93–103.
100. Yoon BS, Ovchinnikov DA, Yoshii I, Mishina Y, Behringer RR, Lyons KM. Bmpr1a and Bmpr1b have overlapping functions and are essential for chondrogenesis in vivo. Proc Natl Acad Sci USA. 2005;102:5062–7.
101. Eames BF, de la Fuente L, Helms JA. Molecular ontogeny of the skeleton. Birth Defects Res C Embryo Today. 2003;69:93–101.
102. Lefebvre V, Behringer RR, de Crombrugghe B. L-Sox5, Sox6 and Sox9 control essential steps of the chondrocyte differentiation pathway. Osteoarthritis Cartilage. 2001;9(Suppl A): S69–75.
103. Ng LJ, Wheatley S, Muscat GE, Conway-Campbell J, Bowles J, Wright E, Bell DM, Tam PP, Cheah KS, Koopman P. SOX9 binds DNA, activates transcription, and coexpresses with type II collagen during chondrogenesis in the mouse. Dev Biol. 1997;183:108–21.
104. Lefebvre V, Li P, de Crombrugghe B. A new long form of Sox5 (L-Sox5), Sox6 and Sox9 are coexpressed in chondrogenesis and cooperatively activate the type II collagen gene. EMBO J. 1998;17:5718–33.
105. Cadigan KM, Nusse R. Wnt signaling: a common theme in animal development. Genes Dev. 1997;11:3286–305.
106. Croce JC, McClay DR. The canonical Wnt pathway in embryonic axis polarity. Semin Cell Dev Biol. 2006;17:168–74.
107. Kleber M, Sommer L. Wnt signaling and the regulation of stem cell function. Curr Opin Cell Biol. 2004;16:681–7.
108. Logan CY, Nusse R. The Wnt signaling pathway in development and disease. Annu Rev Cell Dev Biol. 2004;20:781–810.

109. Luo J, Chen J, Deng ZL, Luo X, Song WX, Sharff KA, Tang N, Haydon RC, Luu HH, He TC. Wnt signaling and human diseases: what are the therapeutic implications? Lab Invest. 2007;87:97–103.
110. Marikawa Y. Wnt/beta-catenin signaling and body plan formation in mouse embryos. Semin Cell Dev Biol. 2006;17:175–84.
111. van Amerongen R, Berns A. Knockout mouse models to study Wnt signal transduction. Trends Genet. 2006;22:678–89.
112. Vincan E. Frizzled/WNT signalling: the insidious promoter of tumour growth and progression. Front Biosci. 2004;9:1023–34.
113. Westendorf JJ, Kahler RA, Schroeder TM. Wnt signaling in osteoblasts and bone diseases. Gene. 2004;341:19–39.
114. Wodarz A, Nusse R. Mechanisms of Wnt signaling in development. Annu Rev Cell Dev Biol. 1998;14:59–88.
115. Clevers H. Wnt/beta-catenin signaling in development and disease. Cell. 2006;127:469–80.
116. Luu HH, Zhang R, Haydon RC, Rayburn E, Kang Q, Si W, Park JK, Wang H, Peng Y, Jiang W, He TC. Wnt/beta-catenin signaling pathway as a novel cancer drug target. Curr Cancer Drug Targets. 2004;4:653–71.
117. Reya T, Clevers H. Wnt signalling in stem cells and cancer. Nature. 2005;434:843–50.
118. Gong Y, Slee RB, Fukai N, Rawadi G, Roman-Roman S, Reginato AM, Wang H, Cundy T, Glorieux FH, Lev D, Zacharin M, Oexle K, Marcelino J, Suwairi W, Heeger S, Sabatakos G, Apte S, Adkins WN, Allgrove J, Arslan-Kirchner M, Batch JA, Beighton P, Black GC, Boles RG, Boon LM, Borrone C, Brunner HG, Carle GF, Dallapiccola B, De Paepe A, Floege B, Halfhide ML, Hall B, Hennekam RC, Hirose T, Jans A, Juppner H, Kim CA, Keppler-Noreuil K, Kohlschuetter A, LaCombe D, Lambert M, Lemyre E, Letteboer T, Peltonen L, Ramesar RS, Romanengo M, Somer H, Steichen-Gersdorf E, Steinmann B, Sullivan B, Superti-Furga A, Swoboda W, van den Boogaard MJ, Van Hul W, Vikkula M, Votruba M, Zabel B, Garcia T, Baron R, Olsen BR, Warman ML. LDL receptor-related protein 5 (LRP5) affects bone accrual and eye development. Cell. 2001;107:513–23.
119. He TC, Chan TA, Vogelstein B, Kinzler KW. PPARdelta is an APC-regulated target of non-steroidal anti-inflammatory drugs. Cell. 1999;99:335–45.
120. He TC, Sparks AB, Rago C, Hermeking H, Zawel L, da Costa LT, Morin PJ, Vogelstein B, Kinzler KW. Identification of c-MYC as a target of the APC pathway. Science. 1998;281: 1509–12.
121. Pennica D, Swanson TA, Welsh JW, Roy MA, Lawrence DA, Lee J, Brush J, Taneyhill LA, Deuel B, Lew M, Watanabe C, Cohen RL, Melhem MF, Finley GG, Quirke P, Goddard AD, Hillan KJ, Gurney AL, Botstein D, Levine AJ. WISP genes are members of the connective tissue growth factor family that are up-regulated in wnt-1-transformed cells and aberrantly expressed in human colon tumors. Proc Natl Acad Sci USA. 1998;95:14717–22.
122. Shtutman M, Zhurinsky J, Simcha I, Albanese C, D'Amico M, Pestell R, Ben-Ze'ev A. The cyclin D1 gene is a target of the beta-catenin/LEF-1 pathway. Proc Natl Acad Sci USA. 1999;96:5522–7.
123. Tetsu O, McCormick F. Beta-catenin regulates expression of cyclin D1 in colon carcinoma cells. Nature. 1999;398:422–6.
124. Xu L, Corcoran RB, Welsh JW, Pennica D, Levine AJ. WISP-1 is a Wnt-1- and beta-catenin-responsive oncogene. Genes Dev. 2000;14:585–95.
125. Bergwitz C, Wendlandt T, Kispert A, Brabant G. Wnts differentially regulate colony growth and differentiation of chondrogenic rat calvaria cells. Biochim Biophys Acta. 2001;1538:129–40.
126. Fischer L, Boland G, Tuan RS. Wnt signaling during BMP-2 stimulation of mesenchymal chondrogenesis. J Cell Biochem. 2002;84:816–31.
127. Gavin BJ, McMahon JA, McMahon AP. Expression of multiple novel Wnt-1/int-1-related genes during fetal and adult mouse development. Genes Dev. 1990;4:2319–32.
128. Gregory CA, Gunn WG, Reyes E, Smolarz AJ, Munoz J, Spees JL, Prockop DJ. How Wnt signaling affects bone repair by mesenchymal stem cells from the bone marrow. Ann N Y Acad Sci. 2005;1049:97–106.

129. Kengaku M, Capdevila J, Rodriguez-Esteban C, De La Pena J, Johnson RL, Izpisua Belmonte JC, Tabin CJ. Distinct WNT pathways regulating AER formation and dorsoventral polarity in the chick limb bud. Science. 1998;280:1274–7.
130. Wang J, Wynshaw-Boris A. The canonical Wnt pathway in early mammalian embryogenesis and stem cell maintenance/differentiation. Curr Opin Genet Dev. 2004;14:533–9.
131. Kang S, Bennett CN, Gerin I, Rapp LA, Hankenson KD, Macdougald OA. Wnt signaling stimulates osteoblastogenesis of mesenchymal precursors by suppressing CCAAT/enhancer-binding protein alpha and peroxisome proliferator-activated receptor gamma. J Biol Chem. 2007;282:14515–24.
132. Takada I, Kouzmenko AP, Kato S. Wnt and PPARgamma signaling in osteoblastogenesis and adipogenesis. Nat Rev Rheumatol. 2009;5:442–7.
133. Glass 2nd DA, Bialek P, Ahn JD, Starbuck M, Patel MS, Clevers H, Taketo MM, Long F, McMahon AP, Lang RA, Karsenty G. Canonical Wnt signaling in differentiated osteoblasts controls osteoclast differentiation. Dev Cell. 2005;8:751–64.
134. Krishnan V, Bryant HU, Macdougald OA. Regulation of bone mass by Wnt signaling. J Clin Invest. 2006;116:1202–9.
135. Bennett CN, Longo KA, Wright WS, Suva LJ, Lane TF, Hankenson KD, MacDougald OA. Regulation of osteoblastogenesis and bone mass by Wnt10b. Proc Natl Acad Sci USA. 2005;102:3324–9.
136. Glass 2nd DA, Karsenty G. In vivo analysis of Wnt signaling in bone. Endocrinology. 2007;148:2630–4.
137. Takada I, Mihara M, Suzawa M, Ohtake F, Kobayashi S, Igarashi M, Youn MY, Takeyama K, Nakamura T, Mezaki Y, Takezawa S, Yogiashi Y, Kitagawa H, Yamada G, Takada S, Minami Y, Shibuya H, Matsumoto K, Kato S. A histone lysine methyltransferase activated by non-canonical Wnt signalling suppresses PPAR-gamma transactivation. Nat Cell Biol. 2007;9:1273–85.
138. Wagner ER, Zhu G, Zhang BQ, Luo Q, Shi Q, Huang E, Gao Y, Gao JL, Kim SH, Rastegar F, Yang K, He BC, Chen L, Zuo GW, Bi Y, Su Y, Luo J, Luo X, Huang J, Deng ZL, Reid RR, Luu HH, Haydon RC, He TC. The therapeutic potential of the Wnt signaling pathway in bone disorders. Curr Mol Pharmacol. 2011;4:14–25.
139. Li X, Liu P, Liu W, Maye P, Zhang J, Zhang Y, Hurley M, Guo C, Boskey A, Sun L, Harris SE, Rowe DW, Ke HZ, Wu D. Dkk2 has a role in terminal osteoblast differentiation and mineralized matrix formation. Nat Genet. 2005;37:945–52.
140. van der Horst G, van der Werf SM, Farih-Sips H, van Bezooijen RL, Lowik CW, Karperien M. Downregulation of Wnt signaling by increased expression of Dickkopf-1 and -2 is a prerequisite for late-stage osteoblast differentiation of KS483 cells. J Bone Miner Res. 2005;20:1867–77.
141. Boyden LM, Mao J, Belsky J, Mitzner L, Farhi A, Mitnick MA, Wu D, Insogna K, Lifton RP. High bone density due to a mutation in LDL-receptor-related protein 5. N Engl J Med. 2002;346:1513–21.
142. Little RD, Carulli JP, Del Mastro RG, Dupuis J, Osborne M, Folz C, Manning SP, Swain PM, Zhao SC, Eustace B, Lappe MM, Spitzer L, Zweier S, Braunschweiger K, Benchekroun Y, Hu X, Adair R, Chee L, FitzGerald MG, Tulig C, Caruso A, Tzellas N, Bawa A, Franklin B, McGuire S, Nogues X, Gong G, Allen KM, Anisowicz A, Morales AJ, Lomedico PT, Recker SM, Van Eerdewegh P, Recker RR, Johnson ML. A mutation in the LDL receptor-related protein 5 gene results in the autosomal dominant high-bone-mass trait. Am J Hum Genet. 2002;70:11–9.
143. Holmen SL, Giambernardi TA, Zylstra CR, Buckner-Berghuis BD, Resau JH, Hess JF, Glatt V, Bouxsein ML, Ai M, Warman ML, Williams BO. Decreased BMD and limb deformities in mice carrying mutations in both Lrp5 and Lrp6. J Bone Miner Res. 2004;19:2033–40.
144. Li X, Zhang Y, Kang H, Liu W, Liu P, Zhang J, Harris SE, Wu D. Sclerostin binds to LRP5/6 and antagonizes canonical Wnt signaling. J Biol Chem. 2005;280:19883–7.
145. Boland GM, Perkins G, Hall DJ, Tuan RS. Wnt 3a promotes proliferation and suppresses osteogenic differentiation of adult human mesenchymal stem cells. J Cell Biochem. 2004;93:1210–30.

146. Holmen SL, Zylstra CR, Mukherjee A, Sigler RE, Faugere MC, Bouxsein ML, Deng L, Clemens TL, Williams BO. Essential role of beta-catenin in postnatal bone acquisition. J Biol Chem. 2005;280:21162–8.
147. Day TF, Guo X, Garrett-Beal L, Yang Y. Wnt/beta-catenin signaling in mesenchymal progenitors controls osteoblast and chondrocyte differentiation during vertebrate skeletogenesis. Dev Cell. 2005;8:739–50.
148. Hill TP, Spater D, Taketo MM, Birchmeier W, Hartmann C. Canonical Wnt/beta-catenin signaling prevents osteoblasts from differentiating into chondrocytes. Dev Cell. 2005;8:727–38.
149. Tu X, Joeng KS, Nakayama KI, Nakayama K, Rajagopal J, Carroll TJ, McMahon AP, Long F. Noncanonical Wnt signaling through G protein-linked PKCdelta activation promotes bone formation. Dev Cell. 2007;12:113–27.
150. Cossu G, Borello U. Wnt signaling and the activation of myogenesis in mammals. EMBO J. 1999;18:6867–72.
151. Hartmann C, Tabin CJ. Dual roles of Wnt signaling during chondrogenesis in the chicken limb. Development. 2000;127:3141–59.
152. Ross SE, Hemati N, Longo KA, Bennett CN, Lucas PC, Erickson RL, MacDougald OA. Inhibition of adipogenesis by Wnt signaling. Science. 2000;289:950–3.
153. Chen Y, Whetstone HC, Lin AC, Nadesan P, Wei Q, Poon R, Alman BA. Beta-catenin signaling plays a disparate role in different phases of fracture repair: implications for therapy to improve bone healing. PLoS Med. 2007;4:e249.
154. Fischer L, Boland G, Tuan RS. Wnt-3A enhances bone morphogenetic protein-2-mediated chondrogenesis of murine C3H10T1/2 mesenchymal cells. J Biol Chem. 2002;277:30870–8.
155. Ogden SK, Ascano Jr M, Stegman MA, Robbins DJ. Regulation of Hedgehog signaling: a complex story. Biochem Pharmacol. 2004;67:805–14.
156. Sciaudone M, Gazzerro E, Priest L, Delany AM, Canalis E. Notch 1 impairs osteoblastic cell differentiation. Endocrinology. 2003;144:5631–9.
157. Almeida M, Han L, Bellido T, Manolagas SC, Kousteni S. Wnt proteins prevent apoptosis of both uncommitted osteoblast progenitors and differentiated osteoblasts by beta-catenin-dependent and -independent signaling cascades involving Src/ERK and phosphatidylinositol 3-kinase/AKT. J Biol Chem. 2005;280:41342–51.
158. Ornitz DM. FGFs, heparan sulfate and FGFRs: complex interactions essential for development. Bioessays. 2000;22:108–12.
159. Ornitz DM, Itoh N. Fibroblast growth factors. Genome Biol. 2001;2:REVIEWS3005.
160. Chen L, Deng CX. Roles of FGF signaling in skeletal development and human genetic diseases. Front Biosci. 2005;10:1961–76.
161. Jackson RA, McDonald MM, Nurcombe V, Little DG, Cool SM. The use of heparan sulfate to augment fracture repair in a rat fracture model. J Orthop Res. 2006;24:636–44.
162. Ornitz DM. FGF signaling in the developing endochondral skeleton. Cytokine Growth Factor Rev. 2005;16:205–13.
163. Ornitz DM, Marie PJ. FGF signaling pathways in endochondral and intramembranous bone development and human genetic disease. Genes Dev. 2002;16:1446–65.
164. Eswarakumar VP, Lax I, Schlessinger J. Cellular signaling by fibroblast growth factor receptors. Cytokine Growth Factor Rev. 2005;16:139–49.
165. Jackson RA, Nurcombe V, Cool SM. Coordinated fibroblast growth factor and heparan sulfate regulation of osteogenesis. Gene. 2006;379:79–91.
166. Woei Ng K, Speicher T, Dombrowski C, Helledie T, Haupt LM, Nurcombe V, Cool SM. Osteogenic differentiation of murine embryonic stem cells is mediated by fibroblast growth factor receptors. Stem Cells Dev. 2007;16:305–18.
167. Valverde-Franco G, Liu H, Davidson D, Chai S, Valderrama-Carvajal H, Goltzman D, Ornitz DM, Henderson JE. Defective bone mineralization and osteopenia in young adult FGFR3–/– mice. Hum Mol Genet. 2004;13:271–84.
168. Chen L, Adar R, Yang X, Monsonego EO, Li C, Hauschka PV, Yayon A, Deng CX. Gly369Cys mutation in mouse FGFR3 causes achondroplasia by affecting both chondrogenesis and osteogenesis. J Clin Invest. 1999;104:1517–25.

169. Sahni M, Ambrosetti DC, Mansukhani A, Gertner R, Levy D, Basilico C. FGF signaling inhibits chondrocyte proliferation and regulates bone development through the STAT-1 pathway. Genes Dev. 1999;13:1361–6.
170. Karaplis AC, Goltzman D. PTH and PTHrP effects on the skeleton. Rev Endocr Metab Disord. 2000;1:331–41.
171. Kronenberg HM. PTHrP and skeletal development. Ann N Y Acad Sci. 2006;1068:1–13.
172. Schipani E, Provot S. PTHrP, PTH, and the PTH/PTHrP receptor in endochondral bone development. Birth Defects Res C Embryo Today. 2003;69:352–62.
173. Kronenberg HM, Lanske B, Kovacs CS, Chung UI, Lee K, Segre GV, Schipani E, Juppner H. Functional analysis of the PTH/PTHrP network of ligands and receptors. Recent Prog Horm Res. 1998;53:283–301. discussion 301–283.
174. Kobayashi T, Chung UI, Schipani E, Starbuck M, Karsenty G, Katagiri T, Goad DL, Lanske B, Kronenberg HM. PTHrP and Indian hedgehog control differentiation of growth plate chondrocytes at multiple steps. Development. 2002;129:2977–86.
175. MacLean HE, Kronenberg HM. Localization of Indian hedgehog and PTH/PTHrP receptor expression in relation to chondrocyte proliferation during mouse bone development. Dev Growth Differ. 2005;47:59–63.
176. Karp SJ, Schipani E, St-Jacques B, Hunzelman J, Kronenberg H, McMahon AP. Indian hedgehog coordinates endochondral bone growth and morphogenesis via parathyroid hormone related-protein-dependent and -independent pathways. Development. 2000;127:543–8.
177. Chiba S. Notch signaling in stem cell systems. Stem Cells. 2006;24:2437–47.
178. Egan SE, St-Pierre B, Leow CC. Notch receptors, partners and regulators: from conserved domains to powerful functions. Curr Top Microbiol Immunol. 1998;228:273–324.
179. Ehebauer M, Hayward P, Arias AM. Notch, a universal arbiter of cell fate decisions. Science. 2006;314:1414–5.
180. Ehebauer M, Hayward P, Martinez-Arias A. Notch signaling pathway. Sci STKE. 2006;2006:cm7.
181. Hurlbut GD, Kankel MW, Lake RJ, Artavanis-Tsakonas S. Crossing paths with Notch in the hyper-network. Curr Opin Cell Biol. 2007;19:166–75.
182. Kanwar R, Fortini ME. Notch signaling: a different sort makes the cut. Curr Biol. 2004;14:R1043–5.
183. Le Borgne R. Regulation of Notch signalling by endocytosis and endosomal sorting. Curr Opin Cell Biol. 2006;18:213–22.
184. Shen J, Bronson RT, Chen DF, Xia W, Selkoe DJ, Tonegawa S. Skeletal and CNS defects in Presenilin-1-deficient mice. Cell. 1997;89:629–39.
185. Iso T, Kedes L, Hamamori Y. HES and HERP families: multiple effectors of the Notch signaling pathway. J Cell Physiol. 2003;194:237–55.
186. Schroeter EH, Kisslinger JA, Kopan R. Notch-1 signalling requires ligand-induced proteolytic release of intracellular domain. Nature. 1998;393:382–6.
187. Song W, Nadeau P, Yuan M, Yang X, Shen J, Yankner BA. Proteolytic release and nuclear translocation of Notch-1 are induced by presenilin-1 and impaired by pathogenic presenilin-1 mutations. Proc Natl Acad Sci USA. 1999;96:6959–63.
188. Swiatek PJ, Lindsell CE, del Amo FF, Weinmaster G, Gridley T. Notch1 is essential for post-implantation development in mice. Genes Dev. 1994;8:707–19.
189. Schnabel M, Fichtel I, Gotzen L, Schlegel J. Differential expression of Notch genes in human osteoblastic cells. Int J Mol Med. 2002;9:229–32.
190. Tezuka K, Yasuda M, Watanabe N, Morimura N, Kuroda K, Miyatani S, Hozumi N. Stimulation of osteoblastic cell differentiation by Notch. J Bone Miner Res. 2002;17:231–9.
191. Deregowski V, Gazzerro E, Priest L, Rydziel S, Canalis E. Notch 1 overexpression inhibits osteoblastogenesis by suppressing Wnt/beta-catenin but not bone morphogenetic protein signaling. J Biol Chem. 2006;281:6203–10.
192. Dahlqvist C, Blokzijl A, Chapman G, Falk A, Dannaeus K, Ibanez CF, Lendahl U. Functional Notch signaling is required for BMP4-induced inhibition of myogenic differentiation. Development. 2003;130:6089–99.

193. Smas CM, Sul HS. Pref-1, a protein containing EGF-like repeats, inhibits adipocyte differentiation. Cell. 1993;73:725–34.
194. Ehlen HW, Buelens LA, Vortkamp A. Hedgehog signaling in skeletal development. Birth Defects Res C Embryo Today. 2006;78:267–79.
195. Hooper JE, Scott MP. Communicating with Hedgehogs. Nat Rev. 2005;6:306–17.
196. Lum L, Beachy PA. The Hedgehog response network: sensors, switches, and routers. Science. 2004;304:1755–9.
197. Nybakken K, Perrimon N. Hedgehog signal transduction: recent findings. Curr Opin Genet Dev. 2002;12:503–11.
198. Riobo NA, Manning DR. Pathways of signal transduction employed by vertebrate Hedgehogs. Biochem J. 2007;403:369–79.
199. Baron MH, Fraser ST. The specification of early hematopoiesis in the mammal. Curr Opin Hematol. 2005;12:217–21.
200. Cohen Jr MM. The hedgehog signaling network. Am J Med Genet. 2003;123A:5–28.
201. Lupo G, Harris WA, Lewis KE. Mechanisms of ventral patterning in the vertebrate nervous system. Nat Rev Neurosci. 2006;7:103–14.
202. McMahon AP, Ingham PW, Tabin CJ. Developmental roles and clinical significance of hedgehog signaling. Curr Top Dev Biol. 2003;53:1–114.
203. Fuchs S, Dohle E, Kirkpatrick CJ. Sonic Hedgehog-mediated synergistic effects guiding angiogenesis and osteogenesis. Vitam Horm. 2012;88:491–506.
204. Stone DM, Hynes M, Armanini M, Swanson TA, Gu Q, Johnson RL, Scott MP, Pennica D, Goddard A, Phillips H, Noll M, Hooper JE, de Sauvage F, Rosenthal A. The tumour-suppressor gene patched encodes a candidate receptor for Sonic hedgehog. Nature. 1996;384:129–34.
205. Marigo V, Scott MP, Johnson RL, Goodrich LV, Tabin CJ. Conservation in hedgehog signaling: induction of a chicken patched homolog by Sonic hedgehog in the developing limb. Development. 1996;122:1225–33.
206. Ingham PW, McMahon AP. Hedgehog signaling in animal development: paradigms and principles. Genes Dev. 2001;15:3059–87.
207. Pola R, Ling LE, Silver M, Corbley MJ, Kearney M, Blake Pepinsky R, Shapiro R, Taylor FR, Baker DP, Asahara T, Isner JM. The morphogen Sonic hedgehog is an indirect angiogenic agent upregulating two families of angiogenic growth factors. Nat Med. 2001;7:706–11.
208. Yuasa T, Kataoka H, Kinto N, Iwamoto M, Enomoto-Iwamoto M, Iemura S, Ueno N, Shibata Y, Kurosawa H, Yamaguchi A. Sonic hedgehog is involved in osteoblast differentiation by cooperating with BMP-2. J Cell Physiol. 2002;193:225–32.
209. Miyaji T, Nakase T, Iwasaki M, Kuriyama K, Tamai N, Higuchi C, Myoui A, Tomita T, Yoshikawa H. Expression and distribution of transcripts for sonic hedgehog in the early phase of fracture repair. Histochem Cell Biol. 2003;119:233–7.
210. Jemtland R, Divieti P, Lee K, Segre GV. Hedgehog promotes primary osteoblast differentiation and increases PTHrP mRNA expression and iPTHrP secretion. Bone. 2003;32:611–20.
211. James AW, Leucht P, Levi B, Carre AL, Xu Y, Helms JA, Longaker MT. Sonic Hedgehog influences the balance of osteogenesis and adipogenesis in mouse adipose-derived stromal cells. Tissue Eng Part A. 2010;16:2605–16.
212. van der Horst G, Farih-Sips H, Lowik CW, Karperien M. Hedgehog stimulates only osteoblastic differentiation of undifferentiated KS483 cells. Bone. 2003;33:899–910.
213. Wang Q, Huang C, Zeng F, Xue M, Zhang X. Activation of the Hh pathway in periosteum-derived mesenchymal stem cells induces bone formation in vivo: implication for postnatal bone repair. Am J Pathol. 2010;177:3100–11.
214. Ito H, Akiyama H, Shigeno C, Iyama K, Matsuoka H, Nakamura T. Hedgehog signaling molecules in bone marrow cells at the initial stage of fracture repair. Biochem Biophys Res Commun. 1999;262:443–51.
215. Iwasaki M, Le AX, Helms JA. Expression of indian hedgehog, bone morphogenetic protein 6 and gli during skeletal morphogenesis. Mech Dev. 1997;69:197–202.
216. Vortkamp A, Pathi S, Peretti GM, Caruso EM, Zaleske DJ, Tabin CJ. Recapitulation of signals regulating embryonic bone formation during postnatal growth and in fracture repair. Mech Dev. 1998;71:65–76.

217. Mueller E, Drori S, Aiyer A, Yie J, Sarraf P, Chen H, Hauser S, Rosen ED, Ge K, Roeder RG, Spiegelman BM. Genetic analysis of adipogenesis through peroxisome proliferator-activated receptor gamma isoforms. J Biol Chem. 2002;277:41925–30.
218. Zhang J, Fu M, Cui T, Xiong C, Xu K, Zhong W, Xiao Y, Floyd D, Liang J, Li E, Song Q, Chen YE. Selective disruption of PPARgamma 2 impairs the development of adipose tissue and insulin sensitivity. Proc Natl Acad Sci USA. 2004;101:10703–8.
219. Tontonoz P, Hu E, Spiegelman BM. Stimulation of adipogenesis in fibroblasts by PPAR gamma 2, a lipid-activated transcription factor. Cell. 1994;79:1147–56.
220. Rosen ED, MacDougald OA. Adipocyte differentiation from the inside out. Nat Rev. 2006;7:885–96.
221. Nakamura T, Shiojima S, Hirai Y, Iwama T, Tsuruzoe N, Hirasawa A, Katsuma S, Tsujimoto G. Temporal gene expression changes during adipogenesis in human mesenchymal stem cells. Biochem Biophys Res Commun. 2003;303:306–12.
222. Allen T, Zhang F, Moodie SA, Clemens LE, Smith A, Gregoire F, Bell A, Muscat GE, Gustafson TA. Halofenate is a selective peroxisome proliferator-activated receptor gamma modulator with antidiabetic activity. Diabetes. 2006;55:2523–33.
223. Rosen ED, Spiegelman BM. PPARgamma: a nuclear regulator of metabolism, differentiation, and cell growth. J Biol Chem. 2001;276:37731–4.
224. Gimble JM, Robinson CE, Wu X, Kelly KA, Rodriguez BR, Kliewer SA, Lehmann JM, Morris DC. Peroxisome proliferator-activated receptor-gamma activation by thiazolidinediones induces adipogenesis in bone marrow stromal cells. Mol Pharmacol. 1996;50:1087–94.
225. Lehmann JM, Moore LB, Smith-Oliver TA, Wilkison WO, Willson TM, Kliewer SA. An antidiabetic thiazolidinedione is a high affinity ligand for peroxisome proliferator-activated receptor gamma (PPAR gamma). J Biol Chem. 1995;270:12953–6.
226. Tzameli I, Fang H, Ollero M, Shi H, Hamm JK, Kievit P, Hollenberg AN, Flier JS. Regulated production of a peroxisome proliferator-activated receptor-gamma ligand during an early phase of adipocyte differentiation in 3T3-L1 adipocytes. J Biol Chem. 2004;279:36093–102.
227. Schopfer FJ, Lin Y, Baker PR, Cui T, Garcia-Barrio M, Zhang J, Chen K, Chen YE, Freeman BA. Nitrolinoleic acid: an endogenous peroxisome proliferator-activated receptor gamma ligand. Proc Natl Acad Sci USA. 2005;102:2340–5.
228. Akune T, Ohba S, Kamekura S, Yamaguchi M, Chung UI, Kubota N, Terauchi Y, Harada Y, Azuma Y, Nakamura K, Kadowaki T, Kawaguchi H. PPARgamma insufficiency enhances osteogenesis through osteoblast formation from bone marrow progenitors. J Clin Invest. 2004;113:846–55.
229. Demay MB, Roth DA, Kronenberg HM. Regions of the rat osteocalcin gene which mediate the effect of 1,25-dihydroxyvitamin D3 on gene transcription. J Biol Chem. 1989; 264:2279–82.
230. Gouveia CH, Schultz JJ, Bianco AC, Brent GA. Thyroid hormone stimulation of osteocalcin gene expression in ROS 17/2.8 cells is mediated by transcriptional and post-transcriptional mechanisms. J Endocrinol. 2001;170:667–75.
231. Morrison NA, Shine J, Fragonas JC, Verkest V, McMenemy ML, Eisman JA. 1,25-dihydroxyvitamin D-responsive element and glucocorticoid repression in the osteocalcin gene. Science. 1989; 246:1158–61.
232. Cao Z, Umek RM, McKnight SL. Regulated expression of three C/EBP isoforms during adipose conversion of 3T3-L1 cells. Genes Dev. 1991;5:1538–52.
233. Lin FT, Lane MD. CCAAT/enhancer binding protein alpha is sufficient to initiate the 3T3-L1 adipocyte differentiation program. Proc Natl Acad Sci USA. 1994;91:8757–61.
234. Freytag SO, Paielli DL, Gilbert JD. Ectopic expression of the CCAAT/enhancer-binding protein alpha promotes the adipogenic program in a variety of mouse fibroblastic cells. Genes Dev. 1994;8:1654–63.
235. Wu Z, Xie Y, Bucher NL, Farmer SR. Conditional ectopic expression of C/EBP beta in NIH-3T3 cells induces PPAR gamma and stimulates adipogenesis. Genes Dev. 1995;9:2350–63.
236. Wu Z, Rosen ED, Brun R, Hauser S, Adelmant G, Troy AE, McKeon C, Darlington GJ, Spiegelman BM. Cross-regulation of C/EBP alpha and PPAR gamma controls the transcriptional pathway of adipogenesis and insulin sensitivity. Mol Cell. 1999;3:151–8.

237. Davis RL, Weintraub H, Lassar AB. Expression of a single transfected cDNA converts fibroblasts to myoblasts. Cell. 1987;51:987–1000.
238. Rudnicki MA, Schnegelsberg PN, Stead RH, Braun T, Arnold HH, Jaenisch R. MyoD or Myf-5 is required for the formation of skeletal muscle. Cell. 1993;75:1351–9.
239. Puri PL, Sartorelli V. Regulation of muscle regulatory factors by DNA-binding, interacting proteins, and post-transcriptional modifications. J Cell Physiol. 2000;185:155–73.
240. Benezra R, Davis RL, Lockshon D, Turner DL, Weintraub H. The protein Id: a negative regulator of helix-loop-helix DNA binding proteins. Cell. 1990;61:49–59.
241. Blais A, Tsikitis M, Acosta-Alvear D, Sharan R, Kluger Y, Dynlacht BD. An initial blueprint for myogenic differentiation. Genes Dev. 2005;19:553–69.

MicroRNAs in Mesenchymal Stem Cells

Mohammad T. Elnakish, Ibrahim A. Alhaider, and Mahmood Khan

Abstract Mesenchymal stem cells (MSC) are adult stem cells that can expand noticeably in culture and are identified in almost every tissue type including early stages of development as well as adult tissues. This wide-ranging distribution of sources along with their genetic stability, compatibility with tissue engineering principles, reproducibility of features, immunoregulatory properties and multilineage potential has led them to be considered as prospective curative candidates for numerous diseases and degenerative purposes. MicroRNAs (miRs) are a class of small non-coding single-stranded RNAs of 19–23 nucleotides that originate from a precursor of approximately 70 nucleotides. Studies with stem cells show a complicated system of miRs regulating main transcription factors and other genes, which sequentially define cell fate. In particular, newly recognized miRs differentiation signatures for MSC designate the presence of distinctive miR patterns in progenitors and terminally differentiated cells, suggesting that such signatures may act as a marker to delineate and track rare cell populations. Recently, several studies have reported the ability of miRs to regulate the differentiation, identity, behavior and self-renewal of MSC. Regulation of MSC homing, reparative and therapeutic efficacies by miRs has been also reported. The overall goal of this chapter is to shed some light on the regulatory role of miRs in different

M.T. Elnakish, Ph.D.
Department of Physiology and Cell Biology, Dorothy M. Davis Heart and Lung Research Institute, The Ohio State University, Columbus, OH, USA

I.A. Alhaider, Ph.D.
Department of Pharmacology, College of Clinical Pharmacy, King Faisal University, Al-Ahsa, KSA, Saudi Arabia

M. Khan, Ph.D. (✉)
Department of Emergency Medicine, Davis Heart and Lung Research Institute, The Ohio State University, Columbus, OH, USA
e-mail: mahmood.khan@osumc.edu

R.C. Zhao (ed.), *Essentials of Mesenchymal Stem Cell Biology and Its Clinical Translation*, DOI 10.1007/978-94-007-6716-4_7,

MSC, as well as the putative role of miRs in regulating the therapeutic efficacy of MSC in the pathophysiology of several diseases.

Keywords MSC • MicroRNA • Differentiation • Self-renewal • Interfacial behavior

Introduction

MSC are adult stem cells with spindle-shaped fibroblast-like appearance that have the ability to expand noticeably in culture. MSC have been identified in almost every tissue type including early stages of development as well as adult tissues [1, 2]. This wide-ranging distribution of sources along with their genetic stability, compatibility with tissue engineering principles, reproducibility of features, immunoregulatory properties and multilineage potential has led them to be considered as prospective curative candidates for numerous diseases and degenerative purposes [3].

Differentiation of MSC into a particular mature cell type is under temporal control of diverse transcription factors, growth factors, and signaling pathways as it has been revealed from global gene expression analysis [4, 5]. Nevertheless, miRs could be projected to regulate mRNA translation and/or stability, so possibly the control of miRs expression patterns corresponds to a novel regulatory network in MSC [6]. In this regard, several studies reported the ability of miRs to regulate the identity, behavior, self-renewal and differentiation of MSC [7–9]. Regulation of MSC homing, reparative and therapeutic efficacies by miRs has been also reported [10, 11]. This chapter focuses on the regulatory role of miRs in MSC as well as the putative involvement of miRs in regulating the therapeutic efficacies of MSC in the future.

MicroRNAs

Definition, Biogenesis and Expression

MiRs are a class of small non-coding single-stranded RNAs of 19–23 nucleotides that originate from a precursor of approximately 70 nucleotides. They can be found in a broad range of organisms, from plants to insects to humans. It has been suggested that there are about 120 miR genes in each invertebrate species and no less than 250 genes in mammals, with some reports described a range of 1,000–10,000 per genome [12–15]. Generally, transcription of miRs genes occurs in the nucleus via RNA polymerase II/III to produce large primary transcripts. Following transcription, precursors of miRs are sliced in the nucleus via Drosha, transported out to cytoplasm via exportin, unwound into a mature single-stranded miR and incorporated into an RNA-induced silencing complex (RISC) after extra slicing via Dicer (Fig. 1) [16].

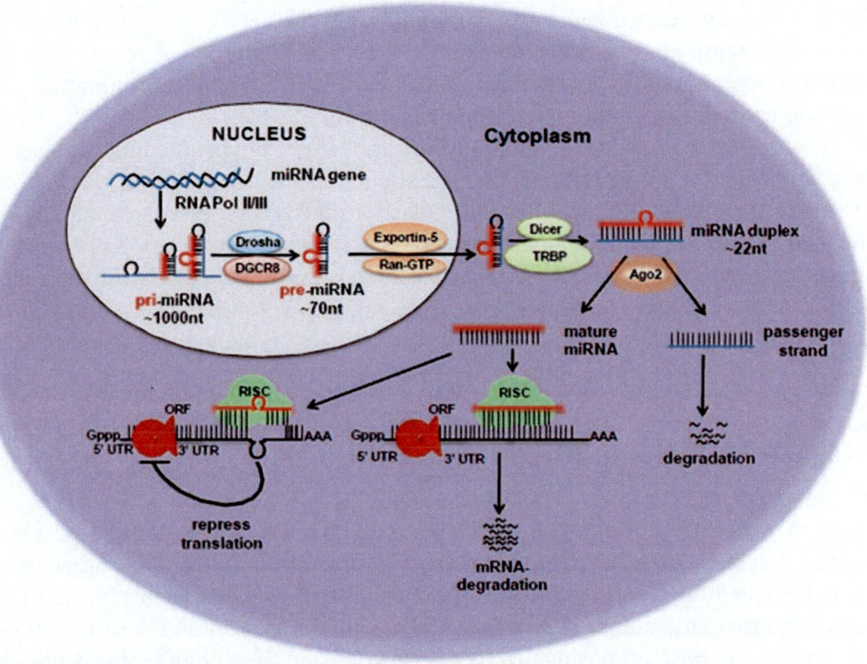

Fig. 1 MicroRNA biogenesis and mechanisms of gene silencing. MicroRNA (*miRNA*) genes are generally transcribed by RNA polymerase II/III in nucleus to form large primary transcript (*pri-miRNA*). These pri-miRNA transcripts are processed to release the ~70-nucleotide hairpin RNA known as precursor-miRNA (*pre-miRNA*), which is transported to the cytoplasm and undergoes another processing to yield a transient ~22-nucleotide RNA duplex. RNA duplex is unwound into a mature single-stranded miRNA, and loaded into RNA-induced silencing complex (*RISC*). miRNA then binds to complementary sites in the 3′-untranslated region (*3′-UTR*) of target mRNA and regulate its expression either by causing degradation of mRNA or repression of their translation, depending on the degree of complementarity between the miRNA and its target (Reproduced from Bhardwaj et al. [17]. Licensed under a Creative Commons Attribution License that allows others to share the work with an acknowledgement of the work's authorship and initial publication in this journal)

Little is known about miRs expression regulation. Frequently, miRs are grouped in the genome, with two or more strongly related miR precursors have been perceived in polycistronic precursors [17, 18]. Further miRs are encoded in the introns of other genes, and some others are edited, forming a dizzying merge of co-expression, expected transcriptional control, and post-transcriptional modifications [19–22].

Targets and Regulation

MiRs operate in an intricate functional system in which each miR possibly manages hundreds of discrete target genes and on the other hand the expression of a particular coding gene can be regulated by many miRs [13, 14]. A list of miRs targets have been currently recognized with important roles in cell growth and apoptosis, homeobox regulation and development [23–26]. This kind of gene regulation signifies a novel regulatory mechanism and is expected to influence numerous vital cellular procedures, counting developmental programs [6].

The difficulty in expecting or confirming the targets of specific miR represents a key confront for integrating miRs regulation into gene expression mechanisms. Trials to map miRs binding sites particularly in the transcriptome of animal cells depend mainly on computational expectations. Several computationally expected targeting interactions are inconsistent between algorithms, and fairly little direct experimental evidence validates these expectations. Yet, it has been proposed that miRs regulate the expression of over 30 % of human's protein-coding genes [6, 27].

Regulatory Roles of MicroRNAs in MSC

Studies with different types of both embryonic and adult stem cells show a complicated system of miRs regulating main transcription factors and other genes, which sequentially define cell fate [6]. Besides, miRs have been reported to be regulated by extracellular signaling pathways that are essential for differentiation into definite tissues, proposing that they are involved in specifying tissue identity [6]. Above all, lately recognized miRs differentiation signatures for MSC designate the presence of distinctive miR patterns in progenitors and terminally differentiated cells, proposing that such signatures may act as a marker to delineate and track rare cell populations [6, 8].

MicroRNAs Regulate MSC Identity

It is worth mentioning that, miRs expression pattern in MSC differs substantially from that in embryonic stem cells (ESC). MSC lack the well-known pluripotent associated markers, such as miRs-302a-d, and miR-200c [28, 29]. Remarkably, in these studies MSC went through substantial consecutive culture expansion, and it is not clear whether the variation in miRs expression between ESC and MSC is a result of intrinsic variation between these two cell types or changes of MSC due to *ex-vivo* expansion [8].

Only a limited number of studies have so far shown the gene expression profiling of MSC compared to their original embryonic stages [7, 30–32]. For instance [31], have recognized extremely expressed genes that encode for membrane proteins that can be used for MSC isolation from differentiating human ESC. On the other hand, Giraud-Triboult et al. [7] have compared the genetic pattern of the pluripotent human ESC with their MSC derivatives, looking for molecular evidence specific for MSC identity. Interestingly, they recognized two miRs contributing to the MSC identity. The decreased expression levels of miR-148a and miR-20b found in MSC

Table 1 MicroRNAs (*miRs*) currently identified as either promoting or suppressing mesenchymal stem cells (*MSC*) differentiation/transdifferentiation into diverse cell lineages

MSC product	Promoting microRNAs	Suppressing microRNAs
Osteocytes	miR-196a, miR-210, miR-2861, miR-148b, let-7, miR-24, miR-29b, miR-15b, and miR-30c	miR-125b⁺, miR-138⁺, miR-133, miR-135, miR-206, miR-204/211, miR-27a, miR-489, miR-26a and miR-31
Chondrocytes	miR-140	miR-199a
Adipocytes	miR-21, miR-200c/141, miR-200b,a/429, miR-204/211 and miR-24	miR-31 and miR-138
Cardiomyocytes	miR-181, miR-206, miR-16 miR-499 and miR-1	miR-124
Neurons	miR-9 and miR-124	Unknown
Hepatocytes	Unknown	let-7 family

+, miR-125b and miR-138 have been predicted by Goff et al. [40] to promote osteogenic differentiation of MSC (*see text for details*)

compared to human ESC would stimulate the over-expression of one of their targets, the transcription regulator *EPAS1* (Endothelial PAS domain 1), which promotes MSC genes expression participating in defining the MSC phenotype [7].

MicroRNAs Regulate Differentiation and Interfacial Behaviors of MSC

Given the multilineage potential of MSC, under proper culture conditions they are capable of differentiating into diverse mesodermal cell lineages such as osteoblasts, chondrocytes, adipocytes and cardiomyocyte-like cells [33]. Additionally, it is evident that MSC can undergo transdifferentiation into non-mesodermal cell lineages including ectodermal cell lineages such as neurons or endodermal cell lineages such as hepatocytes [3]. Through differentiation/transdifferentiation of MSC into particular cell types, consequent miRs are involved (Table 1).

Differentiation of MSC into Osteocytes

Modulation of osteogenic differentiation of MSC by miRs has been reported by several investigators. Numerous miRs have been shown to promote osteogenic differentiation of MSC. MiR-196a was demonstrated to augment osteogenic differentiation probably through its target gene *HOXC8* [34]. Also, miR-210 augmented osteogenic differentiation via reducing ACVR1b in transforming growth factor-β (TGF- β)/activin signaling pathway [35]. Additionally, miR-2861 stimulated bone morphogenetic protein (BMP)-2-mediated ST2 osteogenic differentiation by reducing histone deacetylase 5 expression [36]. Moreover, miR-148b has been detected among a cluster of miRs that were linked to osteogenic differentiation of human MSC in a special 3D substrate [37]. Furthermore, a group of miRNAs (let-7, miR-24, -125b, -138) has been predicted to stimulate osteogenic differentiation of MSC through a platelet-derived growth factor (PDGF) pathway via inhibiting the translation of non-osteogenic target mRNAs to reduce their inhibition of osteogenesis [28]. Similarly, Li et al. [38] demonstrated that miR-29b stimulates osteogenesis by directly down-regulating known inhibitors of osteoblast differentiation, *HDAC4*, *TGFβ3*, *ACVR2A*, *CTNNBIP1*, and *DUSP2* proteins through binding to target 3′ untranslated region (UTR) sequences in their mRNAs [38]. Recently [39], reported three novel over-expressed miRs (miR-30c, -15b, and -130b) in osteo-differentiated human bone marrow (BM)-derived MSC. These miRs were predicted to target genes including *CD29* (MSC marker), stemness-maintaining factor and genes related to cell differentiation. Based on bioinformatics analysis, they suggested that miR-30c and miR-15b which were predicted to target BMP inhibitor genes associated with cartilage formation may be implicated in promoting osteogenic and inhibiting chondrogenic differentiation of MSC.

Conversely, several miRs have been reported to suppress osteogenic differentiation of MSC. At variance with what has been predicted by Goff et al. [28], miR-125b [40] and miR-138 [41] have been shown to suppress osteogenic differentiation of MSC. Mizuno et al. [40] observed that miR-125b was down-regulated in osteoblast-differentiated mouse MSC compared to undifferentiated MSC. They also found that transfection of exogenous miR-125b inhibits osteoblastic differentiation following BMP-4 treatment, signifying a role for miR-125b in suppressing MSC osteogenesis. In this study ERBB2 receptor tyrosine kinase was described as the predicted target gene of miR-125b [40]. Likewise, Eskildsen et al. [41] revealed that miR-138 was down-regulated during osteoblast differentiation of human MSC. Additionally, over-expression of miR-138 prevented *in-vitro* osteoblast differentiation of MSC, while its inhibition stimulated expression of osteoblast-specific genes, alkaline phosphatase activity, and matrix mineralization. The focal adhesion kinase, a kinase implicated in promoting osteoblast differentiation has been identified as a target of miR-138 as evident by target prediction analysis and experimental verification [41]. MiR-133 and miR-135 were also shown to attenuate *RUNX2* and *SMAD5* pathways that synergistically participate in bone formation resulting in a functional inhibition of osteo-progenitors differentiation [42]. MiR-206 reduced osteoblast differentiation of MSC and connexin-43 was proposed to be one of the target genes [43]. MiR-204/211 negatively regulated *RUNX2* transcription factor and served as suppressors of osteoblast differentiation and subsequent mineralization of BM-derived MSC [44]. In addition, miR-489 and -27a have been demonstrated to play an inhibitory role in early osteogenic differentiation of human MSC, partially through the repression of grancalcin [37]. Furthermore, it has been shown that by targeting the *SMAD1* transcription factor and down-regulating bone maker genes, such as alkaline phosphatase, type I collagen, osteocalcin, and osteopontin miR-26a served as a suppressor of osteogenic differentiation during the late stages of human adipose tissue (AT)-derived MSC differentiation toward osteogenic lineage [45]. Recently, Gao et al. [39] recognized four under-expressed miRs (miR-31, -106a, -148a and -424) in osteo-differentiated human BM-derived MSC. These miRs were predicted to target genes linked to bone formation such as *RUNX2*, *CBFB*, and *BMPs*. Functional analysis revealed that inhibition of miR-31 activity promoted osteoblastic differentiation of MSC implying a role for this miR in suppressing osteogenic differentiation of MSC [39].

Differentiation of MSC into Chondrocytes

There is growing evidence that miRs play a key role in regulating chondrogenic differentiation of MSC. Microarray analysis has identified five miRs (miR-130b, -152, -28, -26b, and -193b) to be differentially expressed during chondrogenic differentiation of MSC, and among them only four (miR-130b, -152, -28, -26b) were confirmed by real-time polymerase chain reaction analysis. Potential targets of these confirmed miRs were genes involved in cartilage formation, such as *COL4A1*, *COL2A1*, and *COL6A1* as revealed from bioinformatics analysis [46]. Likewise, microarray analysis in MSC at four different stages of TGF-β3-induced chondrogenic differentiation demonstrated that eight miRs (miR-127, -140, -125b*, -99, -140*, -181a, let-7f and -30a) were significantly up-regulated and five miRs (miR-145, -212, -132, -143 and -125b) were down-regulated. Interestingly, two miRs clusters, miR-143/145 and miR-132/212, kept on down regulation in the process. Predicted target genes known to be involved in chondrogenesis including *SOX6*, *ACVR1B*, *RUNX2* and *ADAMTS5* have been identified by bioinformatics approach [47]. Additionally, MiR-140 has been recognized to promote the differentiation of MSC into chondrocytes. Microarray analysis revealed that miR-140 exhibited the largest expressional difference between chondrocytes and MSC. Enhanced miR-140 expression during chondrogenesis of MSC was in correspondence with expression of *SOX9* and *COL2A1* [48]. Consistently, miR-140 expression during cartilage development was also reported by another group, possibly through inhibition of histone deacetylase 4, an expected co-repressor of *RUNX2* [49]. In contrast, miR-199a was reported to suppress early chondrogenesis by directly targeting the *SMAD1* transcription factor [50].

Differentiation of MSC into Adipocytes

Several miRs have so far been recognized to promote adipogenic differentiation of MSC. For instance, in human AT-derived MSC miR-21 expression was transiently improved following induction of adipogenic differentiation, peaked at 3 days, and returned to the baseline level at 8 days. Over-expression of miR-21 increased adipogenic differentiation and decreased TGFβ-induced inhibition of adipogenic differentiation indicating that miR-21 promote the adipogenic differentiation of MSC via modulating the *TGFβ* signaling [51]. Also, in ST2 mouse MSC mammalian homologues of miR-8 (miR-200c/141 and miR-200b,a/429) have been demonstrated to promote adipogenesis of MSC by inhibiting *Wnt* signaling [52]. Additionally, Huang et al. [44] showed that miR-204/211 act as key endogenous negative regulators of *RUNX2*, which suppress osteogenesis and promote adipogenesis of MSC [44]. Furthermore, miR-24 was found to enhance BMP2-induced G1 arrest as well as BMP2-induced commitment of MSC to adipocytes lineage. Conversely, miR-31 inhibited the cytidine-cytidine-adenosine-adenosine-thymidine (CCAAT) enhancer-binding protein α (an adipocytic differentiation factor) expression at both transcriptional and translational levels and suppressed adipogenic differentiation [52]. An additional adipogenic suppressor that has been shown to be down-regulated during adipogenic differentiation of MSC is miR-138 [53]. Over-expression of miR-138 in AT-derived MSC inhibited expression of key adipogenic transcription factors, CCAAT enhancer binding protein α and peroxisome proliferator-activated receptor gamma-2 as well as a number of other adipogenic markers. Suppressor effects of miR-138 on the adipocyte differentiation of MSCs were partially attributed to repression of early region 1-A-like inhibitor of differentiation-1 (EID-1), a nuclear receptor co-regulator [53].

Differentiation of MSC into Cardiomyocytes

The role of miRs in cardiomyocyte differentiation of MSC has been extensively investigated. Earlier studies showed that cardiomyocyte differentiation of MSC was greatly affected by induction conditions. During differentiation of human MSC into cardiomyocyte, the presence of induction reagent 5-azacytidine resulted in the expression of primary miRs of miR-143 and miR-181, whereas indirect co-culture of human MSC with neonatal rat myocytes enhanced the expression of primary miRs of miR-143, -206, -208, and -181 [54]. Interestingly, both miR-206 and miR-181 are known to promote myogenesis. In particular, miR-181 has been demonstrated to stimulate myoblast differentiation via inhibiting a myogenic suppressor, the homeobox protein *HOX-A11* [55, 56]. Very recently, another group of miRs have been reported to be involved in promoting cardiomyocyte differentiation of MSC. Liu et al. [57] showed that cardiomyocyte differentiation of human BM-derived MSC induced by co-culture with neonatal rat ventricular myocytes was associated with up-regulation of miR-16. Over-expression of miR-16 suppressed the cell cycle related genes, *CDK6*, *CCND1* and *CCND2* resulting in G1 phase arrest and promoted the differentiation of MSC into cardiomyocytes as indicated by enhanced expression of cardiac marker genes, including *GATA4*, *NK2-5*, *MEF2C* and *TNNI3* [57]. Additionally, it has been reported that over-expression of miR-499 in rat BM-derived MSC promotes its differentiation into cardiomyocytes via activating the *Wnt/β-catenin* signaling pathway [58]. Moreover, Huang et al. [44] showed that miR-1 promotes the differentiation of mouse BM-derived MSC into cardiac lineage partially through negative regulation of the downstream target molecular of Notch pathway-Hes-1 [44]. On the contrary, Cai et al. [59] reported that miR-124 was significantly down-regulated during cardiomyocyte differentiation of rat BM-derived MSC induced by co-culture with neonatal rat ventricular myocytes. Over-expression of miR-124 suppressed cardiomyocyte differentiation of MSC-via targeting the 3′ UTR of *STAT3* gene – as indicated by significant decreases in cardiac-specific markers such as ANP, TNT, and α-MHC proteins as well as reduction of cardiac potassium channel currents [59].

Differentiation of MSC into Neurons

It has been reported that specific miRs play a potential role in neurogenesis of MSC, neurotransmitter release by MSC-derived neurons and in synaptic plasticity [10, 60, 61]. Additionally, miR-9 and miR-124 were shown to be regulated by the transcription factor *REST*, which is vital in MSC-mediated dopaminergic neurogenesis [62]. Consistently, miR-9 has been described to promote neuronal differentiation of mouse BM-derived MSC by modulating the Notch signaling pathway [63]. On the other hand, Greco and Rameshwar [60] studied the miR-mediated neurotransmitter regulation in developing neuronal cells. Sixteen different miRs found to be up-regulated in MSC-derived neuronal cells compared to undifferentiated human MSC. Up-regulated miRs have further been analyzed to predict targets of the synaptic transcript, *TAC1* mRNA. MiR-130a, miR-206, and miR-302a binding sites were predicted within the 3′UTR of *TAC1* and target validation confirmed the miR-130a and miR-206 sites thereafter. Specific inhibition of miR-130a and miR-206 in the neuronal cells lead to considerable increases in *TAC1* translation as evident by the enhanced synthesis and release of its encoded protein, neurotransmitter substance-P implying a role for miRs in the regulation of neurotransmitters [60]. In human umbilical cord-derived MSC 11 miRs (miR-206, miR-34a, miR-374, miR-424, miR-100, miR-101, miR-323, miR-368, miR-137, miR-138 and miR-377) were abundantly expressed in trans-differentiated neuronal progenitors. Among them, miR-34a and miR-206 were the only two miRs been coupled to MSC neurogenesis. Specifically, miR-34a was shown to regulate genes associated with cell motility and energy production of neuronal precursor [64].

Differentiation of MSC into Hepatocytes

Little is known about the involvement of miRs in the hepatic commitment of MSC. Koh et al. [65] demonstrated that let-7 family of miRs could indirectly regulate the expression of a known endodermal differentiation marker, hepatic nuclear factor 4 alpha (HNF4A) signifying a role of let-7 in repressing hepatic differentiation of human ESC-derived MSC [65]. Newly, the miRs profile in hepatic differentiation of human umbilical cord-derived MSC has been reported [66]. Dynamic miR profiles

Table 2 MicroRNAs (*miRs*) exhibited the highest levels of over-expression/under-expression during hepatic differentiation of human umbilical cord-derived mesenchymal stem cells (*MSC*) [70]

Over-expressed MiRs		Under-expressed MiRs	
≥4 folds	≥6 folds	≥4 folds	≥6 folds
miR-671-5p	miR-1290	miR-3646	miR-100
miR-542-5p	miR-136	miR-17*	miR-10a
miR-542-3p	miR-424	miR-3679-3p	miR-130b
miR-1185	miR-30a	miR-17	miR-146a
miR-539	miR-148a	miR-155	miR-17
miR-148a	miR-1246	miR-146a	miR-1973
miR-301a			miR-29a
			miR-31
			miR-31*
			miR-762

were recognized that did not overlap or only partly overlapped with miRs described to be implicated in human liver development, hepatocyte regeneration or hepatic differentiation of liver-derived progenitor cells. These miRs also were not enriched in hepatocyte or hepatocellular carcinoma cells and can potentially target liver-enriched transcription factors and genes. A total of 61 miRs exhibited consistent changes and were altered as a minimum twofold between undifferentiated MSC and hepatic differentiated MSC. Among these miRs, 25 miRs were over-expressed and 36 miRs were under-expressed with similar expression pattern. MiRs exhibited the highest levels of over-expression/under-expression during this process are shown in Table 2. Finally, the study suggested that the revelation of miRs profile during the MSC hepatic differentiation presents the foundation for elucidating the role of miRs in hepatic differentiation of MSC and specific miR selection for the conversion of MSC into hepatocytes [66].

MicroRNAs Regulate Interfacial Behaviors of MSC

Cell-substrate interaction is one of the key aspects of tissue engineering. Although, MSC differentiation usually necessitates the use of differentiation factors, substrate topography to which cell shape is restricted alone can also affect stem cell lineage commitments [9]. A recent study described the global marker genes expression and miRs profiling analysis, providing insights of a regulation network into the topography-induced MSC responses. The differentially-expressed miRs combination (miR-140, miR-214, miR-320, miR-351 and miR-674-5p) was shown to promote osteogenesis associated with the topography. The authors suggested that their results help in understanding the mechanism by which microenvironments affect behaviors of progenitor cells via miRs [9]. Figure 2 illustrates the potential downstream protein signaling pathways related to MSC behaviors, including migration, proliferation, apoptosis and differentiation [9].

a

BMSCs					
miRNA	Fold change	miRNA	Fold change	miRNA	Fold change
hsa-miR-935	-14.89	hsa-miR-424*	-2.07	hsa-miR-194	-1.80
hsa-miR-142-3p	-12.74	hsa-miR-381	-2.04	hsa-miR-125a-5p	-1.79
hsa-miR-483-3p	-12.30	hsa-let-7g*	-2.00	hsa-miR-892a	-1.74
hsa-miR-203	-10.15	hsa-miR-500	-1.99	hsa-miR-549	-1.74
hsa-miR-142-5p	-7.30	hsa-miR-148b*	-1.98	hsa-let-7i	-1.72
hsa-miR-335*	-3.62	hsa-miR-24-1*	-1.96	hsa-miR-409-5p	-1.68
hsa-miR-504	-2.96	hsa-miR-574-3p	-1.93	hsa-miR-377*	-1.68
hsa-miR-210	-2.90	hsa-miR-30b	-1.92	hsa-let-7i*	-1.67
hsa-miR-20b*	-2.75	hsa-miR-500*	-1.90	hsa-miR-628-3p	-1.65
hsa-miR-486-5p	-2.73	hsa-miR-99b	-1.90	hsa-miR-502-3p	-1.60
hsa-miR-641	-2.66	hsa-miR-218-2*	-1.90	hsa-miR-188-3p	-1.57
hsa-miR-1208	-2.35	hsa-miR-323-3p	-1.87	hsa-miR-532-5p	-1.56
hsa-miR-663	-2.20	hsa-miR-409-3p	-1.86	hsa-miR-337-3p	-1.48
hsa-miR-181c	-2.20	hsa-miR-376a*	-1.83	hsa-miR-122	1.53
hsa-miR-1247	-2.17	hsa-miR-433	-1.81	hsa-miR-510	1.56

b

ASCs	
miRNA	Fold change
hsa-miR-153	-3.98
hsa-miR-1277	-3.14
hsa-miR-141	-3.07
hsa-miR-33b	-2.87
hsa-miR-33a	-2.74
hsa-miR-299-3p	-2.58
hsa-miR-1295	-2.42
hsa-miR-933	-1.84
hsa-miR-205	-1.75
hsa-miR-18b	-1.74
hsa-let-7g*	-1.68
hsa-miR-519d	-1.55
hsa-miR-452	1.97
hsa-miR-335	4.34

Fig. 2 Age-dependent changes in microRNAs (*miRNA*) profiles of mesenchymal stem cells (*MSC*) derived from adipose stem cells (*ASCs*) and bone marrow stem cells (*BMSCs*). (**a**) Fold regulation of significant miRNA in old versus young BMSCs donors. (**b**) Fold regulation of significant miRNA in old versus young ASCs donors. Upregulated miRNA are denoted in *red*, downregulated miRNAs are *green* (Reproduced from Pandey et al. [86]. Licensed under full BioMed Central Open Access license agreement identical to the Creative Commons Attribution License that allows others to share the work with an acknowledgement of the work's authorship and initial publication in this journal)

MicroRNAs Regulate Survival and Self-Renewal of MSC

Regardless of the benefits of MSC, clinical application of MSC-based therapy is restricted due to the poor viability of the transplanted cells [3]. Additionally, prolonged survival is critical for MSC to be able to transit through the circulation to home to distant injured sites [67]. Also, sufficiently high numbers of MSC obtained via culture expansion are required for cell therapy. Studies showed that MSC exhibit a decreased potential for proliferation after prolonged *in-vitro* culture [68]. Consequently, recognizing the factors associated with MSC survival and proliferation and enhancing these processes in the transplanted MSC could be essential for successful application in cell therapy [69]. Unlike MSC differentiation, only a few studies reported about the implication of miRs in MSC survival and proliferation. Lately, it has been reported that miR-125b did not stimulate cellular differentiation of human MSC but instead had an unpredicted role in improving cell survival in response to withdrawal of cell-matrix adhesion signals; a process that generally activates apoptosis. The ability of MSC to resist apoptosis was attributed to the ability of miR-125b to up-regulate mitogen-activated protein kinase (*MEK*)/extracellular signal-regulated kinase (*ERK*) signaling while down-regulating *p53* expression [67]. Another group has demonstrated alterations in miRs expression in rat BM-derived MSC by hypoxia/serum deprivation; a condition that mimic ischemic environment of injured sites and known to induce apoptosis. They showed that miR-21, miR-23a and miR-210 were up-regulated in response to hypoxia/serum deprivation and they may be involved in protecting MSC against apoptosis. Among them, miR-21 and miR-23a promoted MSC survival partially through inhibiting the decrease in $\Delta\Psi m$; nevertheless, miR-210 may way out to other pathways.

Furthermore, blockage of miR-21, miR-23a or miR-503 aggravated apoptosis, proposing that miR-503 might also be one of the hypoxia related miRs in MSC. Interestingly, it has been reported that miR-146a induced by preconditioning of rat BM-derived MSC with diazoxide, a key regulator of stem cell survival is a powerful and potential target to improve stem cell survival under ischemic condition. Computational analysis established a consensus predicted target site of miR-146a pertinent to apoptosis in the 3′ UTR of *FAS* mRNA [70].

In relevance with MSC proliferation, a deep sequencing analysis of miRs expression profiles in human ESC-derived MSC revealed that let-7 family may be involved in the self-renewal of these cells [65]. Besides, it has been shown that low-level laser irradiation; a known regulator of MSC proliferation increased the proliferation and cell cycle-associated genes in rat BM-derived MSC with differential regulation of subsets of miRs. In particular, miR-193 was the most highly up-regulated miR that has been functionally proven to regulate the proliferation of MSC, but could not affect the apoptosis and differentiation level. Inhibitor of growth family, member 5 (*ING5*) was predicted to be the best target of miR-193 to functionally regulate proliferation and cyclin-dependent kinase 2 (*CDK2*) activity [71]. Similarly, unpublished data from our laboratory demonstrated that transient transfection of MSC with miR-133a-mimic increases cell proliferation by >10-fold, indicating that miRs play a crucial role in stem-cell proliferation.

MicroRNAs Regulate Aging and Replicative Senescence of MSC

Replicative senescence is the process by which cells permanently lose their capacity to divide after carrying out a certain number of divisions but stay viable and metabolically active. Replicative senescence is supposed to be an antitumor mechanism as well as a key player in age-related changes in tissue function [72]. It has been reported that MSC populations are exhausted with age and that decrease in MSC pools participates in human aging and the onset of age-related disease processes [73, 74]. Additionally, Aged MSC has shown to exhibit decreased multipotent differentiation potential and release of useful cytokines for tissue repair [75, 76]. Thus, aged MSC transplantation is expected to result in diminished therapeutic efficacy. Indeed, MSC senescence remains as an unresolved problem and studies are on track to understand the molecular mechanisms of this process [8]. Recent studies have revealed that in MSC replicative senescence miR expression profiling demonstrated up-regulations of miR-371, miR-369-5p, miR-29c, miR-499, and let-7f [77]. Interestingly, DNA-methyl transferase (DNMT) 3A and 3B are known targets of miR-29c in lung cancer tissue [78]. While, DNMT3A and DNMT2 has been predicted to be the targets of miR-371 and miR-499, respectively [8]. It has been also shown that over-expression of miR-486-5p promotes a premature senescence-like phenotype and depresses proliferation as well as adipogenic and osteogenic differentiation of human AT derived-MSC via regulating the expression of silent information regulator 1 (SIRT1), a main regulator of longevity and metabolic disorders [72]. Furthermore, miR expression profiling displayed up-regulations of miR-766 and miR-558 and down-regulations of let-7f, miR-125b, miR-222, miR-199-3p, miR-23a, and miR-221 in old monkey MSC compared to young monkey MSCs implying a role for miRs in MSC aging [79]. Likewise, miR profiling of human AT- and BM-derived MSC from older and younger donors revealed that subsets of miRs are biologically active in human MSC, with the profiles of miRs altering with aging. Interestingly, miRs modulate gene expression linked to a variety of functions, mostly cellular proliferation and inflammation, both of which play an important role in the process of aging. In particular, BM- and AT-derived MSC exhibited more than 95 % (43 miRs) and 85 % (12 miRs) of the notably altered miRs were down-regulated with age, respectively. Conversely, both MSC types had two unique miRs among those screened that were significantly up-regulated in older donors (Fig. 3) [80].

MicroRNAs Regulate MSC-Mediated Cellular Communication

Recently, MSC have been shown to secret microvesicles [81, 82]. Accumulating evidence propose that microvesicles could act as a tool of cellular communication for transferring genetic information or gene products as well as regulating cellular activities [83, 84]. MSC-secreted microvesicles have been demonstrated to harbor a wide array of mRNAs and miRs [82, 83, 85]. Particularly, MSC have been shown to secrete miRs in the precursor rather than the mature form. These pre-miRs were enriched in microvesicles that can be easily transported to adjacent cells, signifying a potential mechanism in regulating activities of other cells. Thus, MSC can potentially apply miR-mediated effects on other cells via secreting pre-miR in microvesicles, which could be beneficial during clinical application [85]. On one side, MSC-secreted microvesicles might reprogram tissue-injured cells via transporting mRNA and/or miR that promotes cell de-differentiation, modulates production of soluble paracrine mediators, and mediates re-entry of cell-cycle, accordingly supports tissue repair. On the other side, phenotypic transfer of MSC to attain tissue-specific cell characteristics can be mediated via delivering mRNAs and/or miRs to MSC by microvesicles released from tissue-injured cells [89]. Examples of miRs enriched in microvesicles secreted from ESC- and BM-derived MSC and might be involved in cellular communication are shown in Fig. 4 [82, 85].

MicroRNAs Regulate Homing and Reparative Phenotype of MSC

One of the most important features of MSC during clinical application is the ability to home to the damaged tissue or inflammatory sites [3]. Identifying factors regulating stem cells homing would be of great benefit in improving therapeutic efficacy of MSC. Recently, miR-34a was shown to regulate genes that were associated with cell motility and energy production during neurogenesis of human umbilical cord-derived MSC. Functional experiments confirmed the ability of miR-34a to inhibit

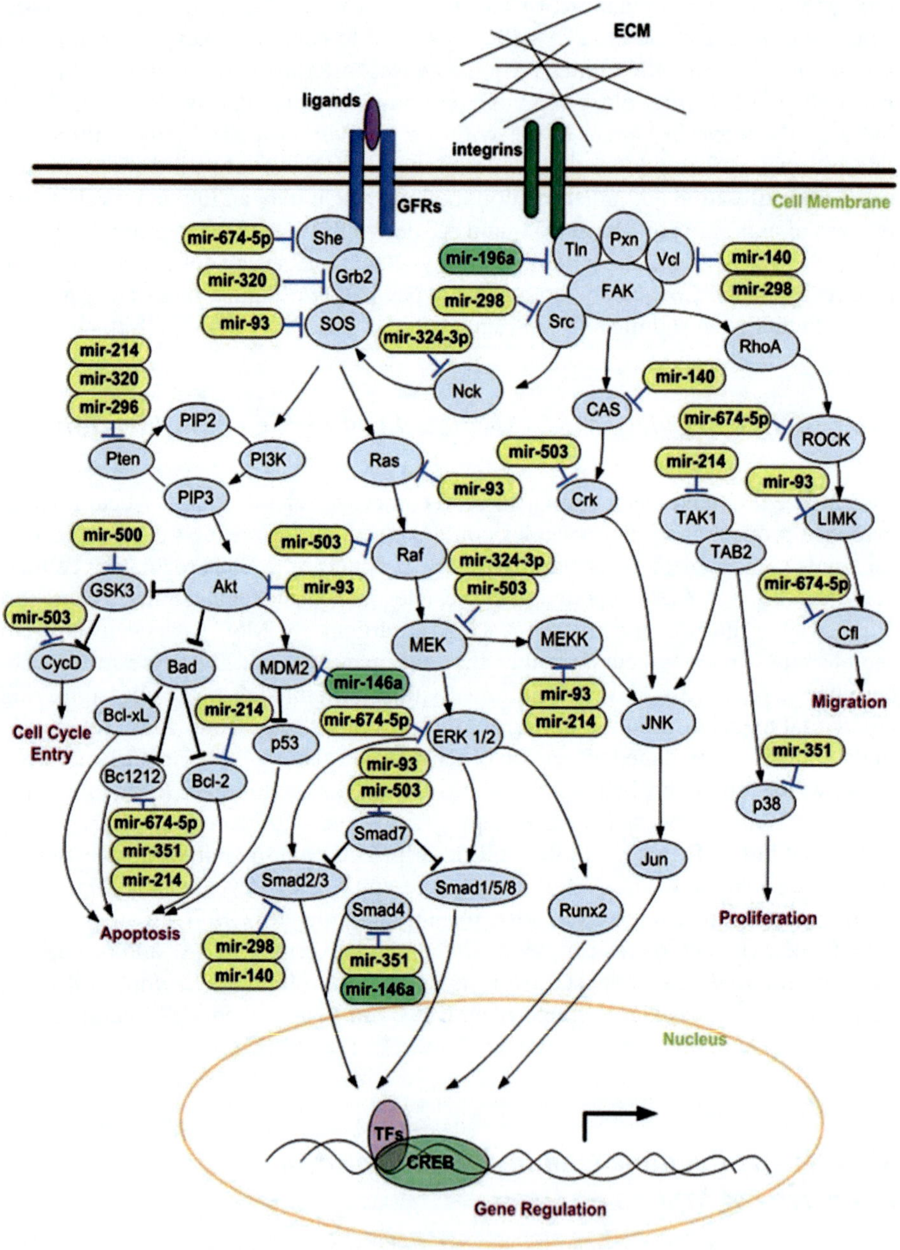

Fig. 3 Proposed signaling networks involving differentially-expressed miRNAs in mesenchymal stem cells (*MSC*) cultured on microgooved substrate. Differentially-expressed miRNAs in MSC cultured on microgooved substrate for migration, proliferation, apoptosis and differentiation could be integrated into complicated signaling networks to control cell behaviors. Integrins are heterodimeric transmembrane receptors composed of eighteen α and eight β subunits that can be non-covalently assembled into 24 combinations connecting extracellular environments with

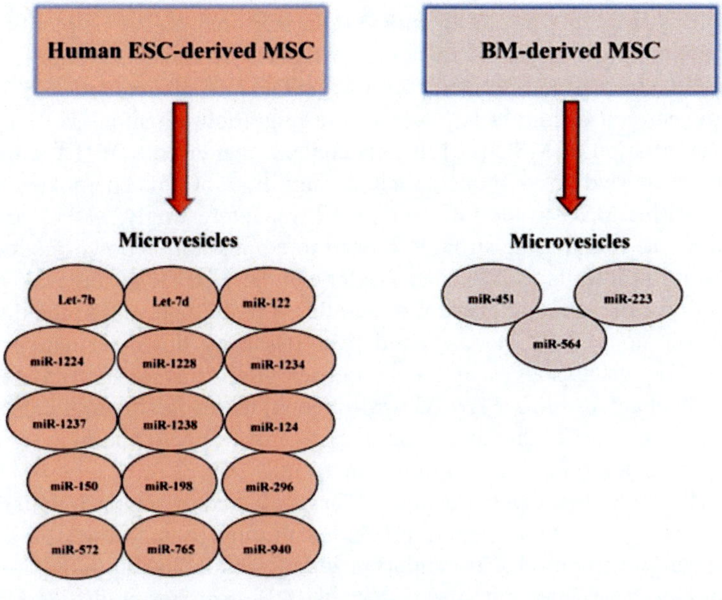

Fig. 4 Examples of microRNAs (*miR*) enriched in microvesicles secreted from mesenchymal stem cells (*MSC*) derived from human embryonic stem cells (*ESC*) and bone marrow (*BM*) [88, 91]

Fig. 3 (continued) cytoskeletons. Integrins can activate several signaling pathways independently. More frequently, they act synergistically with other growth factor receptors (*GFRs*) including receptors of insulin like growth factor (*IGF*), vascular endothelial growth factor (*VEGF*), transforming growth factor beta (*TGF-β receptor*), platelet-derived growth factor beta (*PDGF-β*) and epidermal growth factor (*EGF*). Important pathways including RhoA, Ras, Smad and PI3K etc. are parts of the signaling networks involving differentially expressed miRNAs. *GFR* growth factor receptor, *Shc* shc transforming protein, *Grb2* growth factor receptor-bound protein 2, *SOS* son of sevenless, *FAK* protein-tyrosine kinase, *Tln* Talin, *Pxn* Paxilin, *Vcl* Vinculin, *Src* v-src sarcoma viral oncogene homolog, *Nck* non-catalytic region of tyrosine kinase adaptor protein, *RhoA* ras homolog gene family, member A, *ROCK* Rho kinase, *LIMK* LIM domain kinase, *Cfl* cofilin 1, *CAS* Castor, *Crk* v-crk sarcoma virus CT10 oncogene homolog, *JNK* c-Jun N-terminal kinase, *Jun* Jun oncogene, *TAK1* TGF-beta activated kinase 1/MAP3K7, *TAB2* TGF-beta activated kinase 1/MAP3K7 binding protein 2, *p38* p38 kinase/MAPK14, *Ras* resistance to audiogenic seizures, *Raf* raf kinase, effector of Ras, *MEK* MAP kinase-ERK kinase/MAP2K1, *MEKK* MEK kinase, *ERK* mitogen-activated protein kinase/MAPK1, *Runx2* runt-related transcription factor 2, *Smad* mothers against decapentaplegic Drosophila, homologs, *PTEN* phosphatase and tensin homolog, *PIP* putative proline iminopeptidase protein, *PI3K* phosphatidylinositol 3-kinase, *Akt* v-akt murine thymoma viral oncogene homolog, *GSK3* inosine-guanosine kinase 3, *CycD* cyclin D, *Bad* BCL2-associated agonist of cell death, *Bcl2* B-cell leukemia/lymphoma 2, *Bcl2l2* Bcl-2-like protein 2, *Bcl-xL* B-cell lymphoma-extra large, *MDM2* transformed mouse 3T3 cell double minute 2, *p53* p53 tumor suppressor homolog, *TFs* Transcription factors, *CREB* DNA-binding response regulator. inhibit; d>, promote; miRNAs in green box: down-regulated miRNAs; miRNAs in yellow box: up-regulated miRNAs (Reproduced from Wang et al. [71] with permission from Elsevier, license number: 3005131144084)

MSC motility implying a role for miR-34a in neuronal precursor motility, which may be crucial for stem cells to home to the target sites they should be [64]. Additionally, stromal cell-derived factor-1α (SDF-1α), also called CXCL12, has been demonstrated to play a key role in the migration, chemotaxis, homing and trans-differentiation of MSC [86]. In this context, increased SDF-1α at the site of injury was reported to promote targeted migration of the chemokine receptor CXCR4-positive MSC to the infarcted area [87]. Interestingly, SDF-1 production and release was found to be strongly related to a constitutive down-modulation of miR-886-3p in the cells as it exclusively targets the 3′ UTR of the *SDF-1* mRNA [11]. Most recently, another group reported that miR-27b decreased the directional migration of mouse MSC to damaged liver tissue by down-regulating SDF-1α expression. The group suggested that their observations further contribute to understand the mechanism of *SDF-1α/CXCR4* interactions in modulating cell behavior and may present a novel therapeutic strategy to enhance MSC homing to damaged tissue resulting in more proficient tissue repair [88].

It is well established that resident MSC are activated in response to tissue damage to participate in tissue-repair processes through a multitude of activities, including cell proliferation, differentiation and migration, in addition to the regulation of angiogenesis and immune responses [89]. Another study identified that miRs are involved in regulating the transition between the resting and the reparative phenotypes of human MSC. This study verified the involvement of a specific miR in the coordinated regulation of MSC proliferation, migration and differentiation. Specifically, the study described that miR-335 plays a central role in the gene regulatory network that manages the tissue-repair activities of MSC. Expression miR-335 was high in the undifferentiated MSC in relation to their differentiated cell progeny, and is regulated by key signaling pathways that control MSC biology, including Wnt3a and IFNγ. Predicted miR-335 targets were enriched in genes involved in regulating cellular movement and gene expression, including *RUNX2*. Finally, it has been suggested that the results of this study could help to understand the major molecular mechanisms regulating the therapeutic activity of MSC versus their maintenance in an undifferentiated state, and strongly suggest an important role of miR-335 in tissue homeostasis [89].

Putative Roles of MicroRNAs in the Therapeutic Efficacy of MSC in Selected Pathological Conditions

Over the past decade, clinical application of both miR- and MSC-based therapies have been considered as the two most striking fields in human diseases [90]. Currently, due to their emergent role in regulating MSC biology, miRs are evaluated as promising candidates for regulating the therapeutic efficacy of MSC in various pathological conditions.

Diabetes and Cardiovascular Diseases

One of the current paradigms in the clinical ventures is the application of stem cells, including MSC in the treatment of diabetes (Type 1) and myocardial infarction (MI) [91, 92]. In diabetes and MI, where beta cells and myocardium are damaged, respectively, there is also alteration in the regulation of biological mechanism due to differential expression of miRs which in turn disturb the system promoting the pathological state; the only choice for effective therapy is the use of stem cells and miRs/anti-miRs combinations. On the other hand, exogenous miR or anti-miR can restore the normal biological processes protecting from the deteriorating pathological condition. Consequently, both stem cells and miRs will be potential area for future therapy of diabetes and cardiovascular diseases [93]. So far, stem cells and miRs are individually applied during diabetes and MI therapy. However, an innovative strategy has been suggested, where synergetic approach of stem cell and miR could be developed for rehabilitation. Under these conditions the performance of transplanted stem cells can be directed and firmly regulated by miRs based on the necessity to develop the benefits of stem cell therapy [93]. In particular, the regulatory role of miRs in MSC-induced cardiac repair following MI has been recently reviewed [90]. In this review, the authors described miRs as novel potential regulators in the MSC-based treatment of MI. MiRs were shown to be involved in several MSC-stimulated cardiac processes following MI such as cardiovascular cell differentiation, paracrine effects, anti-arrhythmic effects and others as shown in Fig. 5 [90].

It is worth noting that, full understanding of the mechanisms involved in the adipogenic differentiation of MSC could provide new insights into the pathogenesis of several diseases such as obesity [53], which represents a main risk factor for many other diseases including diabetes and cardiovascular diseases. This will allow the development of appropriate therapeutic approach for these diseases in the future. Newly, evidence of the correlation between one of the AT-derived MSC adipogenic differentiation regulators, miR-21 level and adipocyte number in the white adipose tissue of obese mice has been reported, which presents new insights into the mechanisms of obesity. Briefly, miR-21 was found to be involved in the obesity development through regulating *STAT3* signaling via two different mechanisms: (1) the regulation of the adipocyte precursors proliferation in the early stage and (2) increased adipocyte precursors adipogenic differentiation during the late phase of obesity development [94].

Cancer

The effect of unmodified MSC on tumor progression is still unclear in spite of the extensive investigations that have been done over the past 5 years. Several studies have demonstrated that MSC stimulate tumor progression and metastasis whereas

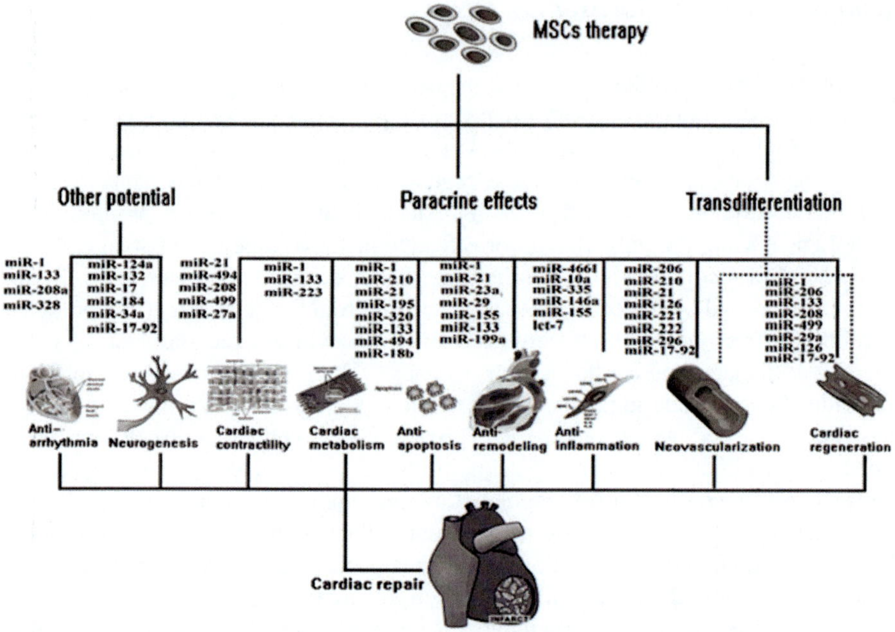

Fig. 5 Summary of putative microRNAs that may be used as important modulators in mesenchymal stem cell (*MSC*)-mediated cardiac repair processes in myocardial infarction. These microRNAs may play central roles in numerous cardiac pathophysiologic processes, such as stem cell differentiation, cardiovascular cell proliferation and migration, myocardial inflammation and apoptosis, cardiac remodelling, contractility, metabolism and arrhythmias, and others. Preconditioning MSC by microRNA modulation may enhance cell capacity to differentiate into cardiovascular cells, enhance anti-arrhythmic and cardiac nerve sprouting potential, and release more paracrine factors and functional microRNAs in a paracrine/autocrine manner, leading to better cardiac repair (Reproduced from Wen et al. [89] with permission from John Wiley and Sons, license number: 3006250868490)

others described MSC as tumor suppressers as reviewed in Klopp et al. [95]. No simple paradigm can explain the contradictory data in the studies of MSC. The option that MSC can stimulate tumor growth and metastasis increases worries about the safety of their clinical use. To date, there is no evidence of tumor formation has been shown in more than 1,000 patients cured with MSC for a number of indications. On the other hand, modified MSC that were engineered to express antitumor cytokines exhibited powerful antitumor effects, proposing that possibly putative MSC-induced tumor promoters could be attenuated by manipulating cytokine expression [95].

Only very little is known about the involvement of miRs in regulating the effects of MSC on tumor progression, so far. One suggestion is that MSC-secreted microvesicles that contain miRs in the precursor form, which represent a subset of miRs found within MSC, may be one of the unknown mechanisms of MSC signaling within the tumor microenvironment [95]. Additionally, comparing the global gene and miR expression profiles between MSC and fibroblasts showed that miR-335 expression was 44-fold higher in MSC than in fibroblasts [96]. One reasonable explanation for this increase has been inferred from a recent study describing miR-335 as a suppressor of breast cancer metastasis [97], signifying that high miR-335 expression might be necessary for the post-transcriptional regulation of metastasis associated genes expressed by MSC [96].

Neurological Disorders

MSC represent a promising therapeutic tool for neurologic disorders such as Parkinson's disease, multiple sclerosis, traumatic brain injury, and spinal cord injury due to their ability to trans-differentiate and stimulate endogenous repair of damaged neural tissue through cytokines and other soluble factors [98–101]. As mentioned above, the role of miRs in neurogenesis from MSC has been confirmed by bioinformatics and functional analyses. MiR-9 and miR-124 were shown to be regulated by the transcription factor *REST*, which is important in MSC-mediated dopaminergic neuron formation that may be helpful in many neurological disorders such as Parkinson's disease [62]. Also, MiR-130a and miR-206 targeting of *TAC1* were found to inhibit production and release of substance P, a key player in many pain and inflammatory processes, from MSC-derived neurons [60]. Furthermore, miR-34a has been described as the main regulator of the de-differentiation-reprogrammed phenotype of MSC that exhibited improved survival and high efficacy in increasing neuronal differentiation and cognitive functions in a neonatal hypoxic–ischemic brain damage rat model [102]. The consequence of these findings is that miRs can potentially regulate the therapeutic efficacy of MSC in neurological disorders and their manipulation could open a new avenue for future treatments. However, it is clear that further research is still needed [10].

Osteoarthritic Diseases, Wound Healing and Preeclampsia

It is apparent that miRs are extensively involved in both osteogenic and chondrogenic differentiation of MSC proposing that the manipulation of miRs expression could provide therapeutic strategies for the treatment of osteoarthritic diseases [103, 104]. For instance, pharmacological inhibition of miR-138 whose over-expression was shown to inhibit osteoblast differentiation of MSC, while its inhibition stimulated expression of osteoblast-specific genes has been suggested to represent a promising therapeutic approach for improving bone formation *in-vivo* [41].

In the wound healing setting, it has been reported that in a skin excision model TGF-β, a key mediator up-regulated in the wound site, promoted the expression of miR-21 in MSC and in the multipotential C3H10T1/2 cells, and stimulated the *in-vitro* proliferation and differentiation of these cells. Delayed healing was consistent with knockdown of miR-21 in the wound bed indicating that miR-21 regulates gene expression and, consequently, the behavior of MSC in wound healing [105]. Moreover, the improvement of the diabetic wound healing impairment with MSC treatment has been reported to be coupled with a considerable elevation in the miR-146a expression and depressed gene expression of its pro-inflammatory target genes [106].

Recently, increased expression of miR-181a in MSC from preeclampsia patients in regard to MSC from normal patients has been reported [107]. Functional experiments showed that miR-181a acts as a suppressor of *TGF-β* signaling pathway and inhibits MSC proliferation. MiR-181a also induced the expression of IL-6, VEGF, and indoleamine 2,3-dioxygenase (IDO) by stimulating the mitogen-activated protein kinases (*MAPK*) pathway and attenuated MSC immunosuppressive properties *in-vitro* and *in-vivo*. The study suggests that miR-181a activity triggers preeclampsia through the down-regulation of *TGF-β* signaling and up-regulation of *MAPK* signaling. This will help understanding the miR function in MSC and could provide the foundation for the development of a potential therapy for preeclampsia [107].

Conclusions and Future Perspectives

It is apparent that miRs play a central role in regulating each and every aspect of MSC biology with possible application to improve therapeutic efficacy of MSC in different pathological condition. However, it is also clear that further studies are needed to fully elucidate the role of miRs in MSC during both regulatory and clinical settings. In regulatory setting, functional experiments are required to identify the exact role of up-regulated and/or down-regulated miRs during MSC differentiation. Additionally, the implication of miRs in other aspects of MSC such as self-renewal, survival, immune-modulation, homing and reparative activity need to be widely elaborated. In clinical setting, miR-modified MSC need to be widely used in defined animal models to recognize the impact of the putative miR on the MSC therapeutic efficacy in specified milieus. Furthermore, explorations into miR-mediated gene expression regulation in cellular networks should consider the interplay among diverse miR targets depending on the varied effects of individual miRs [60, 108]. Similarly, miRs themselves could be regulated in a polygenic manner. It is important to evaluate gene regulation networks as a unit instead of the effects of individual miRs separately. Recognizing miRNA-targeted genes will allow better understanding of MSC biology and facilitate the development of novel MSC-based therapies in the future [8].

Acknowledgements We acknowledge the grant support from AHA SDG (0930181N) to MK.

References

1. Campagnoli C, Roberts IA, Kumar S, Bennett PR, Bellantuono I, Fisk NM. Identification of mesenchymal stem/progenitor cells in human first-trimester fetal blood, liver, and bone marrow. Blood. 2001;98(8):2396–402.
2. da Silva Meirelles L, Chagastelles PC, Nardi NB. Mesenchymal stem cells reside in virtually all post-natal organs and tissues. J Cell Sci. 2006;119(Pt 11):2204–13.
3. Elnakish MT, Hassan F, Dakhlallah D, Marsh CB, Alhaider IA, Khan M. Mesenchymal stem cells for cardiac regeneration: translation to bedside reality. Stem Cells Int. 2012;2012:646038.
4. Doi M, Nagano A, Nakamura Y. Molecular cloning and characterization of a novel gene, EMILIN-5, and its possible involvement in skeletal development. Biochem Biophys Res Commun. 2004;313(4):888–93.
5. Qi H, Aguiar DJ, Williams SM, La Pean A, Pan W, Verfaillie CM. Identification of genes responsible for osteoblast differentiation from human mesodermal progenitor cells. Proc Natl Acad Sci USA. 2003;100(6):3305–10.
6. Lakshmipathy U, Hart RP. Concise review: MicroRNA expression in multipotent mesenchymal stromal cells. Stem Cells. 2008;26(2):356–63.
7. Giraud-Triboult K, Rochon-Beaucourt C, Nissan X, Champon B, Aubert S, Pietu G. Combined mRNA and microRNA profiling reveals that miR-148a and miR-20b control human mesenchymal stem cell phenotype via EPAS1. Physiol Genomics. 2011;43(2):77–86.
8. Guo L, Zhao RC, Wu Y. The role of microRNAs in self-renewal and differentiation of mesenchymal stem cells. Exp Hematol. 2011;39(6):608–16.
9. Wang Y, Jiang XL, Yang SC, Lin X, He Y, Yan C, Wu L, Chen GQ, Wang ZY, Wu Q. MicroRNAs in the regulation of interfacial behaviors of MSCs cultured on microgrooved surface pattern. Biomaterials. 2011;32(35):9207–17.
10. Lim PK, Patel SA, Gregory LA, Rameshwar P. Neurogenesis: role for microRNAs and mesenchymal stem cells in pathological states. Curr Med Chem. 2010;17(20):2159–67.
11. Pillai MM, Yang X, Balakrishnan I, Bemis L, Torok-Storb B. MiR-886-3p down regulates CXCL12 (SDF1) expression in human marrow stromal cells. PLoS One. 2010;5(12):e14304.
12. Bentwich I, Avniel A, Karov Y, Aharonov R, Gilad S, Barad O, Barzilai A, Einat P, Einav U, Meiri E, Sharon E, Spector Y, Bentwich Z. Identification of hundreds of conserved and nonconserved human microRNAs. Nat Genet. 2005;37(7):766–70.
13. Lewis BP, Burge CB, Bartel DP. Conserved seed pairing, often flanked by adenosines, indicates that thousands of human genes are microRNA targets. Cell. 2005;120(1):15–20.
14. Miranda KC, Huynh T, Tay Y, Ang YS, Tam WL, Thomson AM, Lim B, Rigoutsos I. A pattern-based method for the identification of MicroRNA binding sites and their corresponding heteroduplexes. Cell. 2006;126(6):1203–17.
15. Xie X, Lu J, Kulbokas EJ, Golub TR, Mootha V, Lindblad-Toh K, Lander ES, Kellis M. Systematic discovery of regulatory motifs in human promoters and 3′ UTRs by comparison of several mammals. Nature. 2005;434(7031):338–45.
16. Bhardwaj A, Singh S, Singh AP. MicroRNA-based cancer therapeutics: big hope from small RNAs. Mol Cell Pharmacol. 2010;2(5):213–9.
17. He L, Thomson JM, Hemann MT, Hernando-Monge E, Mu D, Goodson S, Powers S, Cordon-Cardo C, Lowe SW, Hannon GJ, Hammond SM. A microRNA polycistron as a potential human oncogene. Nature. 2005;435(7043):828–33.
18. Seitz H, Royo H, Bortolin ML, Lin SP, Ferguson-Smith AC, Cavaille J. A large imprinted microRNA gene cluster at the mouse Dlk1-Gtl2 domain. Genome Res. 2004;14(9):1741–8.
19. Kawahara Y, Zinshteyn B, Sethupathy P, Iizasa H, Hatzigeorgiou AG, Nishikura K. Redirection of silencing targets by adenosine-to-inosine editing of miRNAs. Science. 2007;315(5815):1137–40.
20. Luciano DJ, Mirsky H, Vendetti NJ, Maas S. RNA editing of a miRNA precursor. RNA. 2004;10(8):1174–7.

21. Ruby JG, Jan CH, Bartel DP. Intronic microRNA precursors that bypass Drosha processing. Nature. 2007;448(7149):83–6.
22. Ying SY, Lin SL. Intronic microRNAs. Biochem Biophys Res Commun. 2005;326(3): 515–20.
23. Cheng AM, Byrom MW, Shelton J, Ford LP. Antisense inhibition of human miRNAs and indications for an involvement of miRNA in cell growth and apoptosis. Nucleic Acids Res. 2005;33(4):1290–7.
24. Mansfield JH, Harfe BD, Nissen R, Obenauer J, Srineel J, Chaudhuri A, Farzan-Kashani R, Zuker M, Pasquinelli AE, Ruvkun G, Sharp PA, Tabin CJ, McManus MT. MicroRNA-responsive 'sensor' transgenes uncover Hox-like and other developmentally regulated patterns of vertebrate microRNA expression. Nat Genet. 2004;36(10):1079–83.
25. Nakahara K, Kim K, Sciulli C, Dowd SR, Minden JS, Carthew RW. Targets of microRNA regulation in the Drosophila oocyte proteome. Proc Natl Acad Sci USA. 2005; 102(34):12023–8.
26. O'Donnell KA, Wentzel EA, Zeller KI, Dang CV, Mendell JT. c-Myc-regulated microRNAs modulate E2F1 expression. Nature. 2005;435(7043):839–43.
27. Berezikov E, Guryev V, van de Belt J, Wienholds E, Plasterk RH, Cuppen E. Phylogenetic shadowing and computational identification of human microRNA genes. Cell. 2005;120(1):21–4.
28. Goff LA, Boucher S, Ricupero CL, Fenstermacher S, Swerdel M, Chase LG, Adams CC, Chesnut J, Lakshmipathy U, Hart RP. Differentiating human multipotent mesenchymal stromal cells regulate microRNAs: prediction of microRNA regulation by PDGF during osteogenesis. Exp Hematol. 2008;36(10):1354–69.
29. Lakshmipathy U, Love B, Goff LA, Jornsten R, Graichen R, Hart RP, Chesnut JD. MicroRNA expression pattern of undifferentiated and differentiated human embryonic stem cells. Stem Cells Dev. 2007;16(6):1003–16.
30. de Peppo GM, Svensson S, Lenneras M, Synnergren J, Stenberg J, Strehl R, Hyllner J, Thomsen P, Karlsson C. Human embryonic mesodermal progenitors highly resemble human mesenchymal stem cells and display high potential for tissue engineering applications. Tissue Eng Part A. 2010;16(7):2161–82.
31. Lian Q, Lye E, Suan Yeo K, Khia Way Tan E, Salto-Tellez M, Liu TM, Palanisamy N, El Oakley RM, Lee EH, Lim B, Lim SK. Derivation of clinically compliant MSCs from CD105+, CD24− differentiated human ESCs. Stem Cells. 2007;25(2):425–36.
32. Olivier EN, Rybicki AC, Bouhassira EE. Differentiation of human embryonic stem cells into bipotent mesenchymal stem cells. Stem Cells. 2006;24(8):1914–22.
33. Khan M, Meduru S, Mohan IK, Kuppusamy ML, Wisel S, Kulkarni A, Rivera BK, Hamlin RL, Kuppusamy P. Hyperbaric oxygenation enhances transplanted cell graft and functional recovery in the infarct heart. J Mol Cell Cardiol. 2009;47(2):275–87.
34. Kim YJ, Bae SW, Yu SS, Bae YC, Jung JS. miR-196a regulates proliferation and osteogenic differentiation in mesenchymal stem cells derived from human adipose tissue. J Bone Miner Res. 2009;24(5):816–25.
35. Mizuno Y, Tokuzawa Y, Ninomiya Y, Yagi K, Yatsuka-Kanesaki Y, Suda T, Fukuda T, Katagiri T, Kondoh Y, Amemiya T, Tashiro H, Okazaki Y. miR-210 promotes osteoblastic differentiation through inhibition of AcvR1b. FEBS Lett. 2009;583(13):2263–8.
36. Li H, Xie H, Liu W, Hu R, Huang B, Tan YF, Xu K, Sheng ZF, Zhou HD, Wu XP, Luo XH. A novel microRNA targeting HDAC5 regulates osteoblast differentiation in mice and contributes to primary osteoporosis in humans. J Clin Invest. 2009;119(12):3666–77.
37. Schoolmeesters A, Eklund T, Leake D, Vermeulen A, Smith Q, Force Aldred S, Fedorov Y. Functional profiling reveals critical role for miRNA in differentiation of human mesenchymal stem cells. PLoS One. 2009;4(5):e5605.
38. Li Z, Hassan MQ, Jafferji M, Aqeilan RI, Garzon R, Croce CM, van Wijnen AJ, Stein JL, Stein GS, Lian JB. Biological functions of miR-29b contribute to positive regulation of osteoblast differentiation. J Biol Chem. 2009;284(23):15676–84.

39. Gao J, Yang T, Han J, Yan K, Qiu X, Zhou Y, Fan Q, Ma B. MicroRNA expression during osteogenic differentiation of human multipotent mesenchymal stromal cells from bone marrow. J Cell Biochem. 2011;112(7):1844–56.
40. Mizuno Y, Yagi K, Tokuzawa Y, Kanesaki-Yatsuka Y, Suda T, Katagiri T, Fukuda T, Maruyama M, Okuda A, Amemiya T, Kondoh Y, Tashiro H, Okazaki Y. miR-125b inhibits osteoblastic differentiation by down-regulation of cell proliferation. Biochem Biophys Res Commun. 2008;368(2):267–72.
41. Eskildsen T, Taipaleenmaki H, Stenvang J, Abdallah BM, Ditzel N, Nossent AY, Bak M, Kauppinen S, Kassem M. MicroRNA-138 regulates osteogenic differentiation of human stromal (mesenchymal) stem cells in vivo. Proc Natl Acad Sci USA. 2011;108(15):6139–44.
42. Li Z, Hassan MQ, Volinia S, van Wijnen AJ, Stein JL, Croce CM, Lian JB, Stein GS. A microRNA signature for a BMP2-induced osteoblast lineage commitment program. Proc Natl Acad Sci USA. 2008;105(37):13906–11.
43. Inose H, Ochi H, Kimura A, Fujita K, Xu R, Sato S, Iwasaki M, Sunamura S, Takeuchi Y, Fukumoto S, Saito K, Nakamura T, Siomi H, Ito H, Arai Y, Shinomiya K, Takeda S. A microRNA regulatory mechanism of osteoblast differentiation. Proc Natl Acad Sci USA. 2009;106(49):20794–9.
44. Huang J, Zhao L, Xing L, Chen D. MicroRNA-204 regulates Runx2 protein expression and mesenchymal progenitor cell differentiation. Stem Cells. 2010;28(2):357–64.
45. Luzi E, Marini F, Sala SC, Tognarini I, Galli G, Brandi ML. Osteogenic differentiation of human adipose tissue-derived stem cells is modulated by the miR-26a targeting of the SMAD1 transcription factor. J Bone Miner Res. 2008;23(2):287–95.
46. Han J, Yang T, Gao J, Wu J, Qiu X, Fan Q, Ma B. Specific microRNA expression during chondrogenesis of human mesenchymal stem cells. Int J Mol Med. 2010;25(3):377–84.
47. Yang B, Guo H, Zhang Y, Dong S, Ying D. The microRNA expression profiles of mouse mesenchymal stem cell during chondrogenic differentiation. BMB Rep. 2011;44(1):28–33.
48. Miyaki S, Nakasa T, Otsuki S, Grogan SP, Higashiyama R, Inoue A, Kato Y, Sato T, Lotz MK, Asahara H. MicroRNA-140 is expressed in differentiated human articular chondrocytes and modulates interleukin-1 responses. Arthritis Rheum. 2009;60(9):2723–30.
49. Tuddenham L, Wheeler G, Ntounia-Fousara S, Waters J, Hajihosseini MK, Clark I, Dalmay T. The cartilage specific microRNA-140 targets histone deacetylase 4 in mouse cells. FEBS Lett. 2006;580(17):4214–7.
50. Lin EA, Kong L, Bai XH, Luan Y, Liu CJ. miR-199a, a bone morphogenic protein 2-responsive MicroRNA, regulates chondrogenesis via direct targeting to Smad1. J Biol Chem. 2009;284(17):11326–35.
51. Kim YJ, Hwang SJ, Bae YC, Jung JS. MiR-21 regulates adipogenic differentiation through the modulation of TGF beta signaling in mesenchymal stem cells derived from human adipose tissue. Stem Cells. 2009;27(12):3093–102.
52. Sun F, Wang J, Pan Q, Yu Y, Zhang Y, Wan Y, Li X, Hong A. Characterization of function and regulation of miR-24-1 and miR-31. Biochem Biophys Res Commun. 2009;380(3):660–5.
53. Yang Z, Bian C, Zhou H, Huang S, Wang S, Liao L, Zhao RC. MicroRNA hsa-miR-138 inhibits adipogenic differentiation of human adipose tissue-derived mesenchymal stem cells through adenovirus EID-1. Stem Cells Dev. 2011;20(2):259–67.
54. Shan ZX, Lin QX, Yu XY, Deng CY, Li XH, Zhang XC, Liu XY, Fu YH. MicroRNAs can be expressed in cardiomyocyte-like cells differentiated from human mesenchymal stem cells. Nan Fang Yi Ke Da Xue Xue Bao. 2007;27(12):1813–6.
55. Kim HK, Lee YS, Sivaprasad U, Malhotra A, Dutta A. Muscle-specific microRNA miR-206 promotes muscle differentiation. J Cell Biol. 2006;174(5):677–87.
56. Naguibneva I, Ameyar-Zazoua M, Polesskaya A, Ait-Si-Ali S, Groisman R, Souidi M, Cuvellier S, Harel-Bellan A. The microRNA miR-181 targets the homeobox protein Hox-A11 during mammalian myoblast differentiation. Nat Cell Biol. 2006;8(3):278–84.
57. Liu JL, Jiang L, Lin QX, Deng CY, Mai LP, Zhu JN, Li XH, Yu XY, Lin SG, Shan ZX. MicroRNA 16 enhances differentiation of human bone marrow mesenchymal stem

cells in a cardiac niche toward myogenic phenotypes in vitro. Life Sci. 2012; 90(25–26):1020–6.
58. Zhang LL, Liu JJ, Liu F, Liu WH, Wang YS, Zhu B, Yu B. MiR-499 induces cardiac differentiation of rat mesenchymal stem cells through wnt/beta-catenin signaling pathway. Biochem Biophys Res Commun. 2012;420(4):875–81.
59. Cai B, Li J, Wang J, Luo X, Ai J, Liu Y, Wang N, Liang H, Zhang M, Chen N, Wang G, Xing S, Zhou X, Yang B, Wang X, Lu Y. microRNA-124 regulates cardiomyocyte differentiation of bone marrow-derived mesenchymal stem cells via targeting STAT3 signaling. Stem Cells. 2012;30(8):1746–55.
60. Greco SJ, Rameshwar P. MicroRNAs regulate synthesis of the neurotransmitter substance P in human mesenchymal stem cell-derived neuronal cells. Proc Natl Acad Sci USA. 2007;104(39):15484–9.
61. Hebert SS, Horre K, Nicolai L, Bergmans B, Papadopoulou AS, Delacourte A, De Strooper B. MicroRNA regulation of Alzheimer's Amyloid precursor protein expression. Neurobiol Dis. 2009;33(3):422–8.
62. Trzaska KA, Reddy BY, Munoz JL, Li KY, Ye JH, Rameshwar P. Loss of RE-1 silencing factor in mesenchymal stem cell-derived dopamine progenitors induces functional maturity. Mol Cell Neurosci. 2008;39(2):285–90.
63. Jing L, Jia Y, Lu J, Han R, Li J, Wang S, Peng T. MicroRNA-9 promotes differentiation of mouse bone mesenchymal stem cells into neurons by Notch signaling. Neuroreport. 2011;22(5):206–11.
64. Chang SJ, Weng SL, Hsieh JY, Wang TY, Chang MD, Wang HW. MicroRNA-34a modulates genes involved in cellular motility and oxidative phosphorylation in neural precursors derived from human umbilical cord mesenchymal stem cells. BMC Med Genomics. 2011;4:65.
65. Koh W, Sheng CT, Tan B, Lee QY, Kuznetsov V, Kiang LS, Tanavde V. Analysis of deep sequencing microRNA expression profile from human embryonic stem cells derived mesenchymal stem cells reveals possible role of let-7 microRNA family in downstream targeting of hepatic nuclear factor 4 alpha. BMC Genomics. 2010;11 Suppl 1:S6.
66. Cui L, Zhou X, Li J, Wang L, Wang J, Li Q, Chu J, Zheng L, Wu Q, Han Z, Shi Y, Han Y, Fan D. Dynamic microRNA profiles of hepatic differentiated human umbilical cord lining-derived mesenchymal stem cells. PLoS One. 2012;7(9):e44737.
67. Yu X, Cohen DM, Chen CS. miR-125b Is an adhesion-regulated microRNA that protects mesenchymal stem cells from anoikis. Stem Cells. 2012;30(5):956–64.
68. Baxter MA, Wynn RF, Jowitt SN, Wraith JE, Fairbairn LJ, Bellantuono I. Study of telomere length reveals rapid aging of human marrow stromal cells following in vitro expansion. Stem Cells. 2004;22(5):675–82.
69. Nie Y, Han BM, Liu XB, Yang JJ, Wang F, Cong XF, Chen X. Identification of MicroRNAs involved in hypoxia- and serum deprivation-induced apoptosis in mesenchymal stem cells. Int J Biol Sci. 2011;7(6):762–8.
70. Suzuki Y, Kim HW, Ashraf M, Haider H. Diazoxide potentiates mesenchymal stem cell survival via NF-kappaB-dependent miR-146a expression by targeting Fas. Am J Physiol Heart Circ Physiol. 2010;299(4):H1077–82.
71. Wang J, Huang W, Wu Y, Hou J, Nie Y, Gu H, Li J, Hu S, Zhang H. MicroRNA-193 pro-proliferation effects for bone mesenchymal stem cells after low-level laser irradiation treatment through inhibitor of growth family, member 5. Stem Cells Dev. 2012;21(13): 2508–19.
72. Kim YJ, Hwang SH, Lee SY, Shin KK, Cho HH, Bae YC, Jung JS. miR-486-5p induces replicative senescence of human adipose tissue-derived mesenchymal stem cells and its expression is controlled by high glucose. Stem Cells Dev. 2012;21(10):1749–60.
73. Stolzing A, Jones E, McGonagle D, Scutt A. Age-related changes in human bone marrow-derived mesenchymal stem cells: consequences for cell therapies. Mech Ageing Dev. 2008;129(3):163–73.
74. Van Zant G, Liang Y. The role of stem cells in aging. Exp Hematol. 2003;31(8):659–72.
75. Galderisi U, Helmbold H, Squillaro T, Alessio N, Komm N, Khadang B, Cipollaro M, Bohn W, Giordano A. In vitro senescence of rat mesenchymal stem cells is accompanied

by downregulation of stemness-related and DNA damage repair genes. Stem Cells Dev. 2009;18(7):1033–42.
76. Kretlow JD, Jin YQ, Liu W, Zhang WJ, Hong TH, Zhou G, Baggett LS, Mikos AG, Cao Y. Donor age and cell passage affects differentiation potential of murine bone marrow-derived stem cells. BMC Cell Biol. 2008;9:60.
77. Wagner W, Horn P, Castoldi M, Diehlmann A, Bork S, Saffrich R, Benes V, Blake J, Pfister S, Eckstein V, Ho AD. Replicative senescence of mesenchymal stem cells: a continuous and organized process. PLoS One. 2008;3(5):e2213.
78. Fabbri M, Garzon R, Cimmino A, Liu Z, Zanesi N, Callegari E, Liu S, Alder H, Costinean S, Fernandez-Cymering C, Volinia S, Guler G, Morrison CD, Chan KK, Marcucci G, Calin GA, Huebner K, Croce CM. MicroRNA-29 family reverts aberrant methylation in lung cancer by targeting DNA methyltransferases 3A and 3B. Proc Natl Acad Sci USA. 2007;104(40): 15805–10.
79. Yu JM, Wu X, Gimble JM, Guan X, Freitas MA, Bunnell BA. Age-related changes in mesenchymal stem cells derived from rhesus macaque bone marrow. Aging Cell. 2011; 10(1):66–79.
80. Pandey AC, Semon JA, Kaushal D, O'Sullivan RP, Glowacki J, Gimble JM, Bunnell BA. MicroRNA profiling reveals age-dependent differential expression of nuclear factor kappaB and mitogen-activated protein kinase in adipose and bone marrow-derived human mesenchymal stem cells. Stem Cell Res Ther. 2011;2(6):49.
81. Bruno S, Grange C, Deregibus MC, Calogero RA, Saviozzi S, Collino F, Morando L, Busca A, Falda M, Bussolati B, Tetta C, Camussi G. Mesenchymal stem cell-derived microvesicles protect against acute tubular injury. J Am Soc Nephrol. 2009;20(5):1053–67.
82. Collino F, Deregibus MC, Bruno S, Sterpone L, Aghemo G, Viltono L, Tetta C, Camussi G. Microvesicles derived from adult human bone marrow and tissue specific mesenchymal stem cells shuttle selected pattern of miRNAs. PLoS One. 2010;5(7):e11803.
83. Camussi G, Deregibus MC, Bruno S, Cantaluppi V, Biancone L. Exosomes/microvesicles as a mechanism of cell-to-cell communication. Kidney Int. 2010;78(9):838–48.
84. Lotvall J, Valadi H. Cell to cell signalling via exosomes through esRNA. Cell Adh Migr. 2007;1(3):156–8.
85. Chen TS, Lai RC, Lee MM, Choo AB, Lee CN, Lim SK. Mesenchymal stem cell secretes microparticles enriched in pre-microRNAs. Nucleic Acids Res. 2010;38(1):215–24.
86. Ryu CH, Park SA, Kim SM, Lim JY, Jeong CH, Jun JA, Oh JH, Park SH, Oh WI, Jeun SS. Migration of human umbilical cord blood mesenchymal stem cells mediated by stromal cell-derived factor-1/CXCR4 axis via Akt, ERK, and p38 signal transduction pathways. Biochem Biophys Res Commun. 2010;398(1):105–10.
87. Cheng Z, Ou L, Zhou X, Li F, Jia X, Zhang Y, Liu X, Li Y, Ward CA, Melo LG, Kong D. Targeted migration of mesenchymal stem cells modified with CXCR4 gene to infarcted myocardium improves cardiac performance. Mol Ther. 2008;16(3):571–9.
88. Lu MH, Li CZ, Hu CJ, Fan YH, Wang SM, Wu YY, Liang GP, Yang SM. microRNA-27b suppresses mouse MSC migration to the liver by targeting SDF-1alphain vitro. Biochem Biophys Res Commun. 2012;421(2):389–95.
89. Tome M, Lopez-Romero P, Albo C, Sepulveda JC, Fernandez-Gutierrez B, Dopazo A, Bernad A, Gonzalez MA. miR-335 orchestrates cell proliferation, migration and differentiation in human mesenchymal stem cells. Cell Death Differ. 2011;18(6):985–95.
90. Wen Z, Zheng S, Zhou C, Yuan W, Wang J, Wang T. Bone marrow mesenchymal stem cells for post-myocardial infarction cardiac repair: microRNAs as novel regulators. J Cell Mol Med. 2012;16(4):657–71.
91. Aguayo-Mazzucato C, Bonner-Weir S. Stem cell therapy for type 1 diabetes mellitus. Nat Rev Endocrinol. 2010;6(3):139–48.
92. Elnakish MT, Kuppusamy P, Khan M. Stem cell transplantation as a therapy for cardiac fibrosis. J Pathol. 2013;229(2):347–54.
93. Tyagi AC, Sen U, Mishra PK. Synergy of microRNA and stem cell: a novel therapeutic approach for diabetes mellitus and cardiovascular diseases. Curr Diabetes Rev. 2011; 7(6):367–76.

94. Kim YJ, Hwang SH, Cho HH, Shin KK, Bae YC, Jung JS. MicroRNA 21 regulates the proliferation of human adipose tissue-derived mesenchymal stem cells and high-fat diet-induced obesity alters microRNA 21 expression in white adipose tissues. J Cell Physiol. 2012;227(1):183–93.
95. Klopp AH, Gupta A, Spaeth E, Andreeff M, Marini 3rd F. Concise review: dissecting a discrepancy in the literature: do mesenchymal stem cells support or suppress tumor growth? Stem Cells. 2011;29(1):11–9.
96. Bae S, Ahn JH, Park CW, Son HK, Kim KS, Lim NK, Jeon CJ, Kim H. Gene and microRNA expression signatures of human mesenchymal stromal cells in comparison to fibroblasts. Cell Tissue Res. 2009;335(3):565–73.
97. Tavazoie SF, Alarcon C, Oskarsson T, Padua D, Wang Q, Bos PD, Gerald WL, Massague J. Endogenous human microRNAs that suppress breast cancer metastasis. Nature. 2008;451(7175):147–52.
98. Hardy SA, Maltman DJ, Przyborski SA. Mesenchymal stem cells as mediators of neural differentiation. Curr Stem Cell Res Ther. 2008;3(1):43–52.
99. Keilhoff G, Stang F, Goihl A, Wolf G, Fansa H. Transdifferentiated mesenchymal stem cells as alternative therapy in supporting nerve regeneration and myelination. Cell Mol Neurobiol. 2006;26(7–8):1235–52.
100. Kopen GC, Prockop DJ, Phinney DG. Marrow stromal cells migrate throughout forebrain and cerebellum, and they differentiate into astrocytes after injection into neonatal mouse brains. Proc Natl Acad Sci USA. 1999;96(19):10711–6.
101. Lu P, Jones LL, Tuszynski MH. BDNF-expressing marrow stromal cells support extensive axonal growth at sites of spinal cord injury. Exp Neurol. 2005;191(2):344–60.
102. Liu Y, Jiang X, Zhang X, Chen R, Sun T, Fok KL, Dong J, Tsang LL, Yi S, Ruan Y, Guo J, Yu MK, Tian Y, Chung YW, Yang M, Xu W, Chung CM, Li T, Chan HC. Dedifferentiation-reprogrammed mesenchymal stem cells with improved therapeutic potential. Stem Cells. 2011;29(12):2077–89.
103. Collino F, Bruno S, Deregibus MC, Tetta C, Camussi G. MicroRNAs and mesenchymal stem cells. Vitam Horm. 2011;87:291–320.
104. Hu R, Li H, Liu W, Yang L, Tan YF, Luo XH. Targeting miRNAs in osteoblast differentiation and bone formation. Expert Opin Ther Targets. 2010;14(10):1109–20.
105. Zou Z, Zhang Y, Hao L, Wang F, Liu D, Su Y, Sun H. More insight into mesenchymal stem cells and their effects inside the body. Expert Opin Biol Ther. 2010;10(2):215–30.
106. Xu J, Wu W, Zhang L, Dorset-Martin W, Morris MW, Mitchell ME, Liechty KW. The role of MicroRNA-146a in the pathogenesis of the diabetic wound-healing impairment: correction with mesenchymal stem cell treatment. Diabetes. 2012;61(11):2906–12.
107. Liu L, Wang Y, Fan H, Zhao X, Liu D, Hu Y, Kidd 3rd AR, Bao J, Hou Y. MicroRNA-181a regulates local immune balance by inhibiting proliferation and immunosuppressive properties of mesenchymal stem cells. Stem Cells. 2012;30(8):1756–70.
108. Tran DH, Satou K, Ho TB. Finding microRNA regulatory modules in human genome using rule induction. BMC Bioinformatics. 2008;9 Suppl 12:S5.

Genetic Modification of MSCs for Pharmacological Screening

Jie Qin and Martin Zenke

Abstract Mesenchymal stem cells (MSC) exhibit the potential to differentiate into different cell types, including adipocytes, osteocytes and chondrocytes, and maintain this multipotency in *in vitro* culture. Recent studies demonstrated that MSC differentiation into specific lineages is regulated by different sets of factors. Additionally, there is increasing evidence that those factors are potential targets for therapeutic intervention of diseases, such as type II diabetes, obesity, osteoporosis and osteoarthritis. These traits make MSC a particularly appealing cell source for screening novel drug candidates. Furthermore, genetic modification of MSC by target-specific promoter driving reporter genes vastly enhances the efficiency of high-throughput screening. In this chapter, we will discuss the recent developments of genetic modified MSC in pharmacological screening.

Keywords MSC • Genetic modification • Pharmacological screening

Introduction

After intensive efforts over the past few years we have now a more detailed and complete picture on the biology of mesenchymal stem cells (MSC). MSC reside in nearly every tissue of our body, including bone marrow, adipose tissue and umbilical cord, and are defined as multipotent stem cells [1]. Therefore, they are readily isolated from different tissue samples and expanded in *in vitro* culture. They give

J. Qin • M. Zenke (✉)
Department of Cell Biology, Institute for Biomedical Engineering,
Rhenish-Westphalian Technical University, Aachen University Medical School,
Pauwelsstrasse 30, 52074 Aachen, Germany
e-mail: jie.qin@rwth-aachen.de; martin.zenke@rwth-aachen.de

R.C. Zhao (ed.), *Essentials of Mesenchymal Stem Cell Biology and Its Clinical Translation*, DOI 10.1007/978-94-007-6716-4_8,

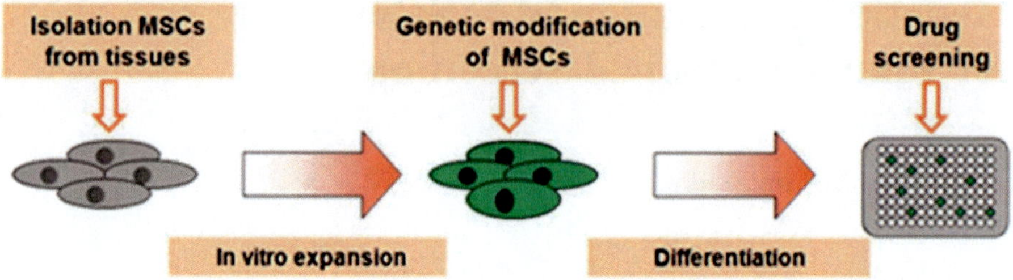

Fig. 1 Scheme of application of genetic modified MSC in drug screening

rise to different progenies, such as adipocytes, osteocytes and chondrocytes, which make them valuable cell candidates both for cellular transplantation and tissue engineering in regenerative therapy [2]. MSC contribute to maintain the homeostasis of different organs. For instance, in bone marrow, they constitute the key component of the niche to regulate proliferation, differentiation and mobilization of hematopoietic stem cells [3]. Most importantly, MSC are recruited to damage sites and participate as immune-modulator in many diseases, for instance, graft-versus-host disease (GVHD), cardiac infarction and Crohn's disease. Currently these activities are being evaluated in an increasing number of clinical trials [4].

MSC can be transduced with different transduction methods (e.g. electroporation, lipofectamine, retrovirus, lentivirus, adenovirus, adeno-associated virus and zinc finger nuclease). Genetic modification of MSC serves as a powerful tool and allows many studies, such as gene therapy and drug screening. MSC are hypo-immunogenic and can migrate into injury sites, which give rise to the notion of utilizing MSC in cellular therapy [5]. But later researches showed that only a small portion of MSC succeeded in homing to the damaged sites after transplantation. Massive inflammation in situ can also affect the efficiency of MSC treatment [6]. To improve the therapeutic outcome, factors have been introduced into MSC, to enhance MSC migration or survival rate, or promote the healing process via paracrine mechanisms [5]. In the field of tissue engineering, where the differentiation capacity of MSC heavily depends on, factors governing differentiation pathways (e.g. BMP2) have been transduced into MSC to improve differentiation efficiency *in vivo* [7]. As a vector in gene therapy, genetic modified MSC gain considerable attention to treat diseases. So far, MSC application in animal models proved efficient in a wide range of diseases, such as skeleton diseases [8], cardiovascular diseases [9] and ischemic damages [10]. Another large application field of genetic modified MSC is in drug development. MSC differentiation into specific lineages is directed by specific sets of factors and increasing evidence indicates that these factors are potential targets for therapeutic intervention of diseases, such as type II diabetes, obesity, osteoporosis and osteoarthritis. Thus, MSC and their differentiated progeny represent particularly attractive cell sources for drug development. Furthermore, genetic modification of MSC by target-specific promoter driving reporter genes vastly enhances the efficiency of high-throughput screening. In this chapter, we summarize the application of MSC in drug development (Fig. 1).

Potential Targets for Therapeutic Drugs

Peroxisome proliferator-activator receptors (PPARs) are a subset of the nuclear hormone receptor superfamily [11]. They comprise three subtypes: PPARα, PPARβ/δ, and PPARγ [12–14]. The three PPAR isoforms have distinct tissue distribution, and function differently. In human, *PPARα* is expressed in heart, kidney, skeletal muscle and large intestine, and its activity correlates with oxidation of fatty acids. *PPARβ/δ* is expressed in all the tissues at rather low levels, and little is known about its function. *PPARγ* expression is predominately in adipose tissue and large intestine [15]. PPARγis a central regulator of adipogenesis [16]. Knockout adipose-specific *PPARγ* result in adipocyte hypocellularity, elevated levels of free fatty acids in plasma and insulin resistance [17]. *PPARγ* expression can be activated by natural and synthetic compounds [18]. Thiazolidinediones (TZDs), a class of insulin sensitizing drugs for type II diabetes, are PPARγ agonists [19]. PPARγ is suggested to be the molecular target for this class of compounds [20]. TZDs can redistribute fat from visceral to subcutaneous adipose tissue, increase uptake of fatty acids in adipose tissue and regulate the secretion of adipocytokines that may improve insulin sensitivity from adipose tissue [21]. However, recent studies identified several side-effects of TZDs, such as liquid retention, congestive heart failure and liver toxicity, due to full agonist of PPARγ [22–26]. Thus, there is clearly the need for developing selective PPARγ agonists with less or no undesirable side-effect. The *Adipocyte protein 2* (*aP2*; *also namely Fatty Acid Binding Protein 4*, *FABP4*) gene is not only one of the specific markers of mature adipocytes, but also one of the PPARγ target genes [27]. Since the interrelation between PPARγ-aP2 axis is well studied, it is supposed that aP2 promoter-driven reporter gene assays represent a rapid and sensitive procedure and particularly versatile tool for identifying novel PPARγ modulators [28]. Indeed, employing aP2 – driven luciferase (luc) reporter cell line, Waki et al. identified one small molecule (Harmine) from 504 compounds, which acts as an agonist of PPARγand improves glucose tolerance in diabetic mice [29].

As for the development of anti-obesity drugs, PPARγis also a candidate target. Currently, there are two types of anti-obesity drugs on the market [30, 31]. One group of compounds is to control food intake by modulating the central nervous system. Sibutramine is a representative from this group. However, Sibutramine may cause insomnia, headache and constipation [32]. The other group of compounds reduces fat intake by inhibiting fat absorption. For example, orlistat inhibits the activity of pancreatic lipase, hence decrease triglyceride digestion. But patients might be affected by gastrointestinal side effects and lose lipid-soluble vitamins and essential fatty acids by steatorrhea [33]. As a result, it is imperative to develop new drugs with higher efficacy and less side-effect. By screening 2,000 natural compounds, Seo et al. showed that Lysimachia foenum-graecum suppresses *PPARγ* and *CCAAT/enhancer binding protein α* (*C/EBPα*) expression, and reduces markedly the weight of white adipose tissue in high-fat diet mice [34].

Today diseases associated with bone loss, such as osteoporosis and Page's disease, represent an enormous burden of the health system. Bone mass homeostasis is critically dependent on maintaining the balance between bone destruction by osteoclasts and bone formation by osteoblasts. In the case of osteoporosis, after the age 40, bone destruction begins to exceed bone formation, which results in local or systemic bone loss [35]. BMP2 plays an essential role in postnatal skeleton formation and is required as a major factor during bone repair [36, 37]. One study identified *BMP2* mutations as a major risk factor in osteoporosis and osteoporotic fractures [38]. As part of the BMP signaling cascade in osteoblasts, *Cbfa1/Runx2* is an osteoblast-specific transcription factor in osteogenic differentiation [39]. Cbfa1/Runx2 knockout mice suffer from maturational arrest of osteoblasts, resulting in a complete block of both intramembranous and endochondral ossification [40]. These findings indicated that BMP2 and Cbfa1/Runx2 might serve as therapeutic targets for bone diseases. Li et al. screened a library of 3,192 compounds to identify 3 candidates, which increases MSC differentiation towards osteolineage, based on BMP2-luc reporter system [41]. Using mouse mesenchymal progenitor cells C3H10T1/2 and Cbfa1/Runx2-luc reporter system, Wu et al. found that purmorphamine 1 can enhance the osteogenic differentiation [42]. Additionally, as a downstream target of Cbfa1/Runx2 [39, 43], *the type I collagen* (*Col1a1*) gene can also be applied in osteogenic compound screening. Hojo et al. used col1a1 promoter driving GFP reporter to identify glabrisoflavone (GI) as an osteogenic compound [44].

Sox9, also known as sex determining region Y–box9, is a key regulator of the chondrocyte lineage [45]. In situ hybridization showed that *Sox9* is expressed in all chondroprogenitor cells during mouse embryogenesis. Heterozygous mutations in the *SOX9* gene may cause severe chondrodysplasia [46]. As a direct target of Sox9, *Collagen type 2 proa1* (*Col2a1*) is an early and abundant marker of chondrocytes, and type II Collagen represents the major extra-cellular matrix in cartilage tissue [45, 47, 48]. Osteoarthritis is one of the common degenerative join diseases. The early pathological changes include the loss of extra-cellular matrix and later lose cartilage [49]. Standard treatments are pain management and join replacement surgery, which however treat symptoms, rather than the disease itself. Therefore alternatives represent the identification of novel drugs, which enhance cartilage regeneration. Sox9-dependent type II Collagen gene promoter driving reporter gene serves as a powerful screening tool. Hojo et al. identified oxytetracycline out of 2,500 natural and synthetic compounds by Col2a1-driven GFP screening system [50].

Why MSC-Based Screening Models?

In current studies, some investigators use murine preadipocytes (3T3-L1), preosteoblasts (KS483, MC3T3-E1) or prechondrocytes (ATDC5), while others use the murine mesenchymal cell line (C3H10T1/2) or human MSC as research model. We discuss here the advantages and disadvantages of using MSC.

First, MSC enable investigators to study the activities of compounds from the cell commitment stage to terminal differentiation stage, while precursor cells only reveal the differentiation process. Therefore, more and more investigators employ first a screening assays based on precursor cell lines, followed by an assay based on MSC to confirm the results obtained [51]. Second, for studying potential side effects of novel drug candidates or developing multifunctional drugs, it is advantageous to use MSC models. For instance, patient with osteoporosis often have obesity problem, and small molecules, such as M-25659, increase osteogenic differentiation while inhibiting adipogenesis, and are therefore suggested as an attractive novel candidate drug [52]. Third, MSC enable researchers to screen on primary human cells. Human cells always serve as a "golden standard" for drug development. After years of intense discussions, researchers established a set of standard protocols for isolation, culturing and differentiation of human MSC [53, 54]. What is more, it is well recognized that cells derived from human MSC *in vitro* share similar characteristics as those *in vivo* in our body [55].

Yet, one major advantage for using murine precursor cell lines is their stability. They can proliferate for quite a long time, without losing their differentiation capacity. Moreover, when introduced with exogenous genes, single cell clones are readily picked and expanded to obtain homogenous cell population. In this respect, MSC (especially human MSC) have some disadvantages. One concern is the aging of MSC. Human MSC undergo aging after several months of *in vitro* culture concomitantly with impaired differentiation [56, 57]. Increasing evidence supports findings that epigenetic modifications play major role during this aging process [58, 59]. Another concern is that it is difficult to generate homogenous genetic modified human MSC, because MSC are difficult to be propagated as single cell, especially after virus infection and antibiotics selection. Immortalization of MSC might provide a solution to its problem. Indeed, it was reported that after immortalizing with HPV16 *E6/E7* genes, human MSC retain a stable MSC phenotype without neoplastic transformation and maintain their multipotent differentiation capacity [60].

The Advantages of Genetically Modified MSC in High-Throughput Screening

High-throughput screening is an approach that allows rapid screening of compounds in large scale to identify putative therapeutic drugs. In general, a complete pharmaceutical screening session involves four steps. The primary screening starts with hundreds of thousands of compounds. Only a few of active compounds are selected, termed "hits". In the secondary screening, biological relevance of those potential compounds is defined, such as cytotoxicity and the half maximal effective concentration (EC_{50}) of the compounds. Those chemicals that pass the secondary screening are named "leads". The follow-up studies are animal experiment and clinical trials (phaseI-V). The challenge lies in maximizing the number of compounds, while simultaneously minimizing the costs and time involved [61, 62].

In the early studies, the primary screening happened in silicon chips, on which particular target proteins or substrates of specific enzymatic reactions were patterned. Those compounds, which bind or stimulate/inhibit the enzymatic reactions, became "hits" and were forwarded to secondary screening in cells. The pitfall of this screening strategy is that it did not put more real candidates into the final screening process. Most of them failed in the secondary assay, due to toxicity to cells and so forth. Thereafter, pharmaceutical companies applied cell based assay as the primary screening to look at the impact on cell physiological events in the first place. In this way, untargeted compounds can be excluded earlier. With the development of miniaturization, cell based primary screening can be put into 3,456-microwell plates, which vastly enhance the screening efficiency and reduce the cost [62].

Conventional read-outs of MSC differentiation include oil red O staining, alkaline phosphatase (ALP) staining and gene expression analysis by RT-PCR, which are time and labor consuming and cost effective. Therefore, for diseases with known molecule targets, reporter gene assays represent the optimal approach in high-throughput screening. Introducing stable cell lines with reporter genes driven by promoters of molecule targets, potential compounds are readily identified. Moreover, it has been proved by many groups that reporter gene expression is in accord with the expression pattern of the target gene analyzed by RT-PCR, indicating that reporter assays have similar sensitivities as RT-PCR [28, 29, 50, 63].

The most frequently employed reporter systems used in industry are enzymatic assays, such as luciferase assays [64]. The luciferase gene comes from the firefly Photinus pyralis. This gene encodes a 61-kDa enzyme that oxidizes D-luciferin in the presence of ATP, oxygen, and Mg2+. The reaction product can be quantified by measuring the released light [65]. Recent studies favor another reporter system, the fluorescence protein assay, which allows measurements with fluorescence spectrophotometer. Shimomura et al. first discovered green fluorescent protein from Aequorea Jellyfish [66]. Thereafter more fluorescent proteins were identified from other organisms [67]. Now a good collection of fluorescent proteins covering every spectral class is available for investigators [68]. The fluorescence protein assay allows real-time and noninvasive observation of samples. Investigators observe the change of individual cells continuously during the screening process without the need for lysing cells and subjecting them to enzymatic assays [69]. It reduces extra time and money spent during the first screening, which clearly represents a strength for high content analysis [70]. Furthermore, unlike luciferase assay, which measures the total amount of luciferase activity of a given cell population in test wells, fluorescence protein assay reveals data on individual cells, which is particularly important when cell populations are not homogeneous [64]. A common concern about the application of fluorescence assay lies in its potential toxic effects on cells, which might probably due to the aggregates formation of fluorescent proteins in cytoplasm [71]. But this potential disadvantage can be met by choosing less toxic fluorescence protein or modulating the copy number of reporter genes in cells [72].

Methods of Genetic Modification of MSC

Several methods have been applied in genetic modification of MSC, including retroviral vector [44, 69], lentiviral vector [63] and zinc finger nuclease vector [73]. For pharmacological application, the aim is to obtain stable genetic modified MSC, while maintaining their multipotency. Therefore, investigators need to chose suitable transduction method and optimize transduction conditions, such as virus titer and MSC viability.

Retrovirus transduction is a particularly widely used technology in stem cell research [74]. It can introduce up to 8 kb of exogenous DNA into dividing cells, but frequently its application is hampered by low transduction efficiency and transgene silencing by epigenetic modifications [75]. Lentivirus, which is a subgroup of retrovirus, can introduce up to 10 kb transgenes into target cells without requirement for cell division. The transduction efficiency of lentivirus is quite high in all mammal cells, even in hematopoietic stem cells (HSC). It was reported that with lentivirus, the infection rate of HSC can reach up to 60 % [76]. However, as lentivirus contains human immunodeficiency virus type 1 (HIV-1) envelope, the safety concern became an issue. The third generation of lentivirus contains a self-inactivating 3′ LTR, which makes the lentivirus unable to produce replication competent virus after integrating into the host chromosome, hence increasing its biosafety [77, 78]. Another concern for lentivirus vector systems is the potential activation of oncogenes due to the promoter insertion. This is particularly relevant since lentiviral vectors integrate into active transcription region of the target cell genome, which might lead to tumor formation [79, 80].

Zinc finger nuclease technique (ZFN) can induce specific integration sites in the host genome. ZFN contains the non-specific cleavage domain of the FokI endonuclease and DNA binding domains of zinc finger proteins. Once the zinc finger proteins bind to predetermined sites of the host genome, FokI endonuclease cuts the target sites into double-stranded DNA. During the following DNA repair process, the exogenous DNA fragments can be integrated into the target sites by homologous recombination [80]. With this technique, investigators can integrate DNA fragments into specific sites of the host genome, such as chemokine [C-C motif] receptor 5 (CCR5), in order to minimize the risk of insertion oncogenesis [73]. However, the ZFN technique also has its pitfalls. For example, ZFN might have cytotoxic effect depending on its concentration [81]. A possible solution is to combine the non-integrated lentivirus with ZFN technique [80]. Non-integrated lentivirus can persist in infected cells as transcriptionally active DNA elements [82]. During cell division, these replication-deficient lentiviral vectors would be diluted, and eventually eliminated from the infected cells [83]. Taking the advantage of high transduction efficiency of lentivirus, ZFN and exogenous DNA fragments can be delivered into target cells by integration-deficient lentivirus. After targeted modifications of the host genome, those ZFN, exogenous DNA fragments and lentivirus would vanish following several cell divisions.

Future Perspectives

The developments in MSC biology over the past few years has allowed to decipher genetic programs that drive lineage commitment and differentiation. The pathogenic mechanism of diseases enables researchers to identify putative therapeutic targets on the molecule basis. As for the screening of therapeutic drugs in adipo-, osteo- and chondrogenic diseases, MSC serve as a particularly attractive study model. With state-of-the-art genetic modification methods and fluorescence reporter gene system, genetic modified MSC with target-specific promoter driving reporter genes have proved to facilitate high-throughput screening. In the future, more and more new therapeutic targets are expected to be discovered and will add to the MSC based screening system.

In pharmacological industry today, 90 % of the potential drugs fail in clinical trials due to insufficient efficacy or unanticipated toxicity [64]. Therefore the challenge remains to develop novel drugs while improving on the screening systems and preclinical models used. Frequently, in basic research different labs screen a number of compounds based on different cell lines and assays. Even though several potential compounds are then selected as potential drugs from these studies, there are rarely follow up studies or clinical trials. Given that in several cases potential molecule targets are known to be associated with specific diseases, a screening protocol should be devised by the joint efforts of investigators and pharmaceutical companies. Such screening protocols should define cell types, screening setups, positive/negative controls and statistics methods. Furthermore, tested compounds should be documented and a database on this information can be established. Such an approach will allow to readily assess the screening results of different groups and to facilitate the cooperation between basic research and pharmaceutical companies.

References

1. da Silva Meirelles L, Chagastelles PC, Nardi NB. Mesenchymal stem cells reside in virtually all post-natal organs and tissues. J Cell Sci. 2006;119:2204–13.
2. Granero-Molto F, Weis JA, Miga MI, Landis B, Myers TJ, O'Rear L, Longobardi L, Jansen ED, Mortlock DP, Spagnoli A. Regenerative effects of transplanted mesenchymal stem cells in fracture healing. Stem Cells. 2009;27:1887–98.
3. Valtieri M, Sorrentino A. The mesenchymal stromal cell contribution to homeostasis. J Cell Physiol. 2008;217:296–300.
4. Sensebe L, Krampera M, Schrezenmeier H, Bourin P, Giordano R. Mesenchymal stem cells for clinical application. Vox Sang. 2010;98:93–107.
5. Myers TJ, Granero-Molto F, Longobardi L, Li T, Yan Y, Spagnoli A. Mesenchymal stem cells at the intersection of cell and gene therapy. Expert Opin Biol Ther. 2010;10:1663–79.
6. Song H, Song BW, Cha MJ, Choi IG, Hwang KC. Modification of mesenchymal stem cells for cardiac regeneration. Expert Opin Biol Ther. 2010;10:309–19.
7. Turgeman G, Pittman DD, Muller R, Kurkalli BG, Zhou S, Pelled G, Peyser A, Zilberman Y, Moutsatsos IK, Gazit D. Engineered human mesenchymal stem cells: a novel platform for skeletal cell mediated gene therapy. J Gene Med. 2001;3:240–51.

8. Pagnotto MR, Wang Z, Karpie JC, Ferretti M, Xiao X, Chu CR. Adeno-associated viral gene transfer of transforming growth factor-beta1 to human mesenchymal stem cells improves cartilage repair. Gene Ther. 2007;14:804–13.
9. Zhang D, Fan GC, Zhou X, Zhao T, Pasha Z, Xu M, Zhu Y, Ashraf M, Wang Y. Over-expression of CXCR4 on mesenchymal stem cells augments myoangiogenesis in the infarcted myocardium. J Mol Cell Cardiol. 2008;44:281–92.
10. Hagiwara M, Shen B, Chao L, Chao J. Kallikrein-modified mesenchymal stem cell implantation provides enhanced protection against acute ischemic kidney injury by inhibiting apoptosis and inflammation. Hum Gene Ther. 2008;19:807–19.
11. Berger J, Moller DE. The mechanisms of action of PPARs. Annu Rev Med. 2002;53:409–35.
12. Schmidt A, Endo N, Rutledge SJ, Vogel R, Shinar D, Rodan GA. Identification of a new member of the steroid hormone receptor superfamily that is activated by a peroxisome proliferator and fatty acids. Mol Endocrinol. 1992;6:1634–41.
13. Issemann I, Green S. Activation of a member of the steroid hormone receptor superfamily by peroxisome proliferators. Nature. 1990;347(6294):645–50.
14. Zhu Y, Qi C, Korenberg JR, Chen XN, Noya D, Rao MS, Reddy JK. Structural organization of mouse peroxisome proliferator-activated receptor gamma (mPPAR gamma) gene: alternative promoter use and different splicing yield two mPPAR gamma isoforms. Proc Natl Acad Sci USA. 1995;92:7921–5.
15. Desvergne B, Wahli W. Peroxisome proliferator-activated receptors: nuclear control of metabolism. Endocr Rev. 1999;20:649–88.
16. Spiegelman BM, Flier JS. Adipogenesis and obesity: rounding out the big picture. Cell. 1996;87:377–89.
17. He W, Barak Y, Hevener A, Olson P, Liao D, Le J, Nelson M, Ong E, Olefsky JM, Evans RM. Adipose-specific peroxisome proliferator-activated receptor gamma knockout causes insulin resistance in fat and liver but not in muscle. Proc Natl Acad Sci USA. 2003;100:15712–7.
18. Fajas L, Auboeuf D, Raspe E, Schoonjans K, Lefebvre AM, Saladin R, Najib J, Laville M, Fruchart JC, Deeb S, Vidal-Puig A, Flier J, Briggs MR, Staels B, Vidal H, Auwerx J. The organization, promoter analysis, and expression of the human PPARgamma gene. J Biol Chem. 1997;272:18779–89.
19. Elte JW, Blickle JF. Thiazolidinediones for the treatment of type 2 diabetes. Eur J Intern Med. 2007;18:18–25.
20. Lehmann JM, Moore LB, Smith-Oliver TA, Wilkison WO, Willson TM, Kliewer SA. An antidiabetic thiazolidinedione is a high affinity ligand for peroxisome proliferator-activated receptor gamma (PPAR gamma). J Biol Chem. 1995;270:12953–6.
21. Semple RK, Chatterjee VK, O'Rahilly S. PPAR gamma and human metabolic disease. J Clin Invest. 2006;116:581–9.
22. Pan HJ, Lin Y, Chen YE, Vance DE, Leiter EH. Adverse hepatic and cardiac responses to rosiglitazone in a new mouse model of type 2 diabetes: relation to dysregulated phosphatidylcholine metabolism. Vascul Pharmacol. 2006;45:65–71.
23. Guan Y, Hao C, Cha DR, Rao R, Lu W, Kohan DE, Magnuson MA, Redha R, Zhang Y, Breyer MD. Thiazolidinediones expand body fluid volume through PPARgamma stimulation of ENaC-mediated renal salt absorption. Nat Med. 2005;11:861–6.
24. Mudaliar S, Chang AR, Henry RR. Thiazolidinediones, peripheral edema, and type 2 diabetes: incidence, pathophysiology, and clinical implications. Endocr Pract. 2003;9:406–16.
25. Page 2nd RL, Gozansky WS, Ruscin JM. Possible heart failure exacerbation associated with rosiglitazone: case report and literature review. Pharmacotherapy. 2003;23:945–54.
26. Berger JP, Petro AE, Macnaul KL, Kelly LJ, Zhang BB, Richards K, Elbrecht A, Johnson BA, Zhou G, Doebber TW, Biswas C, Parikh M, Sharma N, Tanen MR, Thompson GM, Ventre J, Adams AD, Mosley R, Surwit RS, Moller DE. Distinct properties and advantages of a novel peroxisome proliferator-activated protein [gamma] selective modulator. Mol Endocrinol. 2003;17:662–76.
27. Tontonoz P, Hu E, Devine J, Beale EG, Spiegelman BM. PPAR gamma 2 regulates adipose expression of the phosphoenolpyruvate carboxykinase gene. Mol Cell Biol. 1995;15:351–7.

28. Thompson GM, Trainor D, Biswas C, LaCerte C, Berger JP, Kelly LJ. A high-capacity assay for PPARgamma ligand regulation of endogenous aP2 expression in 3T3-L1 cells. Anal Biochem. 2004;330:21–8.
29. Waki H, Park KW, Mitro N, Pei L, Damoiseaux R, Wilpitz DC, Reue K, Saez E, Tontonoz P. The small molecule harmine is an antidiabetic cell-type-specific regulator of PPARgamma expression. Cell Metab. 2007;5:357–70.
30. Bray GA, Tartaglia LA. Medicinal strategies in the treatment of obesity. Nature. 2000; 404:672–7.
31. Korner J, Aronne LJ. Pharmacological approaches to weight reduction: therapeutic targets. J Clin Endocrinol Metab. 2004;89:2616–21.
32. Luque CA, Rey JA. Sibutramine: a serotonin-norepinephrine reuptake-inhibitor for the treatment of obesity. Ann Pharmacother. 1999;33:968–78.
33. Heck AM, Yanovski JA, Calis KA. Orlistat, a new lipase inhibitor for the management of obesity. Pharmacotherapy. 2000;20:270–9.
34. Seo JB, Choe SS, Jeong HW, Park SW, Shin HJ, Choi SM, Park JY, Choi EW, Kim JB, Seen DS, Jeong JY, Lee TG. Anti-obesity effects of Lysimachia foenum-graecum characterized by decreased adipogenesis and regulated lipid metabolism. Exp Mol Med. 2011;43:205–15.
35. Rodan GA, Martin TJ. Therapeutic approaches to bone diseases. Science. 2000;289:1508–14.
36. Bais MV, Wigner N, Young M, Toholka R, Graves DT, Morgan EF, Gerstenfeld LC, Einhorn TA. BMP2 is essential for post natal osteogenesis but not for recruitment of osteogenic stem cells. Bone. 2009;45:254–66.
37. Tsuji K, Bandyopadhyay A, Harfe BD, Cox K, Kakar S, Gerstenfeld L, Einhorn T, Tabin CJ, Rosen V. BMP2 activity, although dispensable for bone formation, is required for the initiation of fracture healing. Nat Genet. 2006;38:1424–9.
38. Styrkarsdottir U, Cazier JB, Kong A, Rolfsson O, Larsen H, Bjarnadottir E, Johannsdottir VD, Sigurdardottir MS, Bagger Y, Christiansen C, Reynisdottir I, Grant SF, Jonasson K, Frigge ML, Gulcher JR, Sigurdsson G, Stefansson K. Linkage of osteoporosis to chromosome 20p12 and association to BMP2. PLoS Biol. 2003;1:E69.
39. Ducy P, Zhang R, Geoffroy V, Ridall AL, Karsenty G. Osf2/Cbfa1: a transcriptional activator of osteoblast differentiation. Cell. 1997;89:747–54.
40. Komori T, Yagi H, Nomura S, Yamaguchi A, Sasaki K, Deguchi K, Shimizu Y, Bronson RT, Gao YH, Inada M, Sato M, Okamoto R, Kitamura Y, Yoshiki S, Kishimoto T. Targeted disruption of Cbfa1 results in a complete lack of bone formation owing to maturational arrest of osteoblasts. Cell. 1997;89:755–64.
41. Li X, Yang J, He X, Yang Z, Ding Y, Zhao P, Liu Z, Shao H, Li Z, Zhang Y, Si S. Identification of upregulators of BMP2 expression via high-throughput screening of a synthetic and natural compound library. J Biomol Screen. 2009;14:1251–6.
42. Wu X, Ding S, Ding Q, Gray NS, Schultz PG. A small molecule with osteogenesis-inducing activity in multipotent mesenchymal progenitor cells. J Am Chem Soc. 2002;124:14520–1.
43. Komori T. Regulation of bone development and extracellular matrix protein genes by RUNX2. Cell Tissue Res. 2010;339:189–95.
44. Hojo H, Igawa K, Ohba S, Yano F, Nakajima K, Komiyama Y, Ikeda T, Lichtler AC, Woo JT, Yonezawa T, Takato T, Chung UI. Development of high-throughput screening system for osteogenic drugs using a cell-based sensor. Biochem Biophys Res Commun. 2008;376:375–9.
45. Bi W, Deng JM, Zhang Z, Behringer RR, de Crombrugghe B. Sox9 is required for cartilage formation. Nat Genet. 1999;22:85–9.
46. Giordano J, Prior HM, Bamforth JS, Walter MA. Genetic study of SOX9 in a case of campomelic dysplasia. Am J Med Genet. 2001;98:176–81.
47. Lefebvre V, Huang W, Harley VR, Goodfellow PN, de Crombrugghe B. SOX9 is a potent activator of the chondrocyte-specific enhancer of the pro alpha1(II) collagen gene. Mol Cell Biol. 1997;17:2336–46.
48. Lefebvre V, Li P, de Crombrugghe B. A new long form of Sox5 (L-Sox5), Sox6 and Sox9 are coexpressed in chondrogenesis and cooperatively activate the type II collagen gene. EMBO J. 1998;17:5718–33.

49. Hollander AP, Pidoux I, Reiner A, Rorabeck C, Bourne R, Poole AR. Damage to type II collagen in aging and osteoarthritis starts at the articular surface, originates around chondrocytes, and extends into the cartilage with progressive degeneration. J Clin Invest. 1995;96:2859–69.
50. Hojo H, Yano F, Ohba S, Igawa K, Nakajima K, Komiyama Y, Kan A, Ikeda T, Yonezawa T, Woo JT, Takato T, Nakamura K, Kawaguchi H, Chung UI. Identification of oxytetracycline as a chondrogenic compound using a cell-based screening system. J Bone Miner Metab. 2010;28:627–33.
51. Noguchi K, Watanabe Y, Fuse T, Takizawa M. A new chondrogenic differentiation initiator with the ability to up-regulate SOX trio expression. J Pharmacol Sci. 2010;112:89–97.
52. Jang EJ, Jeong H, Kang JO, Kim NJ, Kim MS, Choi SH, Yoo SE, Hong JH, Bae MA, Hwang ES. TM-25659 enhances osteogenic differentiation and suppresses adipogenic differentiation by modulating the transcriptional co-activator TAZ. Br J Pharmacol. 2012;165:1584–94.
53. Pittenger MF, Mackay AM, Beck SC, Jaiswal RK, Douglas R, Mosca JD, Moorman MA, Simonetti DW, Craig S, Marshak DR. Multilineage potential of adult human mesenchymal stem cells. Science. 1999;284:143–7.
54. Bruder SP, Jaiswal N, Haynesworth SE. Growth kinetics, self-renewal, and the osteogenic potential of purified human mesenchymal stem cells during extensive subcultivation and following cryopreservation. J Cell Biochem. 1997;64:278–94.
55. Ryden M, Dicker A, Gotherstrom C, Astrom G, Tammik C, Arner P, Le Blanc K. Functional characterization of human mesenchymal stem cell-derived adipocytes. Biochem Biophys Res Commun. 2003;311:391–7.
56. Wagner W, Ho AD, Zenke M. Different facets of aging in human mesenchymal stem cells. Tissue Eng Part B Rev. 2010;16:445–53.
57. Gruber HE, Somayaji S, Riley F, Hoelscher GL, Norton HJ, Ingram J, Hanley Jr EN. Human adipose-derived mesenchymal stem cells: serial passaging, doubling time and cell senescence. Biotech Histochem. 2012;87:303–11.
58. Shibata KR, Aoyama T, Shima Y, Fukiage K, Otsuka S, Furu M, Kohno Y, Ito K, Fujibayashi S, Neo M, Nakayama T, Nakamura T, Toguchida J. Expression of the p16INK4A gene is associated closely with senescence of human mesenchymal stem cells and is potentially silenced by DNA methylation during in vitro expansion. Stem Cells. 2007;25:2371–82.
59. Koch CM, Joussen S, Schellenberg A, Lin Q, Zenke M, Wagner W. Monitoring of cellular senescence by DNA-methylation at specific CpG sites. Aging Cell. 2012;11:366–9.
60. Hung SC, Yang DM, Chang CF, Lin RJ, Wang JS, Low-Tone Ho L, Yang WK. Immortalization without neoplastic transformation of human mesenchymal stem cells by transduction with HPV16 E6/E7 genes. Int J Cancer. 2004;110:313–9.
61. Malo N, Hanley JA, Cerquozzi S, Pelletier J, Nadon R. Statistical practice in high-throughput screening data analysis. Nat Biotechnol. 2006;24:167–75.
62. Dove A. Screening for content – the evolution of high throughput. Nat Biotechnol. 2003;21:859–64.
63. Qin J, Li WQ, Zhang L, Chen F, Liang WH, Mao FF, Zhang XM, Lahn BT, Yu WH, Xiang AP. A stem cell-based tool for small molecule screening in adipogenesis. PLoS One. 2010;5:e13014.
64. Rubin LL. Stem cells and drug discovery: the beginning of a new era? Cell. 2008;132:549–52.
65. Smale ST. Luciferase assay. Cold Spring Harb Protoc. 2010:pdb prot5421. doi: 10.1101/pdb.prot5421
66. Shimomura O, Johnson FH, Saiga Y. Extraction, purification and properties of aequorin, a bioluminescent protein from the luminous hydromedusan, Aequorea. J Cell Comp Physiol. 1962;59:223–39.
67. Shaner NC, Patterson GH, Davidson MW. Advances in fluorescent protein technology. J Cell Sci. 2007;120:4247–60.
68. Shaner NC, Steinbach PA, Tsien RY. A guide to choosing fluorescent proteins. Nat Methods. 2005;2:905–9.
69. Kan A, Ikeda T, Saito T, Yano F, Fukai A, Hojo H, Ogasawara T, Ogata N, Nakamura K, Chung UI, Kawaguchi H. Screening of chondrogenic factors with a real-time fluorescence-monitoring cell line ATDC5-C2ER: identification of sorting nexin 19 as a novel factor. Arthritis Rheum. 2009;60:3314–23.

70. Cardno TS, Poole ES, Mathew SF, Graves R, Tate WP. A homogeneous cell-based bicistronic fluorescence assay for high-throughput identification of drugs that perturb viral gene recoding and read-through of nonsense stop codons. RNA. 2009;15:1614–21.
71. Yanushevich YG, Staroverov DB, Savitsky AP, Fradkov AF, Gurskaya NG, Bulina ME, Lukyanov KA, Lukyanov SA. A strategy for the generation of non-aggregating mutants of Anthozoa fluorescent proteins. FEBS Lett. 2002;511:11–4.
72. Liu HS, Jan MS, Chou CK, Chen PH, Ke NJ. Is green fluorescent protein toxic to the living cells? Biochem Biophys Res Commun. 1999;260:712–7.
73. Benabdallah BF, Allard E, Yao S, Friedman G, Gregory PD, Eliopoulos N, Fradette J, Spees JL, Haddad E, Holmes MC, Beausejour CM. Targeted gene addition to human mesenchymal stromal cells as a cell-based plasma-soluble protein delivery platform. Cytotherapy. 2010;12:394–9.
74. Asahara T, Kalka C, Isner JM. Stem cell therapy and gene transfer for regeneration. Gene Ther. 2000;7:451–7.
75. Miller DG, Adam MA, Miller AD. Gene transfer by retrovirus vectors occurs only in cells that are actively replicating at the time of infection. Mol Cell Biol. 1990;10:4239–42.
76. Evans JT, Kelly PF, O'Neill E, Garcia JV. Human cord blood CD34+ CD38− cell transduction via lentivirus-based gene transfer vectors. Hum Gene Ther. 1999;10:1479–89.
77. Dull T, Zufferey R, Kelly M, Mandel RJ, Nguyen M, Trono D, Naldini L. A third-generation lentivirus vector with a conditional packaging system. J Virol. 1998;72:8463–71.
78. Zufferey R, Dull T, Mandel RJ, Bukovsky A, Quiroz D, Naldini L, Trono D. Self-inactivating lentivirus vector for safe and efficient in vivo gene delivery. J Virol. 1998;72:9873–80.
79. Matrai J, Chuah MK, VandenDriessche T. Recent advances in lentiviral vector development and applications. Mol Ther. 2010;18:477–90.
80. Le Provost F, Lillico S, Passet B, Young R, Whitelaw B, Vilotte JL. Zinc finger nuclease technology heralds a new era in mammalian transgenesis. Trends Biotechnol. 2010;28:134–41.
81. Hoher T, Wallace L, Khan K, Cathomen T, Reichelt J. Highly efficient zinc-finger nuclease-mediated disruption of an eGFP transgene in keratinocyte stem cells without impairment of stem cell properties. Stem Cell Rev. 2012;8:426–34.
82. Philippe S, Sarkis C, Barkats M, Mammeri H, Ladroue C, Petit C, Mallet J, Serguera C. Lentiviral vectors with a defective integrase allow efficient and sustained transgene expression in vitro and in vivo. Proc Natl Acad Sci USA. 2006;103:17684–9.
83. Terskikh AV, Ershler MA, Drize NJ, Nifontova IN, Chertkov JL. Long-term persistence of a non-integrated lentiviral vector in mouse hematopoietic stem cells. Exp Hematol. 2005;33:873–82.

Control of Mesenchymal Stem Cells with Biomaterials

Sandeep M. Nalluri, Michael J. Hill, and Debanjan Sarkar

Abstract *In vivo* micro-environment of mesenchymal stem cells (MSCs), known as stem cell niches, plays an important role in deciding the cell fate which is required to regulate tissue homeostasis. These micro-environmental features include orchestrated interactions between different compartments which are cell-cell, cell-soluble factors, and cell-matrix interactions. In addition to cell-cell interactions and soluble biomolecules present in extra-cellular environment of the niches, stem cell fate is guided through their interactions with extra cellular matrix (ECM) which is tightly regulated at the molecular, cellular and tissue level. These cell-matrix interactions with an artificial matrix should mimic the physicochemical, mechanical and topographical interactions at micro- and nano-scale dimension for recapitulation of stem-cell micro-environment. Therefore, it is important to design biomaterials which can provide the *in vivo* micro-environmental properties of ECM for guiding the fate of the stem cells and potentially can impact tissue regeneration. This chapter discusses the structural and functional characteristics of biomaterials which can guide the fate of MSCs by modulating viability, proliferation, morphology, migration and differentiation. Understanding native stem cell niches and deriving design principles to construct artificial biomaterial based stem-cell microenvironment is important for successful regenerative tissue engineering.

S.M. Nalluri
Department of Chemical and Biological Engineering, University at Buffalo, The State University of New York, Buffalo, NY 14260, USA

M.J. Hill
Department of Biomedical Engineering, University at Buffalo, The State University of New York, Buffalo, NY 14260, USA

D. Sarkar (✉)
Department of Chemical and Biological Engineering, Department of Biomedical Engineering, University at Buffalo, The State University of New York, Buffalo, NY 14260, USA
e-mail: debanjan@buffalo.edu

R.C. Zhao (ed.), *Essentials of Mesenchymal Stem Cell Biology and Its Clinical Translation*, DOI 10.1007/978-94-007-6716-4_9,

Keywords MSC • Cell fate • Biomaterial

Introduction

Regenerative tissue engineering aims to generate new tissues and organs which can replace damaged tissues or enhance the functioning of defective tissues and, thus, represents a therapeutically relevant treatment strategy for various tissue defects and diseases [1, 2]. In principle, biomaterial based tissue engineering strategies involve combining cells and appropriate biomolecules with an artificial matrix essentially derived from a biomaterial with necessary physical, chemical, mechanical and biological signals. Mesenchymal Stem Cells (MSCs) are one of the promising renewable cell sources for tissue engineering. They can be harvested from bone marrow (and other tissues) and can differentiate into osteoblasts (bone) [3], chondrocytes (cartilage) [4], myoblasts (muscle) [5] and adipocytes (fat) [6]. Furthermore, these cells have emerged as a potential candidate cells for tissue engineering applications due to their ability to promote angiogenesis and to modulate immune response [7, 8].

Control of cell fate is extremely important in tissue engineering (and in any regenerative medicine) applications to appropriately replicate tissue structure and function. Fate of MSCs is determined by its ability to maintain stem-cell like character or to differentiate to specific lineages and is regulated by the *in vivo* micro-environment which is called 'niche' [9]. These niches are defined by the presence of other cell types, extra cellular matrix (ECM) characteristics, and soluble factors. These niche components and synchronized interactions between them present specific cues to the cells which controls (stem) cell fate [9–11]. Biomaterial based tissue engineering strategies have, therefore, been developed to control the cell fate in a spatial and temporal manner by presenting appropriate signals. The significance of biomaterials to provide these signals as insoluble matrix cues is immense because cell-matrix interactions constitute a major and critical part of the signaling events. Furthermore, soluble factors, which are conventionally used to differentiate stem cells [12], produce a heterogeneous population of cells [13, 14] and potentially present a challenge for effective tissue organization. Therefore, designing biomaterials to mimic the structural and functional characteristics of native ECM can guide the fate of MSCs efficiently; this will have a major impact in tissue engineering to control cell fate and regulate tissue regeneration.

With recent advances in material synthesis and processing technologies, it is possible to develop biomaterials with a defined set of physicochemical, mechanical and biological properties at macro-, micro-and nano-scale dimensions. For example, with techniques such as photolithography, micro-contact printing and self-assembled monolayers (SAMs) engineers are able to create biomaterials that have distinct properties to guide the stem cell fate [2, 15, 16]. This chapter will discuss the progresses made to control the fate of MSCs by various properties of biomaterials which include physicochemical, mechanical, and biological properties for structural, architectural and functional characteristics at micro- and nano-scales.

Stem Cell Microenvironment

In vivo stem cell niche is a complex multi-dimensional unit which constitutes basic structural and functional micro-environment for the cells. There are three major components of stem cell niche: Cellular, extracellular matrix, and soluble factors as shown in a representative schematic in Fig. 1. Highly coordinated interactions between these niche compartments and the signals derived from the cross-talk between the components guide the cell fate through activation of various genes, which is required for repair or regeneration of tissues [17].

In principle, cell fate is governed by different receptors present on the cell surface and their interaction with different cells, matrix molecules and soluble ligands present in the niche [18]. These receptors, when bound to certain ligands (present on another cell, ECM or in niche milieu) will activate a series of signaling pathways that will guide the cell to activate specific genes which in turn modulate cell fate that include viability, proliferation, differentiation, production of more extracellular matrix proteins or to undergo apoptosis.

Cellular components of MSC niche have been explored and it has been established that cell-cell contact influences the functional characteristics of MSCs in the naïve state. Thus, by utilizing cell-cell interactions, human mesenchymal stem cells (hMSCs) were delivered to *ex vivo* human excisional wound where subpopulations of hMSCs were cultured either in contact with, or physically separated from epidermal keratinocytes. hMSCs which are co-cultured in contact with keratinocytes adopted an epithelial morphology and expressed keratinocyte markers, while those cultured without contact exhibited phenotypes that resembled myofibroblast and early neural lineage, both of which are of dermal origin [19]. Similarly, soluble components of the niche, i.e. soluble growth factors have been shown to modulate the fate of MSCs. For instance, growth factors transforming growth factor-β3 (TGF-β3), bone morphogenic protein-6 (BMP-6), and insulin-like growth factor-1 (IGF-1) are used in different combinations to modulate the proliferation and chondrogenic differentiation of MSCs [20]. Likewise, several other studies have demonstrated the effect of growth factors in controlling MSC fate. Cells integrate these signals derived from different pathways through a myriad of growth factors which modulate their fate [21–23].

In addition to cellular and soluble components, ECM components take various forms in different tissues and in the same tissue during developmental stages. This variance of ECM comes from specific interactions of various molecules of different isoforms and composition, and geometrical arrangements of collageneous proteins, proteoglycans, elastins and, various adhesion proteins which include fibronectins, vitronectins, osteopotin and laminins [11, 18, 24]. ECM contains various cues ranging from micro to nano-scale due to different physical-chemical, mechanical, and biological properties which can guide and regulate

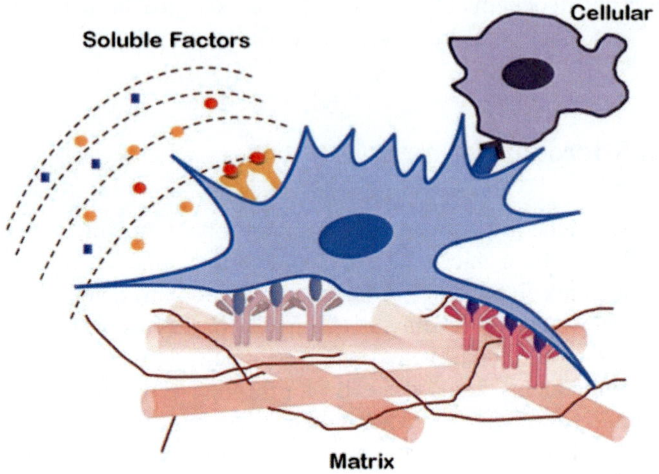

Fig. 1 Schematic presentation of 'Stem Cell Niche'. Stem cell (*blue cell*) interacts with three components of the niche: Soluble factor, cellular and matrix for maintenance stem cell function. Biological, physicochemical and mechanical character of the niche control these interactions to regulate stem cell fate

MSC functions. It is also important to recognize the reciprocal relationship between MSCs and matrix which is underlined by continuous remodeling of matrix molecules by the cells [25]. Physicochemical and mechanical character of the matrix has shown to influence MSCs in an *in vivo* microenvironment. For example, MSCs residing in bone marrow niche experience different matrix character at mineralized bone matrix of the endosteal surface in osteoblastic compartment and in the marrow ECM synthesized by marrow stromal cells in vascular compartment [9, 25, 26]. These two compartments are significantly different in terms of molecular composition and properties and, therefore, are expected to control MSCs through different mechanistic pathways [26–28]. Matrix regulation of MSCs in bone marrow niche has not been studied extensively like hematopoietic stem cells, but nonetheless, matrix component of the niches play a major role in maintaining the undifferentiated state of MSCs or in promoting the differentiation towards a specific lineage.

Thus, stem cell niche integrates cellular and molecular components to act in a highly interactive manner for regulation of stem cell activities. Matrix components of such niches are critical for providing physicochemical, mechanical and biological cues and biomaterials are designed to mimic these features structurally and functionally. Synthetic biomaterials engineered at the molecular level with appropriate architectural features are critical for inducing specific cell-matrix interactions. Recent advancements in biomaterials development and processing have allowed controlling of MSCs at different length scales for tissue engineering applications. In the following sections we summarize the effect of biomaterials on MSC activities in terms of the material's physicochemical, mechanical and structural characteristics.

Physicochemical Effect on Mesenchymal Stem Cells Fate

Controlling MSCs through physical and chemical characteristics of a biomaterial is important for mimicking physicochemical characteristics of ECM. Since ECM is primarily composed of macromolecules, most studies are focused on polymeric biomaterials. However, recent studies have shown that certain small molecules can have potential to modulate MSCs fate. These small molecules can be categorized as soluble factors but their emergence to control MSCs has considerably impacted biomaterials. The precise mechanism through which small molecules modulate the stem cell fate is challenging to unravel and is complicated by many issues, for instance, these molecules may interact with many non-relevant targets that may confound the analysis [29, 30]. The compound 2,6,9-trisubstituted purine, purmorphamine, is found to induce osteogenic differentiation in mouse Mesenchymal progenitor cell line C3H10T1/2. It up-regulates the expression of *Cbfa1/Runx2* (a master regulator of bone development), and also, other bone specific markers such as osteopontin and collagen-I [31]. Some of the other small molecules which are known to modulate stem cell fate are Dexamethasone, Indomethacin, Rosiglitazone, and Vitamin D3. These molecules under carefully defined conditions can induce differentiation in MSCs e.g. dexamethasone and vitamin D3 are known to promote osteogenic differentiation and rosiglitazone and indomethacin are known to promote adipogenic differentiation of MSCs [32–35].

Since ECM is composed of biomacromolecules, polymers are extensively used as biomaterials for MSC based tissue engineering applications to mimic the physicochemical characteristics of ECM. Many polymers, both natural and synthetic ones are used to control MSCs. Synthetic polymers represent a major class of polymers which have been used as artificial matrices for MSCs. For example MSCs' adherence and subsequent osteogenic differentiation on biodegradable poly (lactic-co-glycolic) acid (PLGA) and poly-caprolactone (PCL) was mediated by the polymer character which induced adsorption of distinct serum-derived ECM proteins; PLGA provided a better osteogenic environment than PCL due to this differential protein adsorption [36]. This study indicates molecular composition of polymers can distinctly regulate MSCs. Additionally, studies have shown that incorporation of nano-hydroxyapatite as inorganic mineral phases into the organic polymers matrix enhances the osteogenic potential of MSCs due to better mimic of natural bone matrices [37, 38]. In addition to these polymers, which are classical examples of bioresorbable material, polymers with lesser biodegradability have shown significant control over MSCs. For instance, methacrylate and acrylate monomers are used to synthesize a library of polymers which have different composition, and thus, different properties like contact angle and modulus. It is shown that the composition of the polymer modulates hMSCs adhesion, proliferation, and differentiation, with strong correlation between contact angle, protein adsorption and cell response. However, if cell adhesive peptide sequence Arginine-Glycine-Aspartic acid (RGD) is functionalized on to the surface of the polymers, the polymer composition didn't modulate the cell fate, indicating that, RGD motif overrides the effect of polymer composition on the stem cell fate [39].

This feature indicates the importance of biomimetic approaches in controlling fate of MSCs. Most widely used biomimetic approach is aimed to modulate the adhesion of MSCs to polymeric matrices by using proteins or peptides present on native ECM. Cell adhesion proteins that are present in ECM contain RGD sequence, which the cell recognizes and binds through its transmembrane integrin receptors [40]. This sequence has been identified in many ECM proteins such as Fibronectin, Vitronectin, type I collagen, fibrinogen, von willebrand factor, and osteopontin [41–45]. The presence of RGD sequence incorporated into the microenvironment of hMSCs through the surfaces of biomaterials has been shown to initiate the chondrogenic differentiation, but persistence of RGD for a longer period in ECM has shown to inhibit the chondrogenic differentiation of the MSCs [46]. In a study, to temporally regulate the RGD presence in the ECM, matrix metalloprotease13 (MMP-13) cleavage site has been incorporated into the peptide containing RGD sequence. MMP-13 cleavage site is PENFF (proline-glutamic acid-asparagine-phenylalanine-phenylalanine) which is degraded by MMP-13. MMP-13 is up-regulated within 7–12 days of chondrogenesis in hMSCs [47]. Results have shown that cleavable RGD peptide containing gels, with MMP-13 cleavage site have produced tenfold more glucosaminoglycan compared to un-cleavable RGD peptides containing gels. Similarly, 75 % of cells have stained positive to collagen type II deposition in cleavable RGD gels compared to 19 % in gels where RGD persists. Therefore, temporal regulation of the presence of RGD can modulate the MSCs fate [48]. Similarly, osteogenic differentiation of MSCs was modulated with RGD modified alginate gels where nanoscale presentation of RGD with respect to ligand density and spacing of RGD islands controlled the differentiation of the cells [49, 50]. Apart from RGD, peptide based biomaterials, mainly self-assembled peptides, have emerged as an important biomaterial for regulation of MSCs [51]. In addition, several natural polymers including collagen, hyaluronic acids are also used as biomimetic molecule to control MSCs for specific function and application [52, 53].

Hydrogels are widely used in tissue engineering applications due to their ability to mimic the tissue microenvironment and, therefore, have been an obvious choice to control MSCs fate. Injectable hydrogels (IHs), particularly, can be used to deliver MSCs with minimally invasive surgery and they can easily mold to the shape of the cavity [54]. Hydrogels constructed from natural and synthetic polymers are designed to induce specific interactions with MSCs for effective osteogenic and chondrogenic differentiation. In a study, natural proteins which include collagen type I and collagen type II are mixed in required proportions with alginate to form hydrogels. MSCs are grown in these pure alginate hydrogels and in collagen type I and type II hydrogels, both in serum free medium and medium containing transforming growth factor (TGF) β1. The chondrogenic specific genes like sox9, collagen type II, aggrecan, and COMP were found to be up-regulated more in collagen hydrogels, particularly collagen type II. Presence of TGF β1and dexamethanose in collagen type II hydrogels resulted in providing more favorable conditions for expressing chondrogenic phenotype. This study has shown that collagen type II present in ECM alone has potential to induce chondrogenic differentiation in MSCs [55, 56]. Similarly studies have shown the importance of hydrogels for osteogenic differentiation of

MSCs [55, 57]. hMSCs were encapsulated in a photocrosslinkable, injectable scaffolding system based on poly(ethylene glycol) (PEG) hydrogels. hMSCs differentiated into osteogenic lineage within these PEG hydrogels, when cultured in osteogenic differentiation media, indicating the feasibility of using a PEG-based, photocrosslinkable system to culture and deliver human mesenchymal stem cells for bone tissue regeneration and repair. In another study, MSCs were encapsulated in PEG hydrogels which were functionalized with small chemical groups and the MSCs differentiated into specific lineages depending on the cell-matrix interaction induced by tethered chemical groups [58].

Phosphate- and t-butyl-functionalized PEG hydrogels induced osteogenesis and adipogenesis of MSCs respectively indicating the importance of gel structure and the chemical environment. The results showed that phosphate gels have significant expression of collagen (detected by Masson's trichrome stain) and osteopontin (bone extracellular protein) as osteogenic markers and *t*-butyl gels induced lipid deposition (stained by Oil red) and peroxisome proliferating antigen receptor gamma (PPARG, a critical regulator of adipogenesis) as adipogenic marker. This study also indicated that local hydrophilic/hydrophobic milieu has an impact on MSC behavior and shows the significance of material hydrophilic-hydrophobic character for cell functioning.

Several studies have demonstrated how MSC fate is modulated by hydrophobicity or hydrophilicity of materials. For example, cellular organization and differentiation potential of MSCs were influenced by the matrix hydrophobic character of hydrogels [59]. Results showed that MSCs were able to adhere, migrate and differentiate into specific lineage at an optimum matrix hydrophobicity which induced specific cell-cell and cell-matrix interactions. In another study, silk fibroin protein is modified with diazonium coupling chemistry to tailor its structure and hydrophilicity and its effect on MSCs were examined. When hMSCs were grown on these various surfaces with different hydrophilicity, the cell growth rate and morphology were largely affected, but the osteogenic differentiation capacity didn't change significantly [60]. Another study with acrylic acid grafted on poly(3-hydroxybutyrate-co-3-hydroxyhexnoate) (PHBHHx) membrane, improved the hydrophilic nature of PHBHHx membrane. This hydrophilic property of the membrane affected the morphology of the hMSCs but not the metabolic activity. Cells were more spread on these hydrophilic surfaces and increase in cytoskeleton formation was observed [61].

In addition surface wettability i.e. hydrophobic/hydrophilic character, surface charge has shown to influence MSCs due to diverse electrostatic properties which play key roles in controlling cell functions *in vivo*. In a study, three kinds of water-soluble polymers, positively charged poly(L-lysine) (PLL), negatively charged poly(acrylic acid) (PAAc), and neutral poly(ethylene glycol) (PEG), were compared based on their effects on the adhesion, spread, proliferation and chondrogenic differentiation of human mesenchymal stem cells (MSCs) [62]. MSCs were seeded and cultured in the presence of polymers of different concentrations applied by specific methods. The effects of the water-soluble polymers depended on their electrostatic properties and method of application. A low concentration of PLL promoted MSC adhesion, spread, proliferation and chondrogenic differentiation, while a high concentration of PLL was toxic. The PEG-coated surface facilitated cell aggregation and spheroid formation by inhibiting cell adhesion. A high concentration of mixed PEG promoted cell proliferation in serum-free medium. PAAc showed no obvious effects on MSC adhesion, spread, proliferation, or chondrogenic differentiation. Similar studies with positively charged polyallylamine (PAAm) and negatively charged poly(acrylic acid) (PAAc) were done to study the effect of surface electrostatic properties on osteogenic differentiation of MSCs [63]. Cells adhered, spread, and proliferated somewhat more quickly on the PAAm-modified surface than they did on the PAAc-modified and control surfaces. Additionally, these surface charges acted synergistically with the soluble molecules present to induce osteogenesis of MSCs.

Although biomaterial based regenerative tissue engineering is essentially concentrated on polymeric biomaterials, metals are also known to modulate the stem cell fate. Metallic biomaterials are mainly focused for bone application due to mechanical strength and are primarily investigated for osteogenic differentiation of MSCs. For example, hydroxyapatite (calcium phosphate), titanium, are known for controlling MSC differentiation. Hydroxyapatite is a mineral constituent of natural bone matrix and has been used to improve strength of polymeric matrices and induce differentiation of MSCs into osteogenic lineage [64, 65]. Titanium has shown promise to induce osteogenic differentiation of MSCs which makes titanium an attractive bone regeneration material [66, 67]. MSCs are able to attach and proliferate on titanium surface if presented in different formats which include nanoparticle surfaces and mesh structures for differentiating into osteogenic lineages. In this context, it is important to mention that several metallic nanoparticles have been used to modulate MSC functions. For instance, silver nanoparticles (Ag-Nps) of 100 nm were used to study the effect on proliferation, cytokine release and chemotaxis of hMSCs. The results showed concentration-dependent activation of hMSCs till the Ag-Nps levels of 2.5 μg/ml and cytotoxic cell reactions occurred at concentrations more than 5 μg/ml. With the increasing concentration of Ag-Nps, the cell proliferation and chemotaxis of hMSCs decreased. Different effects were observed on different cytokines with respect to the concentrations of Ag-Nps used [68]. In another instance, it is shown that direct exposure of hMSCs to titanium and zirconium oxide induces apoptosis through increased levels of tumor suppressor proteins P53 and P73 [69].

In summary, mimicking the physical and chemical features of ECM present in MSC niche can be achieved by different biomaterial based approaches. It is important to recognize the essential feature of matrix for inducing MSCs towards a specific direction and presenting it with an artificial biomaterial based matrix.

Physicomechanical Control of Stem Cell Fate

Physicochemical character of biomaterials have been widely used as molecular engineering tool to control MSC fate but recent advancements in our understanding of mechanobiology and mechanotransduction has demonstrated the role of matrix mechanics in cell function and behavior. Thus, biomaterial based researches have

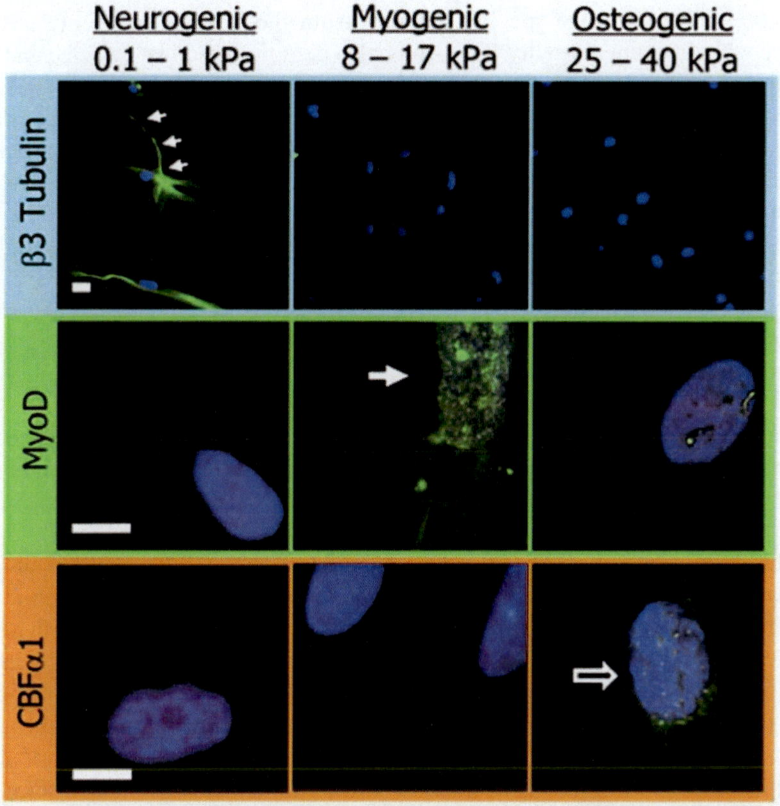

Fig. 2 Expression of lineage specific markers of MSCs as a function of substrate elasticity. MSCs express neurogenic marker **β**3Tubulin on 0.1–10 kPa matrix, myogenic marker MyoD on 8–17 kPa matrix and osteogenic marker CBF**α**1 on 25–40 kPa matrix (Reproduced with permission from Elsevier in Ref. [5])

progressed to capitalize the physicomechanical effect of matrix in regulating MSCs. Adherent cells respond to matrix stiffness through their adhesion complexes/junctions which acts as site for force transduction: cells pull against matrix and feel the resistance to deformation by the adjacent matrix environment [70]. Thus, mechanical response of ECM in physiological and pathological conditions influences the functioning of cells [71]. Matrix stiffness has been shown to alter the behavior of MSCs including proliferation, migration, cytoskeleton arrangement, and differentiation. Effect of matrix elasticity on MSCs was demonstrated by using polyacrylamide gel with different substrate elasticity ranges [72]. On soft matrices (0.1–10 kPa, which mimics the elasticity of brain matrices) MSCs differentiated into neuronal lineage but on stiffer matrices (25–40 kPa, which mimics bone matrices) MSCs differentiated into osteogenic lineage. Whereas, MSCs on intermediate stiffness (8–17 kPa, which mimics muscle elasticity) differentiated into muscle lineage. Figure 2 shows the expression of lineage specific markers from the differentiation of MSCs on surface with different elasticity. This study also demonstrated that cell size and cytoskeleton organization of MSCs correlated to matrix stiffness. This lineage specificity of MSCs due matrix elasticity has shown the role mechanical response of an artificial matrix in controlling stem cell fate. In another study, when MSCs were cultured on polyacrylamide gels of 250 and 7,500 Pa stiffness; 250 Pa gel was chosen based on measurements of bovine marrow and rat adipose tissue, whereas the 7,500 Pa mimicked muscle tissue stiffness. Cells on 250 Pa gels remained rounded and failed to show signs of replication whereas on 7,500 Pa gel, cells spread more [73]. Similar studies performed on these types of system have further emphasized the role of matrix mechanics in controlling MSC behavior [74].

These studies and several others have delved into the biomolecular signaling events resulting from such mechano-sensitive responses of MSCs. It involves proteins in between the ECM and cell which are obviously mechano-sensitive and will undergo deformations that can guide the fate of cells. For instance, the substrate rigidity influences RhoA/ROCK mediated calcium ion oscillations in hMSCs. In turn calcium ion oscillations are known to influence differentiation [75].

Translating these mechano-responses into practically useful biomaterial based approaches will be critical for regenerative tissue engineering application. Researchers have modified alginate gels with cell adhesive RGD sequence in a definitive way to design matrices with tunable elasticity (between 2.5 and 110 kPa) as a function of RGD presentation both in 2-dimensional substrate and in 3-dimensional gel [76]. Results shows MSCs differentiated into specifically different lineages in response to matrix stiffness which was correlated to the traction mediated orientation of adhesion junctions formed at cell-matrix interface. In another study, MSCs were cultured in 3D thixotropic gels designed from polyethylene glycol-silica (PEG-silica) nanocomposites and with varying rheological properties [77]. For these 3D cell cultures in thixotropic gels, the liquefaction stress (minimum shear stress required to liquefy the gel) was used to characterize the matrix stiffness. The highest expressions of neural (ENO2), myogenic (MYOG) and osteogenic (Runx2, OC) transcription factors were obtained for gels with liquefaction stress of 7, 25 and 75 Pa, respectively. Hyaluronic acid based gels with different crosslinking density resulted in gels with different stiffness which influenced the chrondrogenic differentiation of MSCs [78]. Interestingly, matrix mechanics were also used as a factor to maintain the MSCs in a quiescent state by mimicking the bone marrow matrix [73]. In this study 250-Pa polyacrylamide gels coated with collagen type 1 and fibronectin were used as this material mimics the elasticity of bone marrow and fat tissues. hMSCs seeded sparsely on these gels remained quiescent with halted progression through the cell cycle despite the presence of serum.

Collectively, these studies are demonstrating the increasing role of matrix mechanics in controlling MSC fate. As researchers are unraveling the fundamentals of MSC mechano-response, growing trends to mimic the matrix mechanics of a given microenvironment with biomaterial based strategies will improve the tissue regeneration. Adequate control of MSCs with mechanical force to terminally differentiate and function will enhance biomaterial based regenerative strategies.

Micro- and Nano-Structural Control of MSCs Fate

Structural and architectural design of a biomaterial in a given application, both in 2-dimensional substrate and 3-dimensional environment can have significant impact on cell fate. Cellular functions and behavior of MSCs have been influenced by these material features and therefore are an important aspect of biomaterial design. With the advent of photo- and soft-lithographic techniques, there has been a growing interest in fabrication of micro- and nano-scale cues to study fundamental cell-substrate interactions. This is increasingly important because cells, including MSCs, interact with the matrix at different dimensions which translates into specific cell signaling events. Therefore, recent efforts are focused on achieving these topographic and geometric cues from the matrix architecture to control the fate of MSCs.

MSCs alignment, proliferation, migration and differentiation are studied on micro-topographical surfaces [79, 80]. In a study, umbilical cord blood-derived mesenchymal stem cells (UCB-MSCs) were grown on flat polydimethylsiloxane (PDMS) and on grooved PDMS surfaces of different widths (1, 2 and 4 μm). Results showed that proliferation of UCB-MSCs is enhanced on the micro-grooved surfaces compared to flat PDMS. The proliferation on 1 μm width grooved PDMS surface increased twofold compared to that on flat PDMS. But, there was no significant statistical difference in proliferation between 4 μm width grooved PDMS surface and flat PDMS surface. It was also found that UCB-MSCs were differentiated into neural like cells at higher levels on micro grooved PDMS surfaces compared to flat PDMS surface. Particularly, PDMS with 1 and 2 μm induced neuronal-like differentiation from UCB-MSCs drastically, while the differentiation on 4 μm micro grooves is less pronounced, whereas, only half of the cells were found to be differentiated in flat PDMS [80].

In another study, cell alignment and migration on micro-grooved and smooth silicon surfaces was analyzed. Micro grooved surfaces contained step height of 1.6 μm and stripes of 5 μm. Two days after incubation, MSCs aligned along the grooves with rhizomes as shown in Fig. 3a (light microscopy) and b (SEM), whereas, MSCs showed fibroblast-like morphology on the flat silicon surface (Fig. 3c). MSCs exhibited a distinct morphology on micro-grooved surface. They showed a needle-like morphology with two distinct ends. One end is close to the cell body and the other end is elongated and away from the cell body. AFM images showed that cells occupied the grooves rather than the ridges. The images of live MSCs were taken at different time intervals to study the cell migration on both micro-grooved and flat silicon surfaces. Results showed that, one of the edges of the cell starts extending out of the cell, forming a thin, long structure, that eventually stops extending. This thin, long structure leads the direction of migration of cells on the micro-grooved surface. While on the flat silicon surface, the leading edge is the wide edge of the cell and the thin, long extension is trailing. This study shows that micro-grooved surfaces influence the MSCs alignment and migration [79].

MSCs fate can also be influenced by the shape of the microstructures. For instance, a nano-imprint lithography apparatus was used to fabricate square and

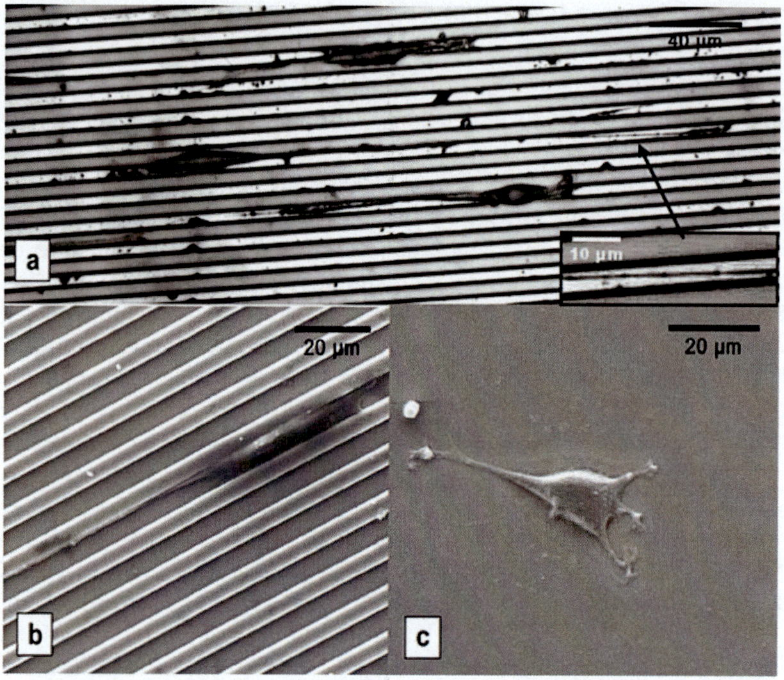

Fig. 3 (**a**) Light microscopy image of MSCs aligned on a Si grid (Rhizomes are marked with a *black arrow* and further enlarged in the *inset*), (**b**) SEM image of a MSC on the micro-patterned silicon surface. A short, crescent-shaped extension and a long, thin extension are visible, (**c**) SEM image of cell on a smooth silicon control surface, exhibiting a fibroblast like morphology (Reproduced with permission from Elsevier in Ref. [79])

round structures on surface of poly (methyl methacrylate) (PMMA). Square structure has width of 2 μm and height of 1 μm, and the rounded structure has width of 50 μm and height of 1 μm. When rat Mesenchymal stem cells (rMSCs) were grown on these structured surfaces and on smooth PMMA, they showed distinct morphologies on these structures. After 3 days of culturing rMSCs on these surfaces, cells grown on round shaped structures have proliferated inside the round shaped wells and spread to the limit of this round structure (Fig. 4a). They showed less focal contacts and actin filament organization compared to cells grown on square structures. Cells grown on square shaped surfaces appeared more stressed and have a star shaped morphology, but, some of the cells also showed spindle, needle-like morphology. Cells are also aligned and many prolongations can be observed with preferential attachment to square structures (Fig. 4b). On flat PMMA surface, cells have various kinds of morphologies and they are randomly oriented (Fig. 4c). However, these structured surfaces didn't influence the proliferation and differentiation of the rMSCs significantly [81]. In another study, the behavior of Mesenchymal Stem Cells was observed on island-patterned (convex) and sunken-patterned (concave) Poly (L-Lacticacid) (PLLA) membranes. Results showed that cell adhesion and

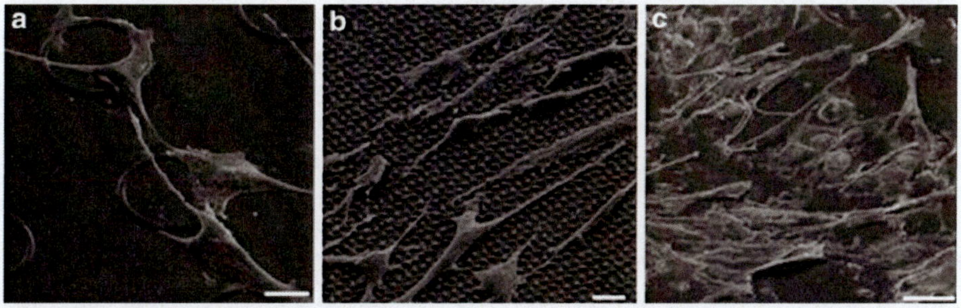

Fig. 4 SEM images of (**a**) rMSCs grown on round structures, they get along these structures, (**b**) rMSCs grown on square structures, they are aligned and several prolongations can be observed that are attached to these structures, (**c**) rMSCS grown on flat PMMA, they have various morphologies and randomly oriented. *Scale bar*: 20 **μ**m for (**a**, **b**) and 50 **μ**m for (**c**) (Reproduced with permission from Elsevier in Ref. [81])

proliferation were enhanced for the MSCs grown on island-patterned surfaces compared to sunken-patterned surfaces. In addition, cells are more biocompatible with island-patterns of 100 μm diameter compared to that of 60 μm diameter [82].

In addition to shape factor, MSCs fate can be modulated by micro-patterning of surfaces. In a study, MSCs differentiation was studied on various geometrically patterned structures. Micro-contact printing technique was used to pattern the shapes of individual cells on substrate. In brief, polydimethylsiloxane (PDMS) stamps were used to pattern the adhesive hydrophobic islands of octadecanethiolate on gold coated glass cover slips. The remaining regions are modified with tri(ethyleneglycol)-terminated monolayer, and then, these cover slips are immersed into a solution of ECM protein fibronectin. These proteins adhere to the above hydrophobic islands to which the MSCs adhere and assume the geometric shape of the underlying island. First, the MSCs were grown in mixed media for 1 week on geometric features of 1,000, 2,500 and 5,000 μm^2 areas to find out if they are biased towards the MSCs fate. Results showed that small islands induce adipocyte characteristics and large islands induce osteoblast fate. MSCs grown on intermediate area (2,500 μm) showed mixed populations of both adipocytes and osteoblasts. Next, this study compared the fate of MSCs on different geometric shape islands having same area but different aspect ratio and curvature. When, MSCs are grown on rectangular surfaces having same area but different aspect ratios of 1:1, 3:2 and 4:1, the yield of osteogenesis increased with aspect ratio. MSCs showed 46, 56 and 61 % osteogenesis on 1:1, 3:2 and 4:1 aspect ratios of rectangles respectively. MSCs are also grown on pentagonal symmetry of same area but with different types of curvatures (convex curved edges, straight line edges, and concave edges with sharp points in the edges). On convex curved edge structures, 62 % MSCs differentiated into adipocytes while remaining differentiated into osteoblasts. On straight line edge structures even distribution of osteoblasts and adipocytes were found, whereas, on the concave edge structures 62 % of MSCS have osteogenic fate. These results show that, not only the area of the micro-structured patterns can modulate the stem cell fate, but also, the small changes in aspect ratio or the curvature of structured patterns for the same area could have a significant impact on the fate of MSCs [6].

Three dimensional micro-porous scaffolds are widely used for tissue regeneration application and can also impact MSC functioning. For instance, hMSCs were grown on coralline hydroxyapatite scaffolds which consisted of pore sizes of 200 and 500 μm. The results showed that proliferation and number of cells accommodated was higher in 500 μm scaffolds. The alkaline phosphate activity assay and reverse transcriptase polymerase chain reaction for 10 osteogenic markers showed that osteogenic differentiation occurred at faster rate in 200 μm scaffold compared to 500 μm scaffolds [83]. In another study, micro-porous scaffold containing pore size of 100–250 μm is made from poly(lactide-co-glycolide) (PLGA) by solution-casting/salt leaching method. When MSCs were seeded, they adhered and proliferated inside these scaffolds. Calcification can be observed within 2 weeks in the ECM and the degree of mineralization has increased with time, which is an indicator that scaffold has induced osteogenic differentiation in MSCs [84]. Particularly studies have shown the importance of pore structures for differentiation of MSCs into bone cells *in vivo* [85].

As the importance of micro-level architectures in controlling MSC function is investigated, studies are increasingly showing the relevance of nano-level features to impact cell functions. The interaction between cells and matrix nano-structures are important because ECM ligands are interactive at this length scale. In general, nanotopography induces changes in focal adhesion, cytoskeleton structure and mechanical properties of MSCs which ultimately controls cell fate through specific cell signaling pathways [86, 87]. For example, nano-grooves are designed to modulate the MSCs fate. In a study, hMSCs are grown on nano-imprinted poly(methyl methacrylate) (PMMA) nano-grooves of 350 nm depth and 350 nm width. Cells and their nuclei are found to be elongated and aligned along the 350 nm nano-grooves while they are not elongated and didn't have any alignment on non-patterned surfaces. It was observed that the neuronal genes up-regulated due to the presence of nano-topography alone, indicating neuronal differentiation of hMSCs. This study also compared the effects of micro-grooves of 1 and 10 μm width with that of 350 nm width nano-grooved surface. The results showed that nano-grooved surfaces influenced the MSCs fate more in terms of proliferation and differentiation than micro-grooved surfaces [88]. However studies have also indicated nanotopography can only act as a guidance to MSCs and should act synergistically with other factors including soluble molecules to control the fate of MSCs, particularly differentiation [89].

Electro-spun nano-fibers can modulate the hMSCs fate. Nano-fibers have been shown to mimic ECM to various degrees in several studies [90–94]. For instance, hMSCs were grown on electrospun nano-fibers made from Poly(D,L-lactide-co-glycolide) (PLGA) beads with a PLA:PGA ratio of 85:15. The fibers have an average diameter of 760 ± 210 nm. hMSCs were proliferating and viable up to 14 days on these scaffolds and they were able to form chondrocytes and osteoblasts when grown in appropriate medium which shows electrospun scaffolds support proliferation and differentiation of hMSCs [95]. In another study, hMSCs were cultured on electrospun type I collagen nano-fibers. Results showed that cells have more flattened and polygonal morphology, and have higher viability compared to control tissue culture polystyrene. Cells showed lower number of vinculin spots and fewer vinculin proteins are recruited for formation of focal adhesion complex compared to control. This might be as a result of nano- fibers resembling pliable ECM. Nano-fibers are also capable of supporting osteogenic differentiation; alkaline phosphate production was similar in both nano-fibers and tissue culture plastic, and gene expression analysis after osteogenic induction in MSCs showed higher or similar levels of RNA transcript production between the nano-fibers and tissue culture polystyrene [96].

Studies have been conducted on growing MSCs on vertical TiO_2 nano-tubes fabricated by metal anodization [3, 97]. In a study, when MSCs are grown on vertical TiO_2 nano-tubes with defined diameters ranging from 15 to 100 nm, the results showed a strong correlation between adhesion, spreading, growth, and differentiation of MSCs and nano-tube diameter. MSCs elicited maximum cell response when grown on tube diameter of 15–30 nm range, approximately corresponding to lateral spacing between integrin receptors in focal contacts on ECM. Cell adhesion, proliferation and spreading were enhanced on these range of tubes (15–30 nm) compared to smooth TiO_2 surface, whereas, cells hardly adhered and proliferated and underwent apoptosis on 50–100 nm nano-tubes. When MSCs were grown for 2 weeks in osteogenic differentiation medium, more calcium phosphate mineralization was observed in 15 nm diameter nano-tubes compared to smooth surface and less mineralization was observed on nano-tube diameters greater than 50 nm compared to the smooth surface. In summary, 15–20 nm nano-tubes may have helped in integrin clustering and formation of focal adhesion complexes which elicit maximum cell response in terms of adhesion, proliferation and differentiation. On the other hand, nano-tubes of diameter greater than 50 nm have significantly impaired cell adhesion and spreading, whereas 100 nm nano-tubes almost completely halted proliferation, differentiation and also induced apoptosis in MSCs [97].

The order and pattern of topographical cues at nano-scale can influence the MSCs behavior. For instance, electron beam lithography (EBL) is used to fabricate nano-pit surfaces with not only highly ordered symmetry but also with random nano-disorder. These nano-pits have 120 nm diameter and depth of 100 nm. This disorder in nano-pits pattern distribution will better mimic the natural cartilage nano-scale topography which is highly random and disordered. The results showed that cellular adhesion and osteogenic differentiation of MSCs on highly ordered nano-pits is low to negligible compared to that on randomly disordered nano-pits. MSCs grown for 14 days on these random nano-pit surfaces exhibited more osteoblast character even though they showed only a slight increase in expression of matrix proteins [98]. In another study, dip pen nanolithography (DPN) was used to pattern gold planar surfaces with nan-dots of 70 nm diameter and spacing between them varying in range of 140–1,000 nm and containing terminal functionalities of simple chemistries that include carboxyl, amino, methyl and hydroxyl. When hMSCs were grown on these patterned surfaces, the cell adhesion and expression of various markers showed dependency on chemistry and also spacing between the chemistry (nano-pits spacing). This DPN technique can be used to generate nano-pit patterns of different size and spacing, and also, with different chemistries. This technique in theory can also be used to generate random, disordered islands which can enhance differentiation capacity of MSCs further [99].

These studies have shown the structural and architectural effects of biomaterials on MSC function. Advanced techniques like photo- and soft-lithography can fabricate both micro- and nano-scale topographies and patterns which mimic ECM in biomaterials and can modulate MSCs fate. These features represent a powerful tool of biomaterial engineering for controlling stem cell fate in tissue regeneration.

Conclusion

Various biomaterials used to mimic the *in vivo* micro-environment of MSCs show promising results to regulate the MSCs fate such as viability, proliferation, morphology, migration and differentiation. Results from 2-dimensional substrates should provide engineering tools to design synthetic niches for stem cells including MSCs. Efforts should be made to fabricate and incorporate various cues into 3-dimensional structures for culturing MSCs because 3-dimensional systems are more relevant for tissue engineering applications. Also, presenting the various cues together with temporal and spatial control is important to mimic the *in vivo* micro-environment which can potentiate the future of biomaterial based tissue regeneration strategies. Furthermore, synergism between MSC biology and biomaterials will provide a strong engineering tool to develop clinically relevant therapeutic strategies. Future studies should foster effective exchange of ideas between stem cell biologists and material scientists with development of high throughput techniques for analysis of large sets of data.

References

1. Griffith LG, Naughton G. Tissue engineering – current challenges and expanding opportunities. Science. 2002;295:1009–14.
2. Khademhosseini A, Langer R, Borenstein J, Vacanti JP. Microscale technologies for tissue engineering and biology. Proc Natl Acad Sci USA. 2006;103:2480–7.
3. Oh S, Brammer KS, Li YSJ, Teng D, Engler AJ, Chien S, Jin S. Stem cell fate dictated solely by altered nanotube dimension. Proc Natl Acad Sci. 2009;106:2130–5.
4. Zandstra PW, Nagy A. Stem cell bioengineering. Annu Rev Biomed Eng. 2001;3:275–305.
5. Engler AJ, Sen S, Sweeney HL, Discher DE. Matrix elasticity directs stem cell lineage specification. Cell (Cambridge, MA, U S). 2006;126:677–89.
6. Kilian KA, Bugarija B, Lahn BT, Mrksich M. Geometric cues for directing the differentiation of mesenchymal stem cells. Proc Natl Acad Sci. 2010;107:4872–7.
7. Duffy GP, Ahsan T, O'Brien T, Barry F, Nerem RM. Bone marrow-derived mesenchymal stem cells promote angiogenic processes in a time- and dose-dependent manner in vitro. Tissue Eng Part A. 2009;15:2459–70.

8. Abdi R, Fiorina P, Adra CN, Atkinson M, Sayegh MH. Immunomodulation by mesenchymal stem cells: a potential therapeutic strategy for type 1 diabetes. Diabetes. 2008;57:1759–67.
9. Kolf CM, Cho E, Tuan RS. Mesenchymal stromal cells. Biology of adult mesenchymal stem cells: regulation of niche, self-renewal and differentiation. Arthritis Res Ther. 2007;9:204.
10. Moore KA, Lemischka IR. Stem cells and their niches. Science. 2006;311:1880–5.
11. Place ES, Evans ND, Stevens MM. Complexity in biomaterials for tissue engineering. Nat Mater. 2009;8:457–70.
12. Pittenger MF, Mackay AM, Beck SC, Jaiswal RK, Douglas R, Mosca JD, Moorman MA, Simonetti DW, Craig S, Marshak DR. Multilineage potential of adult human mesenchymal stem cells. Science. 1999;284:143–7.
13. Hwang NS, Varghese S, Elisseeff J. Controlled differentiation of stem cells. Adv Drug Deliv Rev. 2008;60:199–214.
14. Ding S, Schultz PG. A role for chemistry in stem cell biology. Nat Biotechnol. 2004;22:833–40.
15. Park TH, Shuler ML. Integration of cell culture and microfabrication technology. Biotechnol Prog. 2003;19:243–53.
16. Bettinger CJ, Langer R, Borenstein JT. Engineering substrate topography at the micro- and nanoscale to control cell function. Angew Chem Int Ed. 2009;48:5406–15.
17. Gupta K, Kim DH, Ellison D, Smith C, Kundu A, Tuan J, Suh KY, Levchenko A. Lab-on-a-chip devices as an emerging platform for stem cell biology. Lab Chip. 2010;10:2019–31.
18. Stevens MM, George JH. Exploring and engineering the cell surface interface. Science. 2005;310:1135–8.
19. Sivamani RK, Schwartz MP, Anseth KS, Isseroff RR. Keratinocyte proximity and contact can play a significant role in determining mesenchymal stem cell fate in human tissue. FASEB J. 2011;25:122–31. doi:10.1096/fj.09-148775.
20. Indrawattana N, Chen G, Tadokoro M, Shann LH, Ohgushi H, Tateishi T, Tanaka J, Bunyaratvej A. Growth factor combination for chondrogenic induction from human mesenchymal stem cell. Biochem Biophys Res Commun. 2004;320:914–9.
21. Giancotti FG, Ruoslahti E. Integrin signaling. Science. 1999;285:1028–33.
22. Plopper GE, McNamee HP, Dike LE, Bojanowski K, Ingber DE. Convergence of integrin and growth factor receptor signaling pathways within the focal adhesion complex. Mol Biol Cell. 1995;6:1349.
23. Tran KT, Griffith L, Wells A. Extracellular matrix signaling through growth factor receptors during wound healing. Wound Repair Regen. 2004;12:262–8.
24. Ruoslahti E, Pierschbacher MD. New perspectives in cell adhesion: RGD and integrins. Science. 1987;238:491.
25. Chen X-D. Extracellular matrix provides an optimal niche for the maintenance and propagation of mesenchymal stem cells. Birth Defects Res C Embryo Today. 2010;90:45–54.
26. Shi S, Gronthos S. Perivascular niche of postnatal mesenchymal stem cells in human bone marrow and dental pulp. J Bone Miner Res. 2003;18:696–704.
27. Kuhn NZ, Tuan RS. Regulation of stemness and stem cell niche of mesenchymal stem cells: implications in tumorigenesis and metastasis. J Cell Physiol. 2010;222:268–77.
28. Bianco P, Robey PG, Simmons PJ. Mesenchymal stem cells: revisiting history, concepts, and assays. Cell Stem Cell. 2008;2:313–9.
29. Zhu S, Wurdak H, Schultz PG. Directed embryonic stem cell differentiation with small molecules. Future Med Chem. 2010;2:965–73.
30. Zheng XS, Chan T-F, Zhou HH. Genetic and genomic approaches to identify and study the targets of bioactive small molecules. Chem Biol. 2004;11:609–18.
31. Wu X, Ding S, Ding Q, Gray NS, Schultz PG. A small molecule with osteogenesis-inducing activity in multipotent mesenchymal progenitor cells. J Am Chem Soc. 2002;124:14520–1.
32. Jaiswal N, Haynesworth SE, Caplan AI, Bruder SP. Osteogenic differentiation of purified, culture-expanded human mesenchymal stem cells in vitro. J Cell Biochem. 1997;64:295–312.
33. Grigoriadis AE, Heersche J, Aubin JE. Differentiation of muscle, fat, cartilage, and bone from progenitor cells present in a bone-derived clonal cell population: effect of dexamethasone. J Cell Biol. 1988;106:2139–51.

34. Styner M, Sen B, Xie Z, Case N, Rubin J. Indomethacin promotes adipogenesis of mesenchymal stem cells through a cyclooxygenase independent mechanism. J Cell Biochem. 2010;111: 1042–50.
35. Zhou S, LeBoff MS, Glowacki J. Vitamin D metabolism and action in human bone marrow stromal cells. Endocrinology. 2010;151:14–22.
36. Chastain SR, Kundu AK, Dhar S, Calvert JW, Putnam AJ. Adhesion of mesenchymal stem cells to polymer scaffolds occurs via distinct ECM ligands and controls their osteogenic differentiation. J Biomed Mater Res A. 2006;78A:73–85.
37. Wang H, Li Y, Zuo Y, Li J, Ma S, Cheng L. Biocompatibility and osteogenesis of biomimetic nano-hydroxyapatite/polyamide composite scaffolds for bone tissue engineering. Biomaterials. 2007;28:3338–48.
38. Zhao F, Grayson WL, Ma T, Bunnell B, Lu WW. Effects of hydroxyapatite in 3-D chitosan–gelatin polymer network on human mesenchymal stem cell construct development. Biomaterials. 2006;27:1859–67.
39. Joy A, Cohen DM, Luk A, Anim-Danso E, Chen C, Kohn J. Control of surface chemistry, substrate stiffness, and cell function in a novel terpolymer methacrylate library. Langmuir. 2011;27:1891–9.
40. Bökel C, Brown NH. Integrins in development: moving on, responding to, and sticking to the extracellular matrix. Dev Cell. 2002;3:311.
41. Ruoslahti E, Pierschbacher MD. Arg-Gly-Asp: a versatile cell recognition signal. Cell (Cambridge, Mass). 1986;44:517–8.
42. Hayman EG, Pierschbacher MD, Ruoslahti E. Detachment of cells from culture substrate by soluble fibronectin peptides. J Cell Biol. 1985;100:1948–54.
43. Dedhar S, Ruoslahti E, Pierschbacher MD. A cell surface receptor complex for collagen type I recognizes the Arg-Gly-Asp sequence. J Cell Biol. 1987;104:585–93.
44. Plow EF, Pierschbacher MD, Ruoslahti E, Marguerie GA, Ginsberg MH. The effect of Arg-Gly-Asp-containing peptides on fibrinogen and von Willebrand factor binding to platelets. Proc Natl Acad Sci. 1985;82:8057–61.
45. Oldberg A, Franzén A, Heinegård D. Cloning and sequence analysis of rat bone sialoprotein (osteopontin) cDNA reveals an Arg-Gly-Asp cell-binding sequence. Proc Natl Acad Sci. 1986;83:8819–23.
46. Connelly JT, García AJ, Levenston ME. Inhibition of in vitro chondrogenesis in RGD-modified three-dimensional alginate gels. Biomaterials. 2007;28:1071–83.
47. Sekiya I, Vuoristo JT, Larson BL, Prockop DJ. In vitro cartilage formation by human adult stem cells from bone marrow stroma defines the sequence of cellular and molecular events during chondrogenesis. Proc Natl Acad Sci. 2002;99:4397–402.
48. Salinas CN, Anseth KS. The enhancement of chondrogenic differentiation of human mesenchymal stem cells by enzymatically regulated RGD functionalities. Biomaterials. 2008;29:2370–7.
49. Hsiong SX, Huebsch N, Fischbach C, Kong HJ, Mooney DJ. Integrin-adhesion ligand bond formation of preosteoblasts and stem cells in three-dimensional RGD presenting matrices. Biomacromolecules. 2008;9:1843–51.
50. Hsiong SX, Boontheekul T, Huebsch N, Mooney DJ. Cyclic arginine-glycine-aspartate peptides enhance three-dimensional stem cell osteogenic differentiation. Tissue Eng Part A. 2009;15:263–72.
51. Chow LW, Bitton R, Webber MJ, Carvajal D, Shull KR, Sharma AK, Stupp SI. A bioactive self-assembled membrane to promote angiogenesis. Biomaterials. 2011;32:1574–82.
52. Young RG, Butler DL, Weber W, Caplan AI, Gordon SL, Fink DJ. Use of mesenchymal stem cells in a collagen matrix for achilles tendon repair. J Orthop Res. 1998;16:406–13.
53. Hegewald AA, Ringe J, Bartel J, Krüger I, Notter M, Barnewitz D, Kaps C, Sittinger M. Hyaluronic acid and autologous synovial fluid induce chondrogenic differentiation of equine mesenchymal stem cells: a preliminary study. Tissue Cell. 2004;36:431–8.
54. Bensaïd W, Triffitt JT, Blanchat C, Oudina K, Sedel L, Petite H. A biodegradable fibrin scaffold for mesenchymal stem cell transplantation. Biomaterials. 2003;24:2497–502.
55. Nuttelman CR, Tripodi MC, Anseth KS. In vitro osteogenic differentiation of human mesenchymal stem cells photoencapsulated in PEG hydrogels. J Biomed Mater Res A. 2004;68A: 773–82.

56. Bosnakovski D, Mizuno M, Kim G, Takagi S, Okumura M, Fujinaga T. Chondrogenic differentiation of bovine bone marrow mesenchymal stem cells (MSCs) in different hydrogels: influence of collagen type II extracellular matrix on MSC chondrogenesis. Biotechnol Bioeng. 2006;93:1152–63.
57. Yang F, Williams CG, Wang D-a, Lee H, Manson PN, Elisseeff J. The effect of incorporating RGD adhesive peptide in polyethylene glycol diacrylate hydrogel on osteogenesis of bone marrow stromal cells. Biomaterials. 2005;26:5991–8.
58. Benoit DS, Schwartz MP, Durney AR, Anseth KS. Small functional groups for controlled differentiation of hydrogel-encapsulated human mesenchymal stem cells. Nat Mater. 2008;7:816–23.
59. Ayala R, Zhang C, Yang D, Hwang Y, Aung A, Shroff SS, Arce FT, Lal R, Arya G, Varghese S. Engineering the cell–material interface for controlling stem cell adhesion, migration, and differentiation. Biomaterials. 2011;32:3700–11.
60. Murphy AR, John PS, Kaplan DL. Modification of silk fibroin using diazonium coupling chemistry and the effects on hMSC proliferation and differentiation. Biomaterials. 2008;29:2829–38.
61. Yu B-Y, Hu S-W, Sun Y-M, Lee Y-T, Young T-H. Modulating the activities of human mesenchymal stem cells (hMSCs) and C3A/HepG2 hepatoma cells by modifying the surface characteristics of poly(3-hydroxybutyrate-co-3-hydroxyhexnoate) (PHBHHx). J Biomater Sci Polym Ed. 2009;20:1275–93.
62. Lu H, Guo L, Kawazoe N, Tateishi T, Chen G. Effects of poly(L-lysine), poly(acrylic acid) and poly(ethylene glycol) on the adhesion, proliferation and chondrogenic differentiation of human mesenchymal stem cells. J Biomater Sci Polym Ed. 2009;20:577–89.
63. Guo L, Kawazoe N, Hoshiba T, Tateishi T, Chen G, Zhang X. Osteogenic differentiation of human mesenchymal stem cells on chargeable polymer-modified surfaces. J Biomed Mater Res A. 2008;87A:903–12.
64. Yao J, Radin S, Leboy PS, Ducheyne P. The effect of bioactive glass content on synthesis and bioactivity of composite poly (lactic-co-glycolic acid)/bioactive glass substrate for tissue engineering. Biomaterials. 2005;26:1935–43.
65. Moreau JL, Xu HHK. Mesenchymal stem cell proliferation and differentiation on an injectable calcium phosphate – Chitosan composite scaffold. Biomaterials. 2009;30:2675–82.
66. Stiehler M, Lind M, Mygind T, Baatrup A, Dolatshahi-Pirouz A, Li H, Foss M, Besenbacher F, Kassem M, Bünger C. Morphology, proliferation, and osteogenic differentiation of mesenchymal stem cells cultured on titanium, tantalum, and chromium surfaces. J Biomed Mater Res A. 2008;86A:448–58.
67. Olivares-Navarrete R, Hyzy SL, Park JH, Dunn GR, Haithcock DA, Wasilewski CE, Boyan BD, Schwartz Z. Mediation of osteogenic differentiation of human mesenchymal stem cells on titanium surfaces by a Wnt-integrin feedback loop. Biomaterials. 2011;32:6399–411.
68. Greulich C, Kittler S, Epple M, Muhr G, Köller M. Studies on the biocompatibility and the interaction of silver nanoparticles with human mesenchymal stem cells (hMSCs). Langenbecks Arch Surg. 2009;394:495–502.
69. Wang ML, Tuli R, Manner PA, Sharkey PF, Hall DJ, Tuan RS. Direct and indirect induction of apoptosis in human mesenchymal stem cells in response to titanium particles. J Orthop Res. 2003;21:697–707.
70. Reilly GC, Engler AJ. Intrinsic extracellular matrix properties regulate stem cell differentiation. J Biomech. 2010;43:55–62.
71. Sun Y, Chen CS, Fu J. Forcing stem cells to behave: a biophysical perspective of the cellular microenvironment. Annu Rev Biophys. 2012;41:519–42.
72. Engler AJ, Sen S, Sweeney HL, Discher DE. Matrix elasticity directs stem cell lineage specification. Cell. 2006;126:677–89.
73. Winer JP, Janmey PA, McCormick ME, Funaki M. Bone marrow-derived human mesenchymal stem cells become quiescent on soft substrates but remain responsive to chemical or mechanical stimuli. Tissue Eng Part A. 2009;15:147–54.
74. Guilak F, Cohen DM, Estes BT, Gimble JM, Liedtke W, Chen CS. Control of stem cell fate by physical interactions with the extracellular matrix. Cell Stem Cell. 2009;5:17–26.
75. Kim T-J, Seong J, Ouyang M, Sun J, Lu S, Hong JP, Wang N, Wang Y. Substrate rigidity regulates Ca^{2+} oscillation via RhoA pathway in stem cells. J Cell Physiol. 2009;218:285–93.

76. Huebsch N, Arany PR, Mao AS, Shvartsman D, Ali OA, Bencherif SA, Rivera-Feliciano J, Mooney DJ. Harnessing traction-mediated manipulation of the cell/matrix interface to control stem-cell fate. Nat Mater. 2010;9:518–26.
77. Pek YS, Wan ACA, Ying JY. The effect of matrix stiffness on mesenchymal stem cell differentiation in a 3D thixotropic gel. Biomaterials. 2010;31:385–91.
78. Erickson IE, Huang AH, Sengupta S, Kestle S, Burdick JA, Mauck RL. Macromer density influences mesenchymal stem cell chondrogenesis and maturation in photocrosslinked hyaluronic acid hydrogels. Osteoarthritis Cartilage. 2009;17:1639–48.
79. Zahor D, Radko A, Vago R, Gheber LA. Organization of mesenchymal stem cells is controlled by micropatterned silicon substrates. Mater Sci Eng C. 2007;27:117–21.
80. Kim S-J, Lee JK, Kim JW, Jung J-W, Seo K, Park S-B, Roh K-H, Lee S-R, Hong YH, Kim SJ, Lee Y-S, Kim SJ, Kang K-S. Surface modification of polydimethylsiloxane (PDMS) induced proliferation and neural-like cells differentiation of umbilical cord blood-derived mesenchymal stem cells. J Mater Sci Mater Med. 2008;19:2953–62.
81. Engel E, Martinez E, Mills CA, Funes M, Planell JA, Samitier J. Mesenchymal stem cell differentiation on microstructured poly (methyl methacrylate) substrates. Ann Anat. 2009;191:136–44.
82. Lee IC, Lee Y-T, Yu B-Y, Lai J-Y, Young T-H. The behavior of mesenchymal stem cells on micropatterned PLLA membranes. J Biomed Mater Res A. 2009;91A:929–38.
83. Mygind T, Stiehler M, Baatrup A, Li H, Zou X, Flyvbjerg A, Kassem M, Bünger C. Mesenchymal stem cell ingrowth and differentiation on coralline hydroxyapatite scaffolds. Biomaterials. 2007;28:1036–47.
84. Ren T, Ren J, Jia X, Pan K. The bone formation in vitro and mandibular defect repair using PLGA porous scaffolds. J Biomed Mater Res A. 2005;74A:562–9.
85. Kasten P, Beyen I, Niemeyer P, Luginbühl R, Bohner M, Richter W. Porosity and pore size of β-tricalcium phosphate scaffold can influence protein production and osteogenic differentiation of human mesenchymal stem cells: an in vitro and in vivo study. Acta Biomater. 2008;4:1904–15.
86. Yim EKF, Darling EM, Kulangara K, Guilak F, Leong KW. Nanotopography-induced changes in focal adhesions, cytoskeletal organization, and mechanical properties of human mesenchymal stem cells. Biomaterials. 2010;31:1299–306.
87. Kulangara K, Yang Y, Yang J, Leong KW. Nanotopography as modulator of human mesenchymal stem cell function. Biomaterials. 2012;33:4998–5003.
88. Yim EKF, Pang SW, Leong KW. Synthetic nanostructures inducing differentiation of human mesenchymal stem cells into neuronal lineage. Exp Cell Res. 2007;313:1820–9.
89. You M-H, Kwak MK, Kim D-H, Kim K, Levchenko A, Kim D-Y, Suh K-Y. Synergistically enhanced osteogenic differentiation of human mesenchymal stem cells by culture on nanostructured surfaces with induction media. Biomacromolecules. 2010;11:1856–62.
90. Badami AS, Kreke MR, Thompson MS, Riffle JS, Goldstein AS. Effect of fiber diameter on spreading, proliferation, and differentiation of osteoblastic cells on electrospun poly (lactic acid) substrates. Biomaterials. 2006;27:596.
91. Kim TG, Park TG. Biomimicking extracellular matrix: cell adhesive RGD peptide modified electrospun poly (D, L-lactic-co-glycolic acid) nanofiber mesh. Tissue Eng. 2006;12:221–33.
92. Ma Z, Kotaki M, Inai R, Ramakrishna S. Potential of nanofiber matrix as tissue-engineering scaffolds. Tissue Eng. 2005;11:101–9.
93. Kidoaki S, Kwon IK, Matsuda T. Mesoscopic spatial designs of nano-and microfiber meshes for tissue-engineering matrix and scaffold based on newly devised multilayering and mixing electrospinning techniques. Biomaterials. 2005;26:37–46.
94. Schindler M, Ahmed I, Kamal J. A synthetic nanofibrillar matrix promotes in vivo-like organization and morphogenesis for cells in culture. Biomaterials. 2005;26:5624–31.
95. Xin X, Hussain M, Mao JJ. Continuing differentiation of human mesenchymal stem cells and induced chondrogenic and osteogenic lineages in electrospun PLGA nanofiber scaffold. Biomaterials. 2007;28:316–25.
96. Shih Y-RV Y-RV, Chen C-N, Tsai S-W, Wang YJ, Lee OK. Growth of mesenchymal stem cells on electrospun type I collagen nanofibers. Stem Cells. 2006;24:2391–7.

97. Park J, Bauer S, von der Mark K, Schmuki P. Nanosize and vitality: TiO_2 nanotube diameter directs cell fate. Nano Lett. 2007;7:1686–91.
98. Dalby MJ, Gadegaard N, Tare R, Andar A, Riehle MO, Herzyk P, Wilkinson CDW, Oreffo ROC. The control of human mesenchymal cell differentiation using nanoscale symmetry and disorder. Nat Mater. 2007;6:997–1003.
99. Curran JM, Stokes R, Irvine E, Graham D, Amro NA, Sanedrin RG, Jamil H, Hunt JA. Introducing dip pen nanolithography as a tool for controlling stem cell behaviour: unlocking the potential of the next generation of smart materials in regenerative medicine. Lab Chip. 2010;10:1662–70.

Mesenchymal Stem Cells for Cardiovascular Disease

Wei Wu and Shuyang Zhang

Abstract In recent 10 years, bone marrow mesenchymal stem cells implantation for ischemic heart failure after myocardial infarction was repeatedly inspected in numerous clinical trials. Stem cells can be transferred and proliferate in the infarcted myocardium, differentiate into cardiomyocytes or help cardiac stem cells regeneration by paracrine mechanism. This fascinating strategy has been proved a safe and effective method to help myocardial repair and increasing impaired heart systolic function. However, the efficacy is not so much satisfactory. In clinical trials, we are now informed about the timing, dosage and methods of delivery of mesenchymal stem cells administration. But we are not clear about patient selection (acute or chronic) and improvement of long-term major cardiovascular outcomes. In basic researches, biochemical modifications of mesenchymal stem cells have greatly increased its in vivo retention, proliferation and differentiation abilities. However, these applications have not been test integrated in clinical trials. Further efforts should be done to move this strategy from bench to bedside.

Keywords MSCs • Cardiomyocyte differentiation • Cardiovascular disease

Introduction

Mesenchymal stem cells (MSCs) regeneration strategy in cardiovascular disease focuses mostly on myocardial dysfunction after myocardial infarction (MI) and ischemic heart disease or so-called ischemic cardiomyopathy. Although congestive heart failure, a condition with significant high morbidity and mortality rate, has

W. Wu • S. Zhang (✉)
Department of Cardiology, Peking Union Medical College Hospital,
Chinese Academy of Medical Science and Peking Union Medical College,
No. 1 Shuai Fu Yuan, Dongcheng District, 100730 Beijing, China
e-mail: shuyangzhang103@yahoo.com.cn

R.C. Zhao (ed.), *Essentials of Mesenchymal Stem Cell Biology and Its Clinical Translation*, DOI 10.1007/978-94-007-6716-4_10,

various causes, ischemic aetiology is the most common and important one. Despite advances in medication and device therapies, new approaches to heart failure are eagerly needed since organ shortage makes heart transplantation inevitably unable to cope with the increasing demands. The rationale for cell therapy to be administered after MI is derived from the assumption that given the insufficient regeneration in the injured heart tissue, those cells may be able to replace or repair damaged vascular and cardiac tissue.

Clinical Trials Update

The first clinical trial using bone marrow stem cells transplantation for myocardial infarction is TOPCARE-AMI study which was carried out in Frankfurt in 2001 [1]. Final 5-year results suggesting long-term safety and efficacy was reported in 2011 [2]. TOPCARE-AMI study recruit 59 patients with successfully reperfused acute MI and showed fairly well long-term safety and favorable effects on left ventricular function and functional infarct size. After that, a series of phase I clinical trials using bone marrow mesenchymal stem/progenitor cell therapy for MI further investigated the safety of intervention [3–6]. These trials confirmed the safety of bone marrow MSCs transplantation method and strategy, and also suggested promising improvement in clinical outcomes and cardiac function.

On the basis of Phase I trials, more than 40 randomized controlled trials (RCTs) were carried out worldwide to examine the efficacy of bone marrow stem cells therapy for heart failure after MI. Most trials enrolled a small sample of patients like 20–90 in treatment or control groups. The follow-up duration was 3–6 months mostly [7–12]. For cardiac parameters used to determine changes after stem cells therapy, left ventricular end-diastolic volume (LVEDV), left ventricular end-systolic volume (LVESV), myocardial lesion area, and left ventricular ejection fraction (LVEF) were most often assessed. Interestingly, although echocardiography is more feasible and accessible, magnetic resonance imaging (MRI) became a more preferable method with increasing usage in cardiac imaging.

A meta-analysis provided a systemic assessment of efficacy and safety of bone marrow-derived stem cell (BMSCs) transplantation in treating congestive heart failure after acute myocardial infarction [13]. Thirteen randomized controlled trials with 811 patients were included in the setting of acute MI and the procedure consisted of intervention and any autologous BMSCs freshly isolated. Compared with controls, the infusion of BMSC also reduced LVEDV by 2.47 mL ($P=0.13$), but this difference was not statistically significant. However, BMSC treatment significantly reduced LVESV by 4.74 mL, myocardial lesion area by 3.51 % and improved LVEF by 2.99 %. The improvement is much more subtle than most researchers have expected. Although long term follow-up of all trials have not been reported, differ between bone marrow stem cells treatment and traditional medication treatment maybe remain tiny or disappear.

One explanation for this discontent result is the quick clearance of stem cells from the infarcted myocardium. A consistent finding of cell therapy experimental studies is that very small number of cells could still remain engrafted a few weeks after transplantation. Penicka and associates reported that using nuclear imaging to track the distribution of bone marrow mononuclear cells transplanted by intracoronary artery infusion, most of the transplanted cells had accumulated in the spleen, with only 5 % of the cells detectable in the myocardium at 2 h and 1 % detectable at 18 h after transplantation [14]. Other study confirmed that this percentage will be in the range of 1 % [15], or even lower. Actually, the amount of cardiomyocyte deficit resulting from an infarction large enough to cause congestive heart failure is on the range of one billion cells [16]. Even though one accepts the idea that the benefits of injected cells may not necessarily require that they persist in the tissue over time, provided that they have initially triggered endogenous reparative pathways, at least their initial number should be high enough to effectively exert these paracrine effects [17]. It is thus critical to address this issue of cell transfer, and different strategies are currently being investigated, which include computer-driven injection devices allowing optimized targeting, replacement of injection by cell patches in the case of surgical cell therapy, and techniques that enhance myocardial homing if cells are delivered intravascularly.

There are several other factors affecting the efficacy of improving heart function. One is that some clinical trials use the whole bone marrow mononuclear cells but not purified mesenchymal stem cells which have the ability of proliferation and differentiation. Actually, bone marrow mononuclear cells consist of many types of cells that are not necessary or even may impede stem cell regeneration in the infarcted region. Secondary, whether the patients have been successfully revascularized may be crucial, because sufficient blood flow is important for bone marrow MSCs proliferation and differentiation. Some studies engrafted stem cells by using intramyocardially injection may failed to achieve without reopening occluded coronary arteries. In addition, timing of engraft is also an important aspect. Mesenchymal stem cells may be eliminated by inflammation in the early stage of myocardial infarction or be induced to differentiate into fibroblasts in the late stage of scar formation, which we will discuss later.

Amount of Stem Cells and Strategies for Improving Engraftment

From the experience of literatures, it seems that the improvement of left ventricular function correlated positively with bone marrow MSC dose administered. It suggested that significant effects on LVEF may only be achieved when infusing doses are higher than 10^8 MSC. When lower doses of bone marrow MSC were administered, the mean change in LVEF was statistically significant in favor of no MSC therapy [13]. This is consistent with the idea that myocardial repair or the factors that promote this improvement are dependent on cell number.

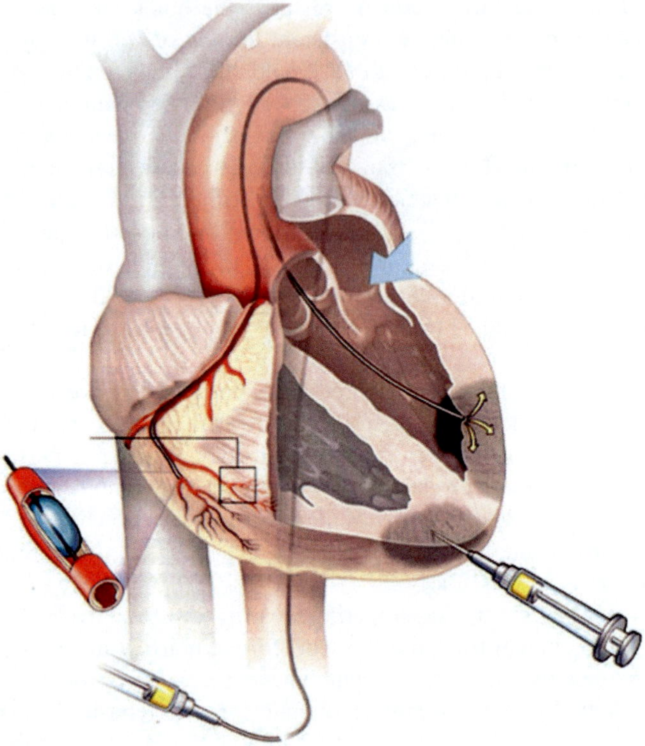

Fig. 1 Three clinical techniques to deliver stem cell. From *right to left*: intracoronary injection; catheter-based transendocardial injection; transepicardial injection requiring surgical access to the heart (Quoted from Ref. [18])

How to increase the survival or retention of implanted stem cells? Researchers were concerned about the stem cells delivery methods. Intravenous infusion is quite a simple and safe method which has lowest cell retention rate to the target infarcted myocardium. Intracoronary artery or bypass graft vessel infusion has moderate retention rate, but may cause microembolism or ischemia during infusion. Transendocardial or surgical intramyocardiual injection allows visualization of the ischemic area of the myocardium and in theory facilitates the delivery of a maximum number of cells to the intended area (Fig. 1) [18]. However, the mechanical leakage of cells at the time of injection has been found to remove about 30 % of the transplanted cells [19]. Even more negative is a report that only 10 % of administered microspheres approximating the size of mesenchymal stem cells were retained within the sites of their injection at 30 min after being injected intramyocardially, as a result of their being "squeezed out" into the nearby vasculature by the mechanical forces of the heart.

Another attempt to improve engraftment and survival is preconditioning or genetic modification of stem cells. The preconditioning methods include hypoxia, ischemia and pharmacological treatment. It was reported that hypoxic preconditioning decreased cell apoptosis, limited the size of myocardial infarcts, and improved neo-angiogenesis in the heart through upregulation of survival signaling pathways involving hypoxia-inducible factor (HIF)-1_ and stromal cell-derived factor (SDF)-1 in animal models of MI [20]. Kim and colleagues [21] showed that the preconditioning of MSC with two 30-min cycles of ischemia plus reoxygenation (I/R) supported their survival through subsequently longer exposures to anoxia and after their engraftment in the infarcted heart. In the other hand, pharmacological treatment such as trimetazidine may have similar effects on mesenchymal stem cells to better stay and survive in infarcted heart tissue [22].

The genetic modification of stem cells is an attractive concept for cellular therapy because of their possible long-term survival in the host. Huang and associates pretreated MSC with corresponding chemokine receptors before intramyocardial injection of the cells into infarcted rat hearts. They were able to demonstrate a significant increase in cell survival and engraftment as compared with that of transplanted cells that were not so pretreated. In addition, there was less fibrosis in areas of infarction treated with the MSC pretreated with chemokine receptors, and the improvement in cardiac function was greater with the pretreated cells, than with unpretreated MSC [23]. Recently, Tao and associates [24] used a porcine model of acute MI to study the benefits of coexpression of VEGF and Ang-1, and investigated the mechanisms underlying their effects. Their results indicated that the coexpression of VEGF and Ang-1 in their porcine model of MI induced angiogenesis, stimulated cardiomyocyte proliferation, and reduced apoptosis. Together, these effects resulted in the improvement of cardiac function.

The preconditioning or modification methods on bone marrow MSCs were not tested in clinical trials. Researchers are concerning about two major problems. One is the inconveniency of cell preparation. Usually, it is a long-time and complicated job that needs high level of laboratory techniques. The other is the increasing possibility of forming neoplasm or arrhythmia by significant proliferation of MSC in engrafted myocardium. Since most phase I clinical trials have proved the safety of bone marrow MSC transplantation, attempts can be made by using preconditioned or modified mesenchymal stem cells in clinical trials.

Timing of Transplantation

The systemic analysis previously mentioned also discussed the correlation between timing of transplantation and left ventricular changes [13]. In this study, the improvement on LVEF was even greater when BMSC were infused later (>7 days) after reperfusion procedure. It was consisted with the results REPAIR-AMI trial which suggested BMSC infusion to be more effective when infused >5 days following reperfusion [12]. In fact, the beneficial effects of BMC infusion on the recovery of contractile function were confined to patients who were treated more than 4 days after infarct reperfusion. BMC infusion on day 5 or later was associated with an absolute increase in LVEF of 5.1 % (P = 0.004), whereas no benefit was observed in patients treated up to day 4 after reperfusion (3.9 ± 5.4 % in the placebo group vs. 4.5 ± 6.8 % in the BMC group; P = 0.62).

It is assumed that inflammation plays the most important role in timing selection of bone marrow mesenchymal stem cell repair. In the very early stage of myocardial infarction, inflammation was significantly drastic that might have negative effect on stem cell survival or proliferation. After adequate time of inflammation cooling down and sufficient medication treatment, local cytokines might promote fibroblast proliferation and scar formation. That might indicate wrong direction of differentiation to fibroblasts but not cardiomyocytes for implanted mesenchymal stem cells. So it is ideal to deliver bone marrow stem cells after falling tide of inflammation and before initiation of fibroblast proliferation.

Recently, new findings on timing of intracoronary administration of autologous bone marrow stem cells after acute myocardial infarction by SWISS-AMI study, which was introduced in 2010 [25], were reported on American Heart Association scientific sessions (2012) in Los Angeles. SWISS-AMI study compared the efficacy of early transplantation (5–7 days) and late transplantation (3–4 weeks) after the initial cardiovascular events. Left ventricular function as well as scar size, transmural extension, and regional wall motion score have been assessed by cardiac magnetic resonance (CMR) studies at baseline and after 4 and 12 months. It was reported that there were no statistical differences on primary and secondary endpoints between early and late administration groups. It was suggested that 1 month was not too late from bone marrow stem cell transplantation. This is good news. If we plan to precondition bone marrow MSCs in vitro before transplantation, there will be enough time for preparation.

Allogeneic Versus Autologous

Mesenchymal stem cells are both immunoprivileged and immunosuppressive, thus bearing the potential to be used as an allograft. To receive autologous bone marrow stem cells transplantation, patients should have normal bone marrow proliferation function and enough period of stem cells preparation. Allogeneic stem cells, however, may solve the problem conveniently. They can be donated by relatively young adults and prepared any time before transplantation. Recently, the POSEIDON randomized trial compared safety and efficacy of allogeneic and autologous bone marrow derived mesenchymal stem cells delivered by transendocardial injection in patients with ischemic cardiomyopathy [26]. Interestingly, In this early-stage study of patients with ischemic heart failure, transendocardial injection of allogeneic and autologous MSCs without a placebo control were both associated with low rates of treatment-emergent severe adverse events (SAEs), including immunologic reactions. In aggregate, MSC injection favorably affected patient functional capacity, quality of life, and ventricular remodeling. The most important point was that, over the duration of the study about 12 months, allogeneic MSCs did not stimulate significant donor-specific alloimmune reactions including long-term adverse events, arrhythmias or major adverse cardiovascular events. This study shows for the first time that allogeneic bone marrow mesenchymal stem cells are comparable in safety and efficacy with the autologous MSCs.

We can imagine that 1 day, bone marrow MSCs could be prepared and stored like conventional medications. Whenever there's patient fit for receiving cell therapy, we can administrate it conveniently.

Perspectives

Although there were several important problems to be answered, bone marrow mesenchymal stem cells transplantation therapy represents a fascinating new approach for management of heart disease. It is the right time of moving this novel strategy from bench to bedside, which is the purpose of translational medicine. Until recently, there have been more than 50 clinical trials focusing the clinical application of bone marrow mesenchymal stem cells in cardiac repair. Its safety and efficacy has been well established. But most randomized clinical trials are phase I or phase II studies, including small sample of patients and relatively short follow-up period. Great differences exist among each trial, such as cell dosage, timing of transplantation, patients selection, aetiology of heart failure and measurement of endpoints. Unlike the conventional efficacy evaluation of medication in treatment using primary endpoints (death, cardiac death or hospitalization), most MSCs trial use the substitution end points such as left ventricular function assessed by echocardiography or cardiac magnetic resonance, as well as infarct size, Minnesota Living with Heart Failure Questionnaire (MLHFQ), and regional wall motion score. These end points were not as rigorous as the conventional "hard" end points. In future, Large randomized multi-center clinical trials with long-term follow up and rigorous end points are needed in future to further prove the efficacy of MSCs in ischemic cardiomyopathy after myocardial infarction.

It still will be a long way to realize the "bench to bedside". Basic researches have great increased the survival, proliferation and differentiation ability of mesenchymal stem cells in vitro or in animals by modification strategies; however, they were not applied in clinical trials. It is not only the technical problem but also the ethic issue. The doubt of increasing neoplasm or arrhythmias may rise in researchers and members of ethic committees. However, reasonable modification of mesenchymal stem cells should be considered in clinical trials in future.

In addition, MSCs regeneration strategy in chronic heart failure was not as exciting as in early stage of heart failure after acute myocardial infarction. But there are much more chronic heart failure patients waiting for better medical care and new strategy. Recently, a study sponsored by the National Heart, Lung, and Blood institute as the Cardiovascular Cell Therapy Research Network (CCTRN) evaluated the safety and efficacy of BMCs in patients with chronic ischemic heart disease and LV dysfunction who have no other revascularization options. This study showed that transendocardial injection of autologous BMCs compared with placebo did not improve LVESV, maximal oxygen consumption, or reversibility on SPECT [27]. It seems that without amelioration of coronary perfusion, BMCs might not improve heart function by themselves. And it also indicates that myocardial repair is not easy in patients with chronic heart failure. Further efforts should be made to better improve the stem cell regeneration treatment in this bigger patient population.

Ten years of basic researches and clinical trials have made it a possible strategy for bone marrow mesenchymal stem cells therapy in ischemic heart failure. Another decade may be needed to optimize the efficacy of this fascinating treatment in cell modification, patients selection and transplantation methods. There is a heavy burden and we need to embark a long road.

References

1. Assmus B, Schachinger V, Teupe C, Britten M, Lehmann R, Dobert N, Grunwald F, Aicher A, Urbich C, Martin H, Hoelzer D, Dimmeler S, Zeiher AM. Transplantation of progenitor cells and regeneration enhancement in acute myocardial infarction (TOPCARE-AMI). Circulation. 2002;106:3009–17.
2. Leistner DM, Fischer-Rasokat U, Honold J, Seeger FH, Schächinger V, Lehmann R, Martin H, Burck I, Urbich C, Dimmeler S, Zeiher AM, Assmus B. Transplantation of progenitor cells and regeneration enhancement in acute myocardial infarction (TOPCARE-AMI): final 5-year results suggest long-term safety and efficacy. Clin Res Cardiol. 2011;100:925–34.
3. Strauer BE, Brehm M, Zeus T, Kostering M, Hernandez A, Sorg RV, Kogler G, Wernet P. Repair of infarcted myocardium by autologous intracoronary mononuclear bone marrow cell transplantation in humans. Circulation. 2002;106:1913–8.
4. Penarrubia MJ, de la Fuente L, Gomez-Bueno M, Cantalapiedra A, Fernandez J, Gutierrez O, Sanchez PL, Hernandez C, Sanz R, Garcia-Sancho J, Sanchez A. Experimental and clinical regenerative capability of human bone marrow cells after myocardial infarction. Circ Res. 2004;95:742–8.
5. Stamm C, Westphal B, Kleine HD, Petzsch M, Kittner C, Klinge H, Schumichen C, Nienaber CA, Freund M, Steinhoff G. Autologous bone-marrow stem-cell transplantation for myocardial regeneration. Lancet. 2003;361:45–6.
6. Wollert KC, Meyer GP, Lotz J, Ringes-Lichtenberg S, Lippolt P, Breidenbach C, Fichtner S, Korte T, Hornig B, Messinger D, Arseniev L, Hertenstein B, Ganser A, Drexler H. Intracoronary autologous bone-marrow cell transfer after myocardial infarction: the BOOST randomised controlled clinical trial. Lancet. 2004;364:141–8.
7. Kang HJ, Lee HY, Na SH, Chang SA, Park KW, Kim HK, Kim SY, Chang HJ, Lee W, Kang WJ, Koo BK, Kim YJ, Lee DS, Sohn DW, Han KS, Oh BH, Park YB, Kim HS. Differential effect of intracoronary infusion of mobilized peripheral blood stem cells by granulocyte colony-stimulating factor on left ventricular function and remodeling in patients with acute myocardial infarction versus old myocardial infarction: the MAGIC Cell-3-DES randomized, controlled trial. Circulation. 2006;114:I145–51.
8. Li ZQ, Zhang M, Jing YZ, Zhang WW, Liu Y, Cui LJ, Yuan L, Liu XZ, Yu X, Hu TS. The clinical study of autologous peripheral blood stem cell transplantation by intracoronary infusion in patients with acute myocardial infarction (AMI). Int J Cardiol. 2007;115:52–6.
9. Penicka M, Horak J, Kobylka P, Pytlik R, Kozak T, Belohlavek O, Lang O, Skalicka H, Simek S, Palecek T, Linhart A, Aschermann M, Widimsky P. Intracoronary injection of autologous bone marrow-derived mononuclear cells in patients with large anterior acute myocardial infarction: a prematurely terminated randomized study. J Am Coll Cardiol. 2007;49:2373–4.

10. Lunde K, Solheim S, Aakhus S, Arnesen H, Abdelnoor M, Egeland T, Endresen K, Ilebekk A, Mangschau A, Fjeld JG, Smith HJ, Taraldsrud E, Grogaard HK, Bjornerheim R, Brekke M, Muller C, Hopp E, Ragnarsson A, Brinchmann JE, Forfang K. Intracoronary injection of mononuclear bone marrow cells in acute myocardial infarction. N Engl J Med. 2006;355:1199–209.
11. Meyer GP, Wollert KC, Lotz J, Steffens J, Lippolt P, Fichtner S, Hecker H, Schaefer A, Arseniev L, Hertenstein B, Ganser A, Drexler H. Intracoronary bone marrow cell transfer after myocardial infarction: eighteen months' follow-up data from the randomized, controlled BOOST (BOne marrOw transfer to enhance ST-elevation infarct regeneration) trial. Circulation. 2006;113:1287–94.
12. Schachinger V, Erbs S, Elsasser A, Haberbosch W, Hambrecht R, Holschermann H, Yu J, Corti R, Mathey DG, Hamm CW, Suselbeck T, Assmus B, Tonn T, Dimmeler S, Zeiher AM. Intracoronary bone marrow-derived progenitor cells in acute myocardial infarction. N Engl J Med. 2006;355:1210–21.
13. Martin-Rendon E, Brunskill SJ, Hyde CJ, Stanworth SJ, Mathur A, Watt SM. Autologous bone marrow stem cells to treat acute myocardial infarction: a systematic review. Eur Heart J. 2008;29:1807–18.
14. Penicka M, Widimsky P, Kobylka P, Kozak T, Lang O. Images in cardiovascular medicine. Early tissue distribution of bone marrow mononuclear cells after transcoronary transplantation in a patient with acute myocardial infarction. Circulation. 2005;112:e63–5.
15. Fukushima S, Varela-Carver A, Coppen SR, Yamahara K, Felkin LE, Lee J, Barton PJ, Terracciano CM, Yacoub MH, Suzuki K. Direct intramyocardial but not intracoronary injection of bone marrow cells induces ventricular arrhythmias in a rat chronic ischemic heart failure model. Circulation. 2007;115:2254–61.
16. Murry CE, Reinecke H, Pabon LM. Regeneration gaps. Observations on stem cells and cardiac repair. J Am Coll Cardiol. 2006;47:1777–85.
17. Menaché P. Current status and future prospects for cell transplantation to prevent congestive heart failure. Semin Thorac Cardiovasc Surg. 2008;20:131–7.
18. Stamm C, Nasseri B, Choi YH, Hetzer R. Cell therapy for heart disease: great expectations, as yet unmet. Heart Lung Circ. 2009;18:245–56.
19. Teng CJ, Luo J, Chiu RC, Shum-Tim D. Massive mechanical loss of microspheres with direct intramyocardial injection in the beating heart: implications for cellular cardiomyoplasty. J Thorac Cardiovasc Surg. 2006;132:628–32.
20. Hu X, Yu SP, Fraser JL, Lu Z, Ogle ME, Wang JA, Wei L. Transplantation of hypoxiapreconditioned mesenchymal stem cells improves infarcted heart function via enhanced survival of implanted cells and angiogenesis. J Thorac Cardiovasc Surg. 2008;135:799–808.
21. Kim HW, Haider HK, Jiang S, Ashraf M. Ischemic preconditioning augments survival of stem cells via miR-210 expression by targeting caspase-8-associated protein 2. J Biol Chem. 2009;284:33161–8.
22. Wisel S, Khan M, Kuppusamy ML, Mohan IK, Chacko SM, Rivera BK, Sun BC, Hideg K, Kuppusamy P. Pharmacological preconditioning of mesenchymal stem cells with trimetazidine (1-[2,3,4-trimethoxybenzyl]piperazine) protects hypoxic cells against oxidative stress and enhances recovery of myocardial function in infarcted heart through Bcl-2 expression. J Pharmacol Exp Ther. 2009;329:543–50.
23. Huang J, Zhang Z, Guo J, Ni A, Deb A, Zhang L, Mirotsou M, Pratt RE, Dzau VJ. Genetic modification of mesenchymal stem cells overexpressing CCR1 increases cell viability, migration, engraftment, and capillary density in the injured myocardium. Circ Res. 2010;106:1753–62.
24. Tao Z, Chen B, Tan X, Zhao Y, Wang L, Zhu T, Cao K, Yang Z, Kan YW, Su H. Coexpression of VEGF and angiopoietin-1 promotes angiogenesis and cardiomyocyte proliferation reduces apoptosis in porcine myocardial infarction (MI) heart. Proc Natl Acad Sci USA. 2011;108:2064–9.
25. Sürder D, Schwitter J, Moccetti T, Astori G, Rufibach K, Plein S, Lo Cicero V, Soncin S, Windecker S, Moschovitis A, Wahl A, Erne P, Jamshidi P, Auf der Maur C, Manka R, Soldati G, Bühler I, Wyss C, Landmesser U, Lüscher TF, Corti R. Cell-based therapy for myocardial repair in patients with acute myocardial infarction: rationale and study design of the SWiss multicenter Intracoronary Stem cells Study in Acute Myocardial Infarction (SWISS-AMI). Am Heart J. 2010;160:58–64.

26. Hare JM, Fishman JE, Gerstenblith G, DiFede Velazquez DL, Zambrano JP, Suncion VY, Tracy M, Ghersin E, Johnston PV, Brinker JA, Breton E, Davis-Sproul J, Schulman IH, Byrnes J, Mendizabal AM, Lowery MH, Rouy D, Altman P, Wong Po Foo C, Ruiz P, Amador A, Da Silva J, McNiece IK, Heldman AW. Comparison of allogeneic vs autologous bone marrow–derived mesenchymal stem cells delivered by transendocardial injection in patients with ischemic cardiomyopathy: the POSEIDON randomized trial. JAMA. 2012;308:2369–79.
27. Perin EC, Willerson JT, Pepine CJ, Henry TD, Ellis SG, Zhao DX, Silva GV, Lai D, Thomas JD, Kronenberg MW, Martin AD, Anderson RD, Traverse JH, Penn MS, Anwaruddin S, Hatzopoulos AK, Gee AP, Taylor DA, Cogle CR, Smith D, Westbrook L, Chen J, Handberg E, Olson RE, Geither C, Bowman S, Francescon J, Baraniuk S, Piller LB, Simpson LM, Loghin C, Aguilar D, Richman S, Zierold C, Bettencourt J, Sayre SL, Vojvodic RW, Skarlatos SI, Gordon DJ, Ebert RF, Kwak M, Moyé LA, Simari RD, Cardiovascular Cell Therapy Research Network (CCTRN). Effect of transendocardial delivery of autologous bone marrow mononuclear cells on functional capacity, left ventricular function, and perfusion in chronic heart failure: the FOCUS-CCTRN trial. JAMA. 2012;307:1717–26.

Mesenchymal Stem Cells as Therapy for Graft Versus Host Disease: What Have We Learned?

Partow Kebriaei, Simon Robinson, Ian McNiece, and Elizabeth Shpall

Abstract Mesenchymal stem cells (MSC) are a population of phenotypically heterogeneous cells that are one component of the supportive, stromal micro-environment. They can be isolated from many readily accessible tissues including bone marrow, umbilical cord, placenta, and adipose tissue, and extensive *ex vivo* and pre-clinical data suggest that subpopulations within MSC contribute to the immunomodulation of the host, without provoking an allo-reactive T cell response. Furthermore, largely through paracrine effects, they contribute to tissue repair. These unique properties make MSC an ideal agent to investigate for the therapy of graft *versus* host disease (G*v*HD). Therapeutic trials with varied MSC dosing schedules and clinical endpoints have shown mixed results. In this chapter, we will review recent preclinical data, and summarize the results of clinical trials utilizing MSC for the treatment of acute and chronic G*v*HD.

Keywords MSC • T cell response • GVHD

Introduction

Mesenchymal stem cells (MSC) are a population of phenotypically heterogeneous cells marked by an absence of hematopoietic markers (CD34, CD45), the expression of CD73, CD90, and CD105 surface markers, and the capacity to differentiate *in vitro* into osteoblasts, chondroblasts, and adipocytes [1–4]. They can be isolated, and expanded *ex vivo* from many readily accessible tissues including bone marrow (BM)

P. Kebriaei (✉) • S. Robinson • I. McNiece • E. Shpall
Department of Stem Cell Transplantation and Cellular Therapy,
The University of Texas MD Anderson Cancer Center,
Unit 423, 1515 Holcombe Blvd, Houston, TX 77030, USA
e-mail: pkebriae@mdanderson.org

R.C. Zhao (ed.), *Essentials of Mesenchymal Stem Cell Biology and Its Clinical Translation*, DOI 10.1007/978-94-007-6716-4_11,

[5–7], adipose tissue [7, 8], umbilical cord [7–10], and placenta [11, 12], where they are a component of the supportive stromal microenvironment. Furthermore, MSC do not express HLA class II histocompatibility antigens, or accessory molecules (CD40, CD80 and CD86), required for immune cell activation, and thus, histocompatibility matching is not required for therapeutic effect [13]. Subpopulations of MSC may contribute directly, and/or via paracrine effects, to immunomodulation [14–17] and tissue repair [18–20]. These unique properties make MSC a rational agent to investigate as therapy for inflammatory disorders, including graft *versus* host disease (G*v*HD), which results in significant morbidity and mortality following allogeneic hematopoietic stem cell transplantation (HSCT). In this review, we will summarize pre-clinical and recent clinical data for MSC as a cellular therapy for G*v*HD.

Pathophysiology and Clinical Manifestations of G*v*HD

The advantage of allogeneic HSCT over chemotherapy alone for the treatment of malignancy is the associated graft *versus* tumor (G*v*T) effect that results from the donor T and natural killer (NK) cells present in the transplant graft. Evidence for G*v*T effect comes from observation that complete remissions can be achieved in patients following donor lymphocyte infusion (DLI) [21, 22]. Unfortunately, this G*v*T effect does not focus specifically on tumor destruction, rather donor T cells can also attack normal host tissue, resulting in G*v*HD [23], although patients who develop G*v*HD following T-cell replete grafts, or DLI, have lower reported rates of disease relapse compared to those who do not. Conversely, patients who receive T-cell depleted grafts have a higher rate of disease relapse compared to T-cell replete grafts [24]. The separation of G*v*T from G*v*HD remains a source of considerable interest amongst hematopoietic transplant specialists. Based on experimental models in mice and dogs, the pathophysiology of acute GVHD can be summarized in three sequential phases. First, the HSCT conditioning regimen damages host tissues, and inflammatory cytokines, such as TNFα, IL-1 and IL-6, are released resulting in activation of host antigen presenting cells (APC). Second, donor T-cells proliferate and differentiate in response to the activated APC, and thirdly, the cellular mediators released by donor cytotoxic T-lymphocytes (CTL) and donor natural killer (NK) cells, as well as soluble inflammatory agents (e.g. TNFα, interferon γ, interleukin 1, and nitric oxide) result in local tissue damage [23]. Classically, G*v*HD that occurs within the first 100 days of HSCT is classified as 'acute' G*v*HD [25], and that occurring beyond day 100 is classified as 'chronic' G*v*HD [26]. This definition has been modified to include an overlap syndrome with manifestations of both acute and chronic G*v*HD [27]. The organs most commonly involved in acute G*v*HD are skin (manifesting as an erythematous maculopapular rash), the gastrointestinal tract (manifesting as nausea, vomiting, and/or diarrhea), and liver dysfunction (manifesting as……). The symptoms of chronic G*v*HD resemble features of auto-immune disorders such as arthritis, scleroderma, and bronchiolitis obliterans and the pathophysiology of chronic G*v*HD is not well characterized.

The incidence of acute GvHD varies from 20 to 70 % following HSCT depending on the extent of human leukocyte antigen (HLA) match, intensity of the conditioning regimen, age of the recipient, and the and stage of the primary disease [28–32]. The standard treatment for acute GvHD remains corticosteroids with reported complete response rates ranging from 18 to 35 % [33, 34]. While the addition of mycophenolate mofetil (MMF) to steroid therapy appeared to improve acute GvHD therapy, with a 60 % CR rate noted at day 28 [35], a subsequent randomized trial of steroids plus MMF *versus* steroids alone, showed no benefit to the addition of MMF, and the trial was closed early. Other attempts to intensify the immunosuppressive therapy as part of the initial treatment of GvHD, have included elevated doses of steroids [36], or the use of antithymocyte globulin (ATG) [37], or daclizumab [38]. However, these approaches did not improve response rates, and some studies reported worse survival rates due to increased disease relapse and GvHD-related mortality compared to corticosteroid use alone [38]. For patients who do not respond to steroid therapy for the treatment of acute GvHD, 1-year survival rates range from 10 to 30 % [39]. Thus, new therapeutic agents that are both safe and effective are urgently needed for the management of acute GvHD.

Biology of MSC as It Pertains to GvHD

The proposed immunomodulatory and reparative characteristics of MSC make them especially suitable for the treatment of various inflammatory disorders, including GvHD. The immunomodulatory properties of MSC are still not fully characterized, but broadly effect cell types of the innate and adaptive immune systems. Within the context of innate immunity, MSC interact with Toll-like receptors (TLR), dendritic cells (DC), and NK cells. Increasing data supports the interaction of TLR and MSC, with "licensing" of MSC toward an 'immunosuppressive', or 'pro-inflammatory' phenotype based on the specific TLR interaction and local micro-environment [40–44]. DC are potent antigen presenting cells (APC) for naïve T-cells, and are critical in the activation of donor T cells during acute GvHD [45]. MSC inhibit differentiation of monocytes to DC, and furthermore, affect DC differentiation, activation, and function [16]. MSC also inhibit NK cell proliferation and cytokine production, and could also potentially modulate DC function through their effects on NK cells [46].

Within the context of adaptive immunity, MSC inhibit allo-reactive T-cell responses via contact-dependent mechanisms and soluble factors [16, 47]. Some studies suggest a shift in T-cell function toward a more regulatory phenotype by the induction of T regulatory cells (T_{reg}) [48]. Importantly, the effects of MSC on T-cells are independent of HLA matching with MSC displaying an immune-privileged phenotype [13]. Sundin and colleagues evaluated the immunogenicity of HLA-mismatched MSC infused after HSCT [49]. Recipient lymphocyte response to MSC and peripheral blood lymphocytes from the MSC or third party donors was measured before and after MSC infusion. Transplant recipients given MSC showed an allo-response to the third-party and MSC donor, but not to the donor MSC themselves, suggesting immune unresponsiveness against the donor MSC, rather than tolerance to the MSC donor. This lack of immune response against the infused donor MSC was sustained after repeat MSC infusions [49].

The role of MSC in tissue repair continues to be studied. When infused intravenously, the majority of MSC are sequestered in the lung, and MSC are not identified in the injured tissues in human studies. The bulk of data suggests that MSC most likely facilitate tissue repair via paracrine effects [50], rather than through direct MSC-mediated tissue repair [51]. For example, in murine models of myocardial infarction [52] and corneal injury[53], human bone marrow-derived MSC trapped in the lung microvasculature are noted to secrete TNFa stimulated gene/protein 6 (TSG-6) which is thought to suppresses the early immune response in both of these settings. Developing a better understanding of the homing capabilities of MSC will also contribute to our understanding of their role in tissue repair. *In vitro*, MSC have the capacity to migrate under the influence of a number of proteins, including complement, growth factors, cytokines, and chemokine receptors such as CXCR4, CXCR5, CXCR6, CCR1, CCR7, and CCR9 [40, 54]. However, *in vivo* data is limited. Better *in vivo* imaging tools will greatly contribute to this field and better reveal the potential for intravenously-infused MSC.

Pre-clinical Data for MSC in Murine Models of G*v*HD

Murine models have been used extensively to investigate the immunomodulatory potential of MSC in ameliorating (preventing and/or treating) G*v*HD, although it must be emphasized that murine and human MSC have distinct and different properties, including a greater propensity for immortalization and transformation [55], lack of IDO-2,3 expression [56], and different *in vitro* immunosuppressive activities [57]. These studies have revealed a mix of results. Some show immunomodulatory efficacy while others do not [57–60]. These conflicting data have highlighted a number of important questions that should be considered. Such questions, once addressed, will likely significantly impact the clinical use of MSC as a cellular therapy for G*v*HD. Such questions include: (i) determining the optimal time at which MSC should be infused for optimal efficacy, (ii) defining the correct dose of MSC for optimal efficacy, and (iii) studying the trafficking and bio-distribution of MSC.

In efforts to determine the optimal schedule for MSC infusion, Polchert and colleagues performed a series of experiments in a murine G*v*HD model generated by the transplant of lethally-irradiated male BALB/c (H-2K^{d}) mice with BM and splenocytes from female C57BL/6 (H-2K^{b}) mice [61]. MSC were infused concurrently with BM, at 2, 20, or 30 days after HSCT. Interestingly, only when MSC were administered 2 or 20 days after transplant was any survival benefit observed [61]. To explain this 'window' of opportunity for effective MSC immunomodulation, the investigators hypothesized that a pro-inflammatory milieu needs time to develop to activate the MSC, and proposed that IFNγ was a key molecule in this process. Several *ex vivo* studies have demonstrated that MSC are activated by IFNγ [62] and that their migration may be driven by an IFNγ-associated upregulation of chemokine receptors expressed by MSC [63, 64]. Interestingly, when Polchert and colleagues transplanted bone marrow and splenocytes into IFNγ knock-out mice, the MSC failed to ameliorate the symptoms of G*v*HD irrespective of the time or dose of administration [61]. Furthermore, in the mice with intact IFNγ function, they demonstrated that serum levels of IFNγ remained low for up to 2 days after transplantation, explaining the time-dependent treatment effect of MSC administration. They postulated that the lack of activity seen for MSC infusion at the late time point (day 30) was due to the presence of overwhelming numbers of activated T-cells by this time point

In efforts to study the effect of dose on MSC function, Joo and colleagues used the mixed lymphocyte reaction (MLR) to determine inhibition of splenocyte proliferation in the presence of MSC [65]. MSC were administered into a murine model of G*v*HD (transplantation of 5×10^6 BM cells and 1×10^6 spleen cells from C3H/he donor mice into lethally-irradiated BALB/c recipient mice). Mice received MSC at 0.5 (low dose), 1.0 (intermediate dose) or 2.0×10^6 (high dose) at the time of spleen cell transplantation to reflect a 0.5:1, 1:1 and 2:1 MSC:splenocyte ratio. They observed significantly improved survival in mice receiving the intermediate and high MSC ratio [65]. They reported an increase in T_{Reg} activity *in vivo* in mice receiving splenocytes and MSC, and consistent with previous reports [66–69], proposed modulation of G*v*HD by MSC through the activation of T_{Reg} [65]. It was hypothesized that this activation was a consequence of factors (e.g. TGFβ) secreted by MSC [16, 47]. Since the amount of any factor liberated by the MSC is proportional to the numbers of MSC, it is likely that a specific dose of MSC might be required to maximally stimulate T_{Reg} proliferation.

Christensen and colleagues also investigated the effect of timing and dose of MSC in 2 murine G*v*HD models mimicking a major histocompatibility complex mismatched (UBI-GFP/BL6[H-2^b] → BALB/c[H-2^d]) and sibling transplant (UBI-GFP/BL6[H-2^b] → BALB.B [H-2^b]) [70]. MSC were administered into mice via intra-peritoneal injections at a dose of 4×10^5 MSC/mouse or 1×10^6 MSC/mouse on days 1, 7, or 14 following HSCT. In contrast to the study by Joo et al., they found the low MSC dose to be the most effective, ameliorating G*v*HD in the sibling model and delaying, but not preventing, G*v*HD in the mismatched model [70]. In accordance with Polchert and colleagues, they noted that MSC treated mice had significantly reduced serum IFN-g, but also noted efficacy for MSC administration at day 1 following HSCT rather than at the later time-points. Furthermore, in MLR assays, although they noted significant reduction in T-cell proliferation with the addition of MSC, they did not observe altered CD4/CD8 T cell ratios or increased in T_{Reg} cells frequencies [70].

Finally, Joo and colleagues utilized bioimaging to study the trafficking of MSC in a murine model of G*v*HD [71]. Recipient BALB/c-nude mice received 500 cGy radiation and 5×10^6 BM cells from normal C57BL/6 donor mice. To induce G*v*HD, 1×10^6 splenocytes from C57BL/6 donor mice expressing the enhanced green fluorescent protein (EGFP) were subsequently injected. Detection of the EGFP signal illustrated the trafficking of splenocytes and identified sites of G*v*HD *in situ*. To study the biodistribution of MSC in this model, MSC were generated from C57BL/6 donor mice expressing red fluorescent protein (RFP). RFP-MSC were transplanted at 1×10^6 MSC/mouse. All cells were injected into the lethally-irradiated BALB/c-nude mice within 24 h of irradiation. Consistent with previous reports [72, 73]. RFP signal associated with MSC was detected in the lungs after 2 days, and EGFP signal, associated with donor splenocytes, was also detected in the lungs. However, after 7 days EGFP (splenocytes) and RFP (MSC) signal intensity reduced in the lungs and increased in the GI tract. After 22–37 days, EGFP and RFP signals co-localized to the liver, skin, and lymph nodes, suggesting that MSC can home to sites of G*v*HD and potentially exert direct cell-cell contact mediated effects, in addition to paracrine effects. However, in contrast to these studies, Sudres and colleagues using an MHC-mismatched model of GVHD (C57BL/6 [H-2 $^{b/b}$] into BALB/c [H-2 $^{d/d}$]), could only detect traces of MSC in G*v*HD target organs, and showed no change in *in vivo* T-cell activation or amelioration of G*v*HD when marrow-derived MSC were infused [57].

More recent work has examined the efficacy of MSC derived from tissue other than the bone marrow. One abundant source is adipose tissue, and Yanez and colleagues investigated the impact of adipose-derived MSC in the murine transplant model: C57BL/6 (B6;H-2$^{b/b}$) (10^7BM cells $= 2 \times 10^7$ splenocytes) → B6D2FI (H-2$^{b/d}$) [74]. Repeated infusions of 5×10^4 adipose-derived MSC were administered on days 0, 7, and 14 or on days 14, 21, and 28. Similar to Polchert's findings, the efficacy of the MSC were timedependent. Only mice infused with adipose tissue-derived MSC at the early time-points survived G*v*HD [74]. In conclusion, while there is much discrepancy among the various murine studies, there is a trend that the timing and dose of MSC administration may be critical for the demonstration of anti-G*v*HD efficacy.

Results of Clinical Trials Utilizing MSC for the Treatment of G*v*HD

Acute GvHD

As illustrated in Table 1, MSC have been most extensively studied in the setting of steroid-refractory acute G*v*HD, following the initial, dramatic response of haplo-identical MSC used to treat a young patient with advanced, steroid-refractory liver and gut G*v*HD [89]. This observation prompted a phase II study by the European Blood and Marrow Transplant MSC consortium, using a shared expansion protocol for manufacturing MSC, to treat 25 pediatric and 30 adult patients with sibling HLA-identical, haplo-identical, or third-party mismatched, bone marrow-derived MSC for steroid-refractory G*v*HD [78] (Table 1). A single MSC infusion, with

Table 1 Results of clinical trials utilizing MSC for therapy of acute and chronic GVHD

Study	Indication	N	Med age (range)	Episodes GVHD	MSC regimen (M, 10^6 MSC)	Results
Kebriaei et al., 2009 [75]	De novo acute GVHD	32	52 (34–67)	Grade II: 21 Grade III: 8 Grade IV: 3	2 or 8 M/kg at 1 and 3 days after GVHD + steroids; mismatched MSC	94 % initial response (77 % CR, 16 % PR), 61 % sustained CR; No difference b/w high/low MSC dose; No infusional toxicity
Ringden et al., 2006 [76]	Steroid refractory, acute GVHD	8	56 (8–61)	Grade II: 2 Grade III: 5 Grade IV:1	1 M/kg (range 0.7–9) 1 dose (range 1–2); mismatched/sib/haplo MSC	6/8 CR; 5/8 survival No infusional toxicity
Fang et al., 2007 [77]	Steroid refractory, acute GVHD	6	39 (22–49)	Grade III: 2 Grade IV: 4	1 M/kg adipose MSC 1 dose (range 1–2); mismatched/haplo MSC	5/6 CR, 4/6 survival No infusional toxicity
Le Blanc et al., 2008 [78]	Steroid refractory, acute GVHD	55	22 (.5–64)	Grade II: 5 Grade III: 25 Grade IV: 25	1.4 M/kg (range 0.4–9) 1 dose (range 1–5); mismatched/sib/haplo MSC	71 % initial response (55 % CR, 16 % PR); 2-year survival benefit for CR, 53 % vs. 16 % No infusional toxicity
von Bonin et al., 2009 [79]	Steroid refractory, acute GVHD	13	58 (21–69)	Grade III: 2 Grade IV: 11	0.9 M/kg (range 0.6–1.1) 2 doses; mismatched MSC expanded in platelet lysate-containing medium	2/13 CR, 5/13 mixed response 4/13 alive at median 257 days No infusional toxicity

(continued)

Table 1 (continued)

Study	Indication	N	Med age (range)	Episodes GVHD	MSC regimen (M, 10^6 MSC)	Results
Kurtzberg et al., 2010 [80]	Steroid refractory, acute GVHD	59	8	Grade II: 6 Grade III: 20 Grade IV: 33	2 M/kg; 8 bi-weekly × 4 week, followed by 4 infusions weekly × 4 if PR; mismatched MSC	64 % ORR at day 28 76 % vs. 9 % survival at day 100 No infusional toxicity
Martin et al., 2010 [81]	Steroid refractory, acute GVHD	244	44 MSC; 40 control	MSC vs. control II: 38 vs. 23 III: 88 vs. 50 IV: 47 vs. 14	2 M/kg; 8 bi-weekly × 4 weeks, followed by 4 infusions weekly × 4 if PR; mismatched MSC	No diff in durable CR b/w MSC and control; liver, GI GVHD significantly better response 81 % vs. 68 %, p = .035
Perez-Simon et al., 2011 [82]	Steroid refractory acute GVHD	10	43 (21–63)	Grade II: 2 Grade III/IV: 8	0.7–2.9 M/kg; 1–4 infusions Platelet lysate medium	1/10 CR, 6/10 PR; 1 disease relapse
Martino Introna et al., 2012 [83]	Steroid refractory acute GVHD, chronic GVHD	47	26 (1–67)	Grade II: 15 Grade III: 23 Grade IV: 9 10 cases cGVHD	1.5 M/kg 3 dose (range 1–8) Platelet lysate medium	64 % ORR 28 % CR, 36 % PR No acute or late toxicity GVHD biomarkers
te Boome et al., 2012 [84]	Steroid refractory acute GVHD	42	52 (1–66)	Grade II: 11 Grade III: 28 Grade IV: 3	2 M/kg, 3 dose (range 1–4) Platelet lysate medium	56 % CR OS significantly better for responders; GVHD biomarkers
Ringden et al., 2013 [12]	Steroid refractory, acute GVHD	9	57 (10 months–64)	Grade III: 5 Grade IV: 4	0.9–2.8 M/kg; 15 infusions Placental-derived MSC	2/8 CR, 4/8 PR; one patient seizure
Muller et al., 2008 [85]	Acute/chronic GVHD	5	14 (4–17)	2 acute 3 chronic	0.4–3 M/kg 1 dose (range 1–3) Haplo MSC	2/2 acute GVHD did not progress 1/3 chronic GVHD improvement No infusional toxicity

Zhou et al., 2010 [86]	Chronic GVHD	4	42 (38–43)	4 sclerodermal chronic GVHD	1–2 × 10^7 MSC/kg; 4–8 intra-BM injections per patient; mismatched MSC	4/4 significant improvement No infusional toxicity
Weng et al., 2010 [87]	Chronic GVHD	19	29 (18–39)	Extensive chronic GVHD	0.6 M/kg (range 0.2–1.4) 1–5 doses	74 % ORR (4CR, 10PR), five patients able to stop immunosuppression, 2 year OS 78 %; *In vivo* immunomodulation noted in responsive group

Adapted from Kebriaei and Robinson [88]

median dose 1.4×10^6 MSC/kg, was infused into 27 patients and the remaining patients were treated with two or more infusions. The initial response rate was 70 % (30 CR, 9 PR). The median time from infusion of first MSC to response was 18 days, with 19 patients having sustained CR at 6 weeks following infusion. Patients with CR had a statistically significant lower treatment-related mortality (TRM) at 1-year and overall survival (OS) at 2-years following transplant compared to non-responders, 37 % *versus* 72 %, $p=0.002$, and 53 % *versus* 16 %, $p=0.018$, respectively [78]. The infusions were well tolerated with no significant adverse events noted. There was a trend for better response in the pediatric patients, with a statistically better survival. The majority of patients received third-party donors, precluding an efficacy analysis for MSC match grade. Of note, the authors recently published a retrospective comparison of a subset of adult patients treated on this study (n = 15) and compared them to a group of similar patients who did not receive MSC (n = 13) during the period 2002–2006. Within the limitations of a retrospective comparison, they found no difference in survival or non-relapse mortality (NRM) between the two groups [90], underscoring the necessity of prospective, randomized studies to better study the effect of MSC.

Positive findings were also noted in a large, pediatric phase II study of third-party, mismatched MSC (Prochymal®, Osiris therapeutics, Inc.) for steroid-refractory acute G*v*HD. Fifty-nine patients, with median age 8 years received 8 bi-weekly infusions of 2×10^6 MSC/kg for 4 weeks, followed by additional four infusions weekly as "maintenance" in patients with partial remission (PR). The majority of patients had severe gut and liver G*v*HD, and had progressed through a median of 3.2 lines of prior therapy for G*v*HD. At day 28, the overall response rate, defined as organ improvement of at least one stage without worsening in any other, was 64 %. These patients had a significantly better survival at 100 days compared to patients who did not achieve a response at day 28, 76 % *versus* 9 % [80].

Lucchini and colleagues treated ten patients with steroid refractory G*v*HD with 2–5 infusions of third-party, bone marrow-derived MSC cultured in platelet lysate at a median dose of 1.5×10^6/kg [91]. They noted an overall response rate of 70 % with a CR rate of 30 %. Importantly, they measured two validated serum biomarkers of G*v*HD in ten patients (IL-2Ra and TNFRI) [92], and demonstrated changes in biomarker levels after treatment with MSC that correlated with response [93]. Furthermore, they investigated the effect of MSC infusions on circulating lymphocytes in the peripheral blood, and demonstrated a shift in the ratios of pro-inflammatory T cells subsets (TH1 and TH17) and anti-inflammatory T_{Reg} population to support an anti-inflammatory environment after MSC infusions [94–96]. However, as the authors correctly comment, these changes are not specific to MSC and any change, such as infection, can skew this balance. The multicenter Italian trial was recently updated at the American Society of Hematology meeting in December 2012. Introna and colleagues reported the outcomes for 47 patients treated to date, and saw similar response rates, and similar patterns in G*v*HD biomarker profiles [83]. Prior to the MSC infusion, patients received only steroids (n = 22), pentostatin (n = 12), or other immunosuppressant (n = 13). Notably, the patients who received a prior dose of pentostatin had a better response and survival than the other two groups, highlighting the difficulty of ascribing response specifically to the MSC [83].

Te Boome and colleagues also reported on the use of third-party, bone marrow-derived MSC expanded in platelet lysate [84]. Fifty patients received a median of 3 MSC infusions for steroid-refractory G*v*HD. A complete remission (CR) was noted in 56 % of patients, with responding patients having a significantly better overall survival. The authors also noted corroborating changes in G*v*HD biomarkers following response to MSC [84]. Similar positive findings were noted in the remaining, smaller patient series using MSC for steroid refractory G*v*HD, with transiently higher response rates noted than compared with historical data, and no significant adverse effects noted with MSC infusion (Table 1) [77, 79, 85, 97].

However, in contrast to these findings, preliminary results from the only completed, randomized, Phase III clinical trial for steroid-refractory acute G*v*HD failed to show any benefit to MSC administration, as compared to placebo (September 8, 2009, http://investor.osiris.com/releasedetail.cfm?releaseid=407404). Bi-weekly MSC (Prochymal®) were administered for 4 weeks with individual dosing at 2×10^6 MSC/kg. The trial did not reach the primary endpoint of durable CR≧28 days. However, select patients with either steroid-refractory liver or gastrointestinal G*v*HD were reported to have significantly improved response rates (81 % *versus* 68 %, p=.035), but they were not part of the intent-to-treat population. No significant difference was noted with respect to toxicity or recurrent malignancy rates [81]. To better study the effect of MSC specifically in liver and gut G*v*HD, the Dutch cooperative study group HOVON has recently initiated a multicenter, randomized study of MSC or placebo plus mycophenolate mofetil for patients with early, steroid refractory liver and/or gut G*v*HD. Patients receive two doses of 2×10^6 MSC/kg and G*v*HD biomarkers will also be assessed (clinical trials.gov).

Only one large, multicenter trial has been reported for *de novo* acute G*v*HD. Thirty-two adult patients received two treatments of MSC (Prochymal®) at a dose of either 2 or 8×10^6 MSC/kg in combination with a conventional corticosteroid regimen. Thirty-one patients were evaluable; the initial response rate was 94 % (24 CR, 5 PR) with 79 % of CR patients maintaining CR for at least 90 days. No infusional toxicities or ectopic tissue formation were noted. Although the trial was not designed to detect a difference between the two different MSC doses, no obvious differences were observed [75]. Again, however, preliminary results from the multicenter, randomized, Phase III clinical trials for *de novo* acute GVHD failed to show a benefit for MSC, as compared to placebo (September 8, 2009, http://investor.osiris.com/releasedetail.cfm?releaseid=407404).

Chronic GvHD

Investigations with MSC for the treatment of chronic G*v*HD are more limited, and difficult to interpret. The studies are essentially descriptive, with less than ten patients treated in each series with a variety of treatment schedules. The largest series has been reported by Lucchini and colleagues. Six pediatric patients received unrelated, HLA-disparate, bone marrow-derived MSC, expanded in platelet-lysate medium, for chronic GvHD [91]. The median dose was 1.2×10^6/kg (range 0.7–2.8×10^6/kg) infused as a single dose at a median of 5 months following HSCT (range 1–10 months) in all patients except for one, who received four doses of MSC at 0.7×10^6/kg. The majority of patients had chronic GvHD of skin and mucosa. A transient benefit was noted, with three partial and one complete response that subsequently reflared [91]. In contrast, significant improvement following repeated intra-bone marrow injections of MSC at a dose of $1–2 \times 10^7$ MSC/kg was reported for patients with sclerodermal-type chronic GvHD [86]. Four patients with extensive skin changes and ulcers received a range of 4–8 MSC injections from the same MSC donor; correspondingly, the ratio of helper T-lymphocyte (Th)1 cells to Th2 cells dramatically reversed following MSC infusion, with an increase in Th1 and decrease in Th2 cells [86].

Conclusion

In conclusion, the clinical experience with MSC for the treatment of GVHD is intriguing, but incomplete. More than 15 years have passed since MSC were first infused in a clinical trial, and the data thus far suggest that their use is safe [98]. A meta-analysis of 8 randomized clinical trials that enrolled 321 clinical trial participants did not detect any association between MSC and acute infusional toxicity, organ system complications, infections, death, or malignancy [99]. However, their efficacy for GvHD remains to be established. First, data from preclinical murine models of GvHD suggest that the schedule and dose of MSC administration are critical to its effects, but the optimal treatment schedule remains to be defined for patients. The bulk of the studies, mainly in the steroid refractory setting would suggest that repeated MSC infusions around a dose of 2×10^6/kg are needed. However, when to optimally initiate infusions remains an important question. Furthermore, animal models suggest that a local inflammatory milieu is needed to activate MSC, and thus the route of administration may be important. Currently, all trials, except for a pilot trial in chronic GvHD which used in intra-bone marrow injections, infuse MSC systemically, and the bulk of MSC get trapped in the lungs. Improved labeling and *in vivo* imaging techniques feasible for human clinical trials would yield significant data regarding MSC trafficking and homing. Second, consensus regarding the optimal culture and manufacturing conditions has not been established. In efforts to minimize auto-antibody formation against MSC cultured in fetal bovine serum (FBS), recent trials have used platelet lysate, which in phase 2 studies appears effective. Finally, vigilant long-term follow-up of patients on current clinical trials is necessary to determine if any late toxicity are associated with MSC use. Carefully planned studies with well-defined endpoints are necessary to the continued understand of the therapeutic potential of MSC for GvHD.

References

1. Jiang Y, Jahagirdar BN, Reinhardt RL, Schwartz RE, Keene CD, Ortiz-Gonzalez XR, et al. Pluripotency of mesenchymal stem cells derived from adult marrow. Nature. 2002;418(6893): 41–9. Epub 2002/06/22.
2. Dominici M, Le Blanc K, Mueller I, Slaper-Cortenbach I, Marini F, Krause D, et al. Minimal criteria for defining multipotent mesenchymal stromal cells. The International Society for Cellular Therapy position statement. Cytotherapy. 2006;8(4):315–7.
3. Horwitz EM, Le Blanc K, Dominici M, Mueller I, Slaper-Cortenbach I, Marini FC, et al. Clarification of the nomenclature for MSC: The International Society for Cellular Therapy position statement. Cytotherapy. 2005;7(5):393–5.
4. Pittenger MF, Mackay AM, Beck SC, Jaiswal RK, Douglas R, Mosca JD, et al. Multilineage potential of adult human mesenchymal stem cells. Science. 1999;284(5411):143–7.
5. Robinson SN, Ng J, Niu T, Yang H, McMannis JD, Karandish S, et al. Superior ex vivo cord blood expansion following co-culture with bone marrow-derived mesenchymal stem cells. Bone Marrow Transplant. 2006;37(4):359–66. Epub 2006/01/10.
6. McNiece I, Harrington J, Turney J, Kellner J, Shpall EJ. Ex vivo expansion of cord blood mononuclear cells on mesenchymal stem cells. Cytotherapy. 2004;6(4):311–7. Epub 2005/09/09.
7. Najar M, Raicevic G, Boufker HI, Fayyad Kazan H, De Bruyn C, Meuleman N, et al. Mesenchymal stromal cells use PGE2 to modulate activation and proliferation of lymphocyte subsets: combined comparison of adipose tissue, Wharton's jelly and bone marrow sources. Cell Immunol. 2010;264(2):171–9. Epub 2010/07/14.
8. Najar M, Raicevic G, Boufker HI, Fayyad-Kazan H, De Bruyn C, Meuleman N, et al. Adipose-tissue-derived and Wharton's jelly-derived mesenchymal stromal cells suppress lymphocyte responses by secreting leukemia inhibitory factor. Tissue Eng Part A. 2010;16(11):3537–46. Epub 2010/07/06.
9. Majore I, Moretti P, Stahl F, Hass R, Kasper C. Growth and differentiation properties of mesenchymal stromal cell populations derived from whole human umbilical cord. Stem Cell Rev. 2011;7(1):17–31. Epub 2010/07/03.
10. Tong CK, Vellasamy S, Tan BC, Abdullah M, Vidyadaran S, Seow HF, et al. Generation of mesenchymal stem cell from human umbilical cord tissue using combination of enzymatic and mechanical disassociation method. Cell Biol Int. 2011;33(3):211–6. Epub 2010/10/16.
11. Zhang Y, Li C, Jiang X, Zhang S, Wu Y, Liu B, et al. Human placenta-derived mesenchymal progenitor cells support culture expansion of long-term culture-initiating cells from cord blood CD34+ cells. Exp Hematol. 2004;32(7):657–64. Epub 2004/07/13.
12. Ringden O, Erkers T, Nava S, Uzunel M, Iwarsson E, Conrad R, et al. Fetal membrane cells for treatment of steroid-refractory acute graft-versus-host disease. Stem Cells. 2013;31(3): 592–601. Epub 2013/01/12.
13. Le Blanc K, Tammik L, Sundberg B, Haynesworth SE, Ringden O. Mesenchymal stem cells inhibit and stimulate mixed lymphocyte cultures and mitogenic responses independently of the major histocompatibility complex. Scand J Immunol. 2003;57(1):11–20. Epub 2003/01/25.
14. Di Nicola M, Carlo-Stella C, Magni M, Milanesi M, Longoni PD, Matteucci P, et al. Human bone marrow stromal cells suppress T-lymphocyte proliferation induced by cellular or nonspecific mitogenic stimuli. Blood. 2002;99(10):3838–43. Epub 2002/05/03.
15. Aggarwal S, Pittenger MF. Human mesenchymal stem cells modulate allogeneic immune cell responses. Blood. 2005;105(4):1815–22. Epub 2004/10/21.
16. Uccelli A, Moretta L, Pistoia V. Mesenchymal stem cells in health and disease. Nat Rev Immunol. 2008;8(9):726–36. Epub 2009/01/28.
17. Ramasamy R, Tong CK, Seow HF, Vidyadaran S, Dazzi F. The immunosuppressive effects of human bone marrow-derived mesenchymal stem cells target T cell proliferation but not its effector function. Cell Immunol. 2008;251(2):131–6. Epub 2008/05/27.
18. Deans RJ, Moseley AB. Mesenchymal stem cells: biology and potential clinical uses. Exp Hematol. 2000;28(8):875–84. Epub 2000/09/16.

19. Toma C, Pittenger MF, Cahill KS, Byrne BJ, Kessler PD. Human mesenchymal stem cells differentiate to a cardiomyocyte phenotype in the adult murine heart. Circulation. 2002;105(1):93–8. Epub 2002/01/05.
20. Horwitz EM, Gordon PL, Koo WK, Marx JC, Neel MD, McNall RY, et al. Isolated allogeneic bone marrow-derived mesenchymal cells engraft and stimulate growth in children with osteogenesis imperfecta: implications for cell therapy of bone. Proc Natl Acad Sci USA. 2002;99(13):8932–7. Epub 2002/06/27.
21. Marks DI, Lush R, Cavenagh J, Milligan DW, Schey S, Parker A, et al. The toxicity and efficacy of donor lymphocyte infusions given after reduced-intensity conditioning allogeneic stem cell transplantation. Blood. 2002;100(9):3108–14. Epub 2002/10/18.
22. Roddie C, Peggs KS. Donor lymphocyte infusion following allogeneic hematopoietic stem cell transplantation. Expert Opin Biol Ther. 2011;11(4):473–87. Epub 2011/01/29.
23. Ferrara JL, Levine JE, Reddy P, Holler E. Graft-versus-host disease. Lancet. 2009;373(9674): 1550–61. Epub 2009/03/14.
24. Appelbaum FR. Haematopoietic cell transplantation as immunotherapy. Nature. 2001;411 (6835):385–9. Epub 2001/05/18.
25. Przepiorka D, Weisdorf D, Martin P, Klingemann HG, Beatty P, Hows J, et al. 1994 consensus conference on acute GVHD grading. Bone Marrow Transplant. 1995;15(6):825–8. Epub 1995/06/01.
26. Sullivan KM, Shulman HM, Storb R, Weiden PL, Witherspoon RP, McDonald GB, et al. Chronic graft-versus-host disease in 52 patients: adverse natural course and successful treatment with combination immunosuppression. Blood. 1981;57(2):267–76. Epub 1981/02/01.
27. Filipovich AH, Weisdorf D, Pavletic S, Socie G, Wingard JR, Lee SJ, et al. National Institutes of Health consensus development project on criteria for clinical trials in chronic graft-versus-host disease: I. Diagnosis and staging working group report. Biol Blood Marrow Transplant. 2005;11(12):945–56. Epub 2005/12/13.
28. Storb R, Deeg HJ, Whitehead J, Appelbaum F, Beatty P, Bensinger W, et al. Methotrexate and cyclosporine compared with cyclosporine alone for prophylaxis of acute graft versus host disease after marrow transplantation for leukemia. N Engl J Med. 1986;314(12):729–35.
29. Storb R, Deeg HJ, Fisher L, Appelbaum F, Buckner CD, Bensinger W, et al. Cyclosporine v methotrexate for graft-v-host disease prevention in patients given marrow grafts for leukemia: long-term follow-up of three controlled trials. Blood. 1988;71(2):293–8.
30. Ratanatharathorn V, Nash RA, Przepiorka D, Devine SM, Klein JL, Weisdorf D, et al. Phase III study comparing methotrexate and tacrolimus (prograf, FK506) with methotrexate and cyclosporine for graft-versus-host disease prophylaxis after HLA-identical sibling bone marrow transplantation. Blood. 1998;92(7):2303–14.
31. Nash RA, Antin JH, Karanes C, Fay JW, Avalos BR, Yeager AM, et al. Phase 3 study comparing methotrexate and tacrolimus with methotrexate and cyclosporine for prophylaxis of acute graft-versus-host disease after marrow transplantation from unrelated donors. Blood. 2000; 96(6):2062–8.
32. Bacigalupo A. Management of acute graft-versus-host disease. Br J Haematol. 2007;137(2): 87–98.
33. Martin PJ, Schoch G, Fisher L, Byers V, Anasetti C, Appelbaum FR, et al. A retrospective analysis of therapy for acute graft-versus-host disease: initial treatment. Blood. 1990;76(8):1464–72.
34. MacMillan ML, Weisdorf DJ, Wagner JE, DeFor TE, Burns LJ, Ramsay NK, et al. Response of 443 patients to steroids as primary therapy for acute graft-versus-host disease: comparison of grading systems. Biol Blood Marrow Transplant. 2002;8(7):387–94.
35. Alousi AM, Weisdorf DJ, Logan BR, Bolanos-Meade J, Carter S, Difronzo N, et al. Etanercept, mycophenolate, denileukin, or pentostatin plus corticosteroids for acute graft-versus-host disease: a randomized phase 2 trial from the Blood and Marrow Transplant Clinical Trials Network. Blood. 2009;114(3):511–7. Epub 2009/05/16.
36. Van Lint MT, Uderzo C, Locasciulli A, Majolino I, Scime R, Locatelli F, et al. Early treatment of acute graft-versus-host disease with high- or low-dose 6-methylprednisolone: a multicenter randomized trial from the Italian Group for Bone Marrow Transplantation. Blood. 1998;92(7): 2288–93. Epub 1998/09/25.

37. Cragg L, Blazar BR, Defor T, Kolatker N, Miller W, Kersey J, et al. A randomized trial comparing prednisone with antithymocyte globulin/prednisone as an initial systemic therapy for moderately severe acute graft-versus-host disease. Biol Blood Marrow Transplant. 2000; 6(4A):441–7.
38. Lee SJ, Zahrieh D, Agura E, MacMillan ML, Maziarz RT, McCarthy Jr PL, et al. Effect of upfront daclizumab when combined with steroids for the treatment of acute graft-versus-host disease: results of a randomized trial. Blood. 2004;104(5):1559–64.
39. Deeg HJ. How I, treat refractory acute GVHD. Blood. 2007;109(10):4119–26. Epub 2007/01/20.
40. English K. Mechanisms of mesenchymal stromal cell immunomodulation. Immunol Cell Biol. 2013;91(1):19–26. Epub 2012/10/24.
41. Waterman RS, Tomchuck SL, Henkle SL, Betancourt AM. A new mesenchymal stem cell (MSC) paradigm: polarization into a pro-inflammatory MSC1 or an Immunosuppressive MSC2 phenotype. PLoS One. 2010;5(4):e10088. Epub 2010/05/04.
42. Opitz CA, Litzenburger UM, Lutz C, Lanz TV, Tritschler I, Koppel A, et al. Toll-like receptor engagement enhances the immunosuppressive properties of human bone marrow-derived mesenchymal stem cells by inducing indoleamine-2,3-dioxygenase-1 via interferon-beta and protein kinase R. Stem Cells. 2009;27(4):909–19. Epub 2009/04/09.
43. Liotta F, Angeli R, Cosmi L, Fili L, Manuelli C, Frosali F, et al. Toll-like receptors 3 and 4 are expressed by human bone marrow-derived mesenchymal stem cells and can inhibit their T-cell modulatory activity by impairing Notch signaling. Stem Cells. 2008;26(1):279–89. Epub 2007/10/27.
44. Romieu-Mourez R, Francois M, Boivin MN, Bouchentouf M, Spaner DE, Galipeau J. Cytokine modulation of TLR expression and activation in mesenchymal stromal cells leads to a proinflammatory phenotype. J Immunol. 2009;182(12):7963–73. Epub 2009/06/06.
45. Shlomchik WD. Graft-versus-host disease. Nat Rev Immunol. 2007;7(5):340–52. Epub 2007/04/18.
46. Spaggiari GM, Capobianco A, Becchetti S, Mingari MC, Moretta L. Mesenchymal stem cell-natural killer cell interactions: evidence that activated NK cells are capable of killing MSCs, whereas MSCs can inhibit IL-2-induced NK-cell proliferation. Blood. 2006;107(4):1484–90. Epub 2005/10/22.
47. Keating A. How do mesenchymal stromal cells suppress T cells? Cell Stem Cell. 2008;2(2): 106–8. Epub 2008/03/29.
48. Prevosto C, Zancolli M, Canevali P, Zocchi MR, Poggi A. Generation of CD4+ or CD8+ regulatory T cells upon mesenchymal stem cell-lymphocyte interaction. Haematologica. 2007;92(7): 881–8. Epub 2007/07/04.
49. Sundin M, Barrett AJ, Ringden O, Uzunel M, Lonnies H, Dackland AL, et al. HSCT recipients have specific tolerance to MSC but not to the MSC donor. J Immunother. 2009;32(7):755–64. Epub 2009/06/30.
50. Prockop DJ, Olson SD. Clinical trials with adult stem/progenitor cells for tissue repair: let's not overlook some essential precautions. Blood. 2007;109(8):3147–51. Epub 2006/12/16.
51. Ball L, Bredius R, Lankester A, Schweizer J, van den Heuvel-Eibrink M, Escher H, et al. Third party mesenchymal stromal cell infusions fail to induce tissue repair despite successful control of severe grade IV acute graft-versus-host disease in a child with juvenile myelo-monocytic leukemia. Leukemia. 2008;22(6):1256–7. Epub 2007/11/02.
52. Lee MJ, Kim J, Kim MY, Bae YS, Ryu SH, Lee TG, et al. Proteomic analysis of tumor necrosis factor-alpha-induced secretome of human adipose tissue-derived mesenchymal stem cells. J Proteome Res. 2010;9(4):1754–62. Epub 2010/02/27.
53. Roddy GW, Oh JY, Lee RH, Bartosh TJ, Ylostalo J, Coble K, et al. Action at a distance: systemically administered adult stem/progenitor cells (MSCs) reduce inflammatory damage to the cornea without engraftment and primarily by secretion of TNF-alpha stimulated gene/protein 6. Stem Cells. 2011;29(10):1572–9. Epub 2011/08/13.
54. Honczarenko M, Le Y, Swierkowski M, Ghiran I, Glodek AM, Silberstein LE. Human bone marrow stromal cells express a distinct set of biologically functional chemokine receptors. Stem Cells. 2006;24(4):1030–41. Epub 2005/10/29.

55. Miura M, Miura Y, Padilla-Nash HM, Molinolo AA, Fu B, Patel V, et al. Accumulated chromosomal instability in murine bone marrow mesenchymal stem cells leads to malignant transformation. Stem Cells. 2006;24(4):1095–103. Epub 2005/11/12.
56. Meisel R, Brockers S, Heseler K, Degistirici O, Bulle H, Woite C, et al. Human but not murine multipotent mesenchymal stromal cells exhibit broad-spectrum antimicrobial effector function mediated by indoleamine 2,3-dioxygenase. Leukemia. 2011;25(4):648–54. Epub 2011/01/19.
57. Sudres M, Norol F, Trenado A, Gregoire S, Charlotte F, Levacher B, et al. Bone marrow mesenchymal stem cells suppress lymphocyte proliferation in vitro but fail to prevent graft-versus-host disease in mice. J Immunol. 2006;176(12):7761–7. Epub 2006/06/06.
58. Chung NG, Jeong DC, Park SJ, Choi BO, Cho B, Kim HK, et al. Cotransplantation of marrow stromal cells may prevent lethal graft-versus-host disease in major histocompatibility complex mismatched murine hematopoietic stem cell transplantation. Int J Hematol. 2004;80(4):370–6. Epub 2004/12/24.
59. Tisato V, Naresh K, Girdlestone J, Navarrete C, Dazzi F. Mesenchymal stem cells of cord blood origin are effective at preventing but not treating graft-versus-host disease. Leukemia. 2007;21(9):1992–9. Epub 2007/07/13.
60. Min CK, Kim BG, Park G, Cho B, Oh IH. IL-10-transduced bone marrow mesenchymal stem cells can attenuate the severity of acute graft-versus-host disease after experimental allogeneic stem cell transplantation. Bone Marrow Transplant. 2007;39(10):637–45. Epub 2007/03/21.
61. Polchert D, Sobinsky J, Douglas G, Kidd M, Moadsiri A, Reina E, et al. IFN-gamma activation of mesenchymal stem cells for treatment and prevention of graft versus host disease. Eur J Immunol. 2008;38(6):1745–55. Epub 2008/05/22.
62. Croitoru-Lamoury J, Lamoury FM, Zaunders JJ, Veas LA, Brew BJ. Human mesenchymal stem cells constitutively express chemokines and chemokine receptors that can be upregulated by cytokines, IFN-beta, and Copaxone. J Interferon Cytokine Res. 2007;27(1):53–64. Epub 2007/02/03.
63. New JY, Li B, Koh WP, Ng HK, Tan SY, Yap EH, et al. T cell infiltration and chemokine expression: relevance to the disease localization in murine graft-versus-host disease. Bone Marrow Transplant. 2002;29(12):979–86. Epub 2002/07/05.
64. Wang L, Li Y, Chen X, Chen J, Gautam SC, Xu Y, et al. MCP-1, MIP-1, IL-8 and ischemic cerebral tissue enhance human bone marrow stromal cell migration in interface culture. Hematology. 2002;7(2):113–7. Epub 2002/08/21.
65. Joo SY, Cho KA, Jung YJ, Kim HS, Park SY, Choi YB, et al. Mesenchymal stromal cells inhibit graft-versus-host disease of mice in a dose-dependent manner. Cytotherapy. 2010;12(3):361–70. Epub 2010/01/19.
66. Ye Z, Wang Y, Xie HY, Zheng SS. Immunosuppressive effects of rat mesenchymal stem cells: involvement of CD4+ CD25+ regulatory T cells. Hepatobiliary Pancreat Dis Int. 2008;7(6): 608–14. Epub 2008/12/17.
67. Gonzalez-Rey E, Gonzalez MA, Varela N, O'Valle F, Hernandez-Cortes P, Rico L, et al. Human adipose-derived mesenchymal stem cells reduce inflammatory and T cell responses and induce regulatory T cells in vitro in rheumatoid arthritis. Ann Rheum Dis. 2010;69(1):241–8. Epub 2009/01/07.
68. Casiraghi F, Azzollini N, Cassis P, Imberti B, Morigi M, Cugini D, et al. Pretransplant infusion of mesenchymal stem cells prolongs the survival of a semiallogeneic heart transplant through the generation of regulatory T cells. J Immunol. 2008;181(6):3933–46. Epub 2008/09/05.
69. Di Ianni M, Del Papa B, De Ioanni M, Moretti L, Bonifacio E, Cecchini D, et al. Mesenchymal cells recruit and regulate T regulatory cells. Exp Hematol. 2008;36(3):309–18. Epub 2008/02/19.
70. Christensen ME, Turner BE, Sinfield LJ, Kollar K, Cullup H, Waterhouse NJ, et al. Mesenchymal stromal cells transiently alter the inflammatory milieu post-transplant to delay graft-versus-host disease. Haematologica. 2010;95(12):2102–10. Epub 2010/08/31.
71. Joo SY, Cho KA, Jung YJ, Kim HS, Park SY, Choi YB, et al. Bioimaging to monitor the in vivo distribution of infused mesenchymal stem cells: in a mouse model of graft versus host disease. Cell Biol Int. 2011;35(4):417–21. Epub 2010/11/11.
72. Gao J, Dennis JE, Muzic RF, Lundberg M, Caplan AI. The dynamic in vivo distribution of bone marrow-derived mesenchymal stem cells after infusion. Cells Tissues Organs. 2001;169(1): 12–20. Epub 2001/05/08.

73. Lee RH, Pulin AA, Seo MJ, Kota DJ, Ylostalo J, Larson BL, et al. Intravenous hMSCs improve myocardial infarction in mice because cells embolized in lung are activated to secrete the anti-inflammatory protein TSG-6. Cell Stem Cell. 2009;5(1):54–63. Epub 2009/07/03.
74. Yanez R, Lamana ML, Garcia-Castro J, Colmenero I, Ramirez M, Bueren JA. Adipose tissue-derived mesenchymal stem cells have in vivo immunosuppressive properties applicable for the control of the graft-versus-host disease. Stem Cells. 2006;24(11):2582–91. Epub 2006/07/29.
75. Kebriaei P, Isola L, Bahceci E, Holland K, Rowley S, McGuirk J, et al. Adult human mesenchymal stem cells added to corticosteroid therapy for the treatment of acute graft-versus-host disease. Biol Blood Marrow Transplant. 2009;15(7):804–11. Epub 2009/06/23.
76. Ringden O, Uzunel M, Rasmusson I, Remberger M, Sundberg B, Lonnies H, et al. Mesenchymal stem cells for treatment of therapy-resistant graft-versus-host disease. Transplantation. 2006;81(10):1390–7.
77. Fang B, Song Y, Zhao RC, Han Q, Lin Q. Using human adipose tissue-derived mesenchymal stem cells as salvage therapy for hepatic graft-versus-host disease resembling acute hepatitis. Transplant Proc. 2007;39(5):1710–3. Epub 2007/06/21.
78. Le Blanc K, Frassoni F, Ball L, Locatelli F, Roelofs H, Lewis I, et al. Mesenchymal stem cells for treatment of steroid-resistant, severe, acute graft-versus-host disease: a phase II study. Lancet. 2008;371(9624):1579–86. Epub 2008/05/13.
79. von Bonin M, Stolzel F, Goedecke A, Richter K, Wuschek N, Holig K, et al. Treatment of refractory acute GVHD with third-party MSC expanded in platelet lysate-containing medium. Bone Marrow Transplant. 2009;43(3):245–51. Epub 2008/09/30.
80. Kurtzberg J, Prasad V, Grimley MS, Horn B, Carpenter PA, Jacobsen D, Prockop S, editors. Allogeneic human mesenchymal stem cell therapy (prochymal) as a rescue agent for severe treatment resistant GVHD in pediatric patients. Orlando: American Society of Blood and Marrow Transplantation; 2009.
81. Martin BJ, Uberti JP, Soiffer RJ, Klingemann H, Waller EK, Daly AS, Herrmann RP, Kebriaei P, editors. Prochymal improves response rates in patients with steroid-refractory acute graft versus host disease (SR-GVHD) involving the liver and gut: results of a randomized, placebo-controlled, multicenter phase III trial in GVHD. Orlando: American Society of Bone Marrow Transplantation; 2010.
82. Perez-Simon JA, Lopez-Villar O, Andreu EJ, Rifon J, Muntion S, Campelo MD, et al. Mesenchymal stem cells expanded in vitro with human serum for the treatment of acute and chronic graft-versus-host disease: results of a phase I/II clinical trial. Haematologica. 2011;96(7):1072–6. Epub 2011/03/12.
83. Martino Introna GL, Dander E, Rovelli A, Balduzzi A, Daniela L, Pavan F, Masciochi F, Algarotti A, Micò C, Grassi A, Cavattoni I, Deola S, Gaipa G, Belotti D, Perseghin P, Parma M, Pogliani EM, Golay J, Gotti E, Capelli C, Cortelazzo S, D'Amico G, Biondi A, Rambaldi A, Biagi E, editors. Safe and effective treatment of graft versus host disease with platelet lysate-expanded human mesenchymal stromal cells: a phase 1 study on 47 adult and pediatric patients. Atlanta: American Society of Hematology; 2012.
84. te Boome CMM L, Lindemans CA, Van der Wagen LE, Cuijpers M, Slaper-Cortenbach I, Rozemuller H, Petersen EJ, Spierings E, Bierings M, Boelens JJ, Wulffraat N, Kuball JH, editors. Treatment of steroid resistant grade II to IV acute GVHD by infusion of mesenchymal stroma cells expanded with human plasma and platelet lysate – a phase I/II study. Atlanta: American Society of Hematology; 2012.
85. Muller I, Kordowich S, Holzwarth C, Isensee G, Lang P, Neunhoeffer F, et al. Application of multipotent mesenchymal stromal cells in pediatric patients following allogeneic stem cell transplantation. Blood Cells Mol Dis. 2008;40(1):25–32. Epub 2007/09/18.
86. Zhou H, Guo M, Bian C, Sun Z, Yang Z, Zeng Y, et al. Efficacy of bone marrow-derived mesenchymal stem cells in the treatment of sclerodermatous chronic graft-versus-host disease: clinical report. Biol Blood Marrow Transplant. 2010;16(3):403–12. Epub 2009/11/21.
87. Weng JY, Du X, Geng SX, Peng YW, Wang Z, Lu ZS, et al. Mesenchymal stem cell as salvage treatment for refractory chronic GVHD. Bone Marrow Transplant. 2010;45(12):1732–40. Epub 2010/09/08.

88. Kebriaei P, Robinson S. Treatment of graft-versus-host-disease with mesenchymal stromal cells. Cytotherapy. 2011;13(3):262–8.
89. Le Blanc K, Rasmusson I, Sundberg B, Gotherstrom C, Hassan M, Uzunel M, et al. Treatment of severe acute graft-versus-host disease with third party haploidentical mesenchymal stem cells. Lancet. 2004;363(9419):1439–41.
90. Remberger M, Ringden O. Treatment of severe acute graft-versus-host disease with mesenchymal stromal cells: a comparison with non-MSC treated patients. Int J Hematol. 2012;96(6): 822–4. Epub 2012/11/13.
91. Lucchini G, Introna M, Dander E, Rovelli A, Balduzzi A, Bonanomi S, et al. Platelet-lysate-expanded mesenchymal stromal cells as a salvage therapy for severe resistant graft-versus-host disease in a pediatric population. Biol Blood Marrow Transplant. 2010;16(9):1293–301. Epub 2010/03/31.
92. Paczesny S, Krijanovski OI, Braun TM, Choi SW, Clouthier SG, Kuick R, et al. A biomarker panel for acute graft-versus-host disease. Blood. 2009;113(2):273–8. Epub 2008/10/04.
93. Dander E, Lucchini G, Vinci P, Introna M, Masciocchi F, Perseghin P, et al. Mesenchymal stromal cells for the treatment of graft-versus-host disease: understanding the in vivo biological effect through patient immune monitoring. Leukemia. 2012;26(7):1681–4. Epub 2012/02/01.
94. Dander E, Balduzzi A, Zappa G, Lucchini G, Perseghin P, Andre V, et al. Interleukin-17-producing T-helper cells as new potential player mediating graft-versus-host disease in patients undergoing allogeneic stem-cell transplantation. Transplantation. 2009;88(11):1261–72. Epub 2009/12/10.
95. Zhao Q, Xiao X, Wu Y, Wei Y, Zhu LY, Zhou J, et al. Interleukin-17-educated monocytes suppress cytotoxic T-cell function through B7-H1 in hepatocellular carcinoma patients. Eur J Immunol. 2011;41(8):2314–22. Epub 2011/06/16.
96. Zhao XY, Xu LL, Lu SY, Huang XJ. IL-17-producing T cells contribute to acute graft-versus-host disease in patients undergoing unmanipulated blood and marrow transplantation. Eur J Immunol. 2011;41(2):514–26. Epub 2011/01/27.
97. Prasad VK, Lucas KG, Kleiner GI, Talano JA, Jacobsohn D, Broadwater G, et al. Efficacy and safety of ex-vivo cultured adult human mesenchymal stem cells (prochymal(TM)) in pediatric patients with severe refractory acute graft-versus-host disease in a compassionate use study. Biol Blood Marrow Transplant. 2011;17(4):534–41. Epub 2010/05/12.
98. Lazarus HM, Haynesworth SE, Gerson SL, Rosenthal NS, Caplan AI. Ex vivo expansion and subsequent infusion of human bone marrow-derived stromal progenitor cells (mesenchymal progenitor cells): implications for therapeutic use. Bone Marrow Transplant. 1995;16(4): 557–64. Epub 1995/10/01.
99. Lalu MM, McIntyre L, Pugliese C, Fergusson D, Winston BW, Marshall JC, et al. Safety of cell therapy with mesenchymal stromal cells (SafeCell): a systematic review and meta-analysis of clinical trials. PLoS One. 2012;7(10):e47559. Epub 2012/11/08.

Mesenchymal Stem Cells for Liver Disease

Feng-chun Zhang

Abstract Mesenchymal stem cells (MSCs), represent an attractive tool for the establishment of a successful stem-cell-based therapy of liver diseases. A number of different mechanisms contribute to the therapeutic effects exerted by MSCs, since these cells can differentiate into functional hepatic cells and can also produce a series of growth factors and cytokines which are able to suppress inflammatory responses, reduce hepatocyte apoptosis, mitigate liver fibrosis, and improve hepatic function. To date, the infusion of MSCs has shown encouraging results in the treatment of fulminant hepatic failure and in end-stage liver disease in experimental studies. However, the application of MSCs in clinical trials was severely hampered by some issues under debate. The short-term efficacy of MSCs was favorable, but long-term outcomes remain controversial. In addition, MSCs therapy in clinic is also severely hampered by impossible monitoring of transplanted cells in patients and lack of standardized clinical protocols. Further studies should be taken to achieve a better understanding of the potential benefits and risks of MSCs in clinic.

Keywords MSC • Growth factors and cytokines • Liver disease • Fibrosis

Introduction

Liver diseases are major causes of human mortality and morbidity worldwide. Acute injury or chronic liver damage can be caused by different diseases, such as virus infection, drug-induced hepatitis, autoimmune hepatitis etc. Liver cirrhosis is generally considered to be an irreversible process and represents a frequent cause of

---

F.-c. Zhang (✉)
Department of Rheumatology, Peking Union Medical College Hospital,
Chinese Academy of Medical Sciences, Beijing 100032, China
e-mail: zhangfccra@yahoo.com.cn

R.C. Zhao (ed.), *Essentials of Mesenchymal Stem Cell Biology and Its Clinical Translation*, DOI 10.1007/978-94-007-6716-4_12,

death worldwide. Liver transplantation offers a definitive cure for many end stage liver disease. However, the complex invasive procedure and paucity of donor livergraft organs limits clinical applicability. To compensate for this, development of new regenerative therapies for liver cirrhosis is an urgent task. Mesenchymal stem cell transplantation is a new way for liver regenerative therapies. Embryonic and adult stem cells can be differentiated into hepatocytes [1]. Mesenchymal stem cells (MSCs) being the most potent component of bone marrow cells in hepatic differentiation. Several animal studies have revealed the benefit of MSCs for the treatment of liver failure [2–4]. Some clinical trials have been performed in patients with end-stage liver disease caused by hepatitis B, hepatitis C, and alcoholic fibrosis, which with satisfactory tolerability and clinically relevant effects. We present an overview of the current status of clinical trials and future prospects for liver regeneration therapies using MSCs.

MSCs Based Treatments of Liver Diseases

In 1999, Petersen et al. [5] first showed that liver stem cells might be derived from bone marrow(BM), in a rat model of liver injury, and it was suggested that BM could contribute to the mature hepatocyte population. Several studies have demonstrated that MSCs can differentiate in vitro along the hepatogenic lineage [2–4]. Studies on animal models reported the beneficial effect of MSCs in promoting hepatic tissue regeneration. Sun et al. [2] established a PBC animal model by injecting C57BL/6 mice with polyinosinic-polycytidylic acid sodium (polyI:C) to investigate the therapeutic effect of bone marrow-derived mesenchymal stem cells (BM-MSC) on this model. After 6 weeks of MSCs infusion, serum aminotransferase and autoimmune antibodies declined, and histological examination by hematoxylin and eosin staining showed significant amelioration of monocytes infiltration around bile ducts of mice treated with BM-MSC. Kuo et al. [6] have showed that both mesenchymal stem cell-derived hepatocytes and mesenchymal stem cells, transplanted by either intrasplenic or intravenous route, engrafted recipient liver, differentiated into functional hepatocytes, and rescued liver failure. Intravenous transplantation was more effective in rescuing liver failure than intrasplenic transplantation. Patrizia Burra et al. [7] evaluate the therapeutic potential of mesenchymal stem cells from human umbilical cord (UCMSCs), a readily available source of mesenchymal stem cells, in the CCl4-induced acute liver injury model. The experiment show that UCMSCs can be reliably isolated, have hepatogenic properties and following systemic administration are able to accelerate the resolution of an acute liver injury without any differentiation and manipulation. These features make UCMSCs strong candidates for future application in regenerative medicine for human acute liver disease.

To date, only a few clinical trials have been performed in patients with end-stage liver disease caused by hepatitis B, hepatitis C, and alcoholic fibrosis. Eight patients (four hepatitis B, one hepatitis C, one alcoholic, and two cryptogenic) with end-stage liver disease having Model for End-Stage Liver Disease score>or =10 were included. Autologous MSCs were taken from iliac crest. Approximately, 30–50 million MSCs were proliferated and injected into peripheral or the portal vein. Liver function and clinical features were evaluated at baseline and 1, 2, 4, 8, and 24 weeks after injection. Treatment was well-tolerated by all patients. Liver function improved as verified by the Model for End-Stage Liver Disease score [8]. Peng [9] investigated the short-term efficacy and long-term prognosis of patients with liver failure who are caused by hepatitis B after a single transplantation with autologous marrow mesenchymal stem cells (MMSCs). A total of 527 inpatients with liver failure caused by hepatitis B were recruited and received the same medical treatments, among whom 53 patients underwent a single transplantation with autologous MMSCs. A total of 105 patients matched. The MMSC suspension was slowly transfused into the liver through the proper hepatic artery. The success rate of transplantation was 100 %, without serious side effects or complications. There were no dramatic differences in incidence of hepatocellular carcinoma or mortality between the two groups. Levels of ALB, TBIL, and PT and MELD score of patients in the transplantation group were markedly improved from 2 to 3 weeks after transplantation. Short-term efficacy was favorable, but long-term outcomes were not markedly improved. Fu-Sheng Wang [15] have recruited 30 patients with hepatitis B. They are all have liver cirrhosis and receiving UC-MSC transfusion. At the same time, 15 patients received saline as the control. There was a significant reduction in the volume of ascites in patients treated with UC-MSC transfusion compared with controls. UC-MSC therapy also significantly improved liver function, as indicated by the increase of serum albumin levels, decrease in total serum bilirubin levels, and decrease in the sodium model for end-stage liver disease scores. The results of these studies have shown that MSCs injection can be used for the treatment of end-stage liver diseases, with satisfactory tolerability and clinically relevant effects. In these clinical trials, the success rate of transplantation was 100 %, without serious side effects or complications. Liver function including MELD score and some serum biochemical index (Album, bilirubin, prothrombin time etc.) improved. Short-term efficacy was favorable, but long-term outcomes were not markedly improved. Most of the patients have autologous MSCs transplantation, UC-MSCs was also be used. They have similar number of cells. Nonetheless, no studies have provided definitive evidence that MSCs have a capability to differentiate into functional hepatocytes in vivo. Improvements could be attributed to the secretion of growth factors by MSCs rather than to their transdifferentiation into hepatocytes. MSCs demonstrate multi-potentiality and can promote liver regeneration, secrete cytokines/growth factors, inhibit inflammation, inhibit activation of liver astrocytes, block the production of extracellular matrix (ECM), and facilitate the degradation of excessive ECM, leading to improvement of chronic hepatitis B, impediment of liver fibrosis, and repair of injured liver tissues. Great progress has been made in the treatment of liver diseases with the use of autologous MSC transplantation and has included basic research and clinical studies. But there are still a number of problems requiring resolution in clinical practice, the mechanism of MSCs in vivo needs to be studied. In the same time, the long-term prognosis of these patients still need to be observed.

There are three ways of human MSCs transplantation for human liver disease: hepatic artery, peripheral veins and portal vein. Different number of MSCs transfusion is needed, but no exact evidence showed that which way may achieve more favorable outcomes. Peripheral veins injection is simple and easy to operate, which need more cells and some animal experiments show that most of the MSCs gathered in lung tissue and may lead to pulmonary infarction. The portal vein is usually accessed percutaneously with ultrasound or X ray guidance by puncturing through the liver or via a transjugular route from the neck. Arteries are accessed via femoral artery puncture through the groin. The technical challenges include bleeding, especially patients with liver disease often have abnormal clotting function. These operations are not as easy as peripheral veins injection and brings the risks of puncture risk. After infusion into the portal circulation, the transplanted cells move along the portal tracts into the sinusoids and engraft into the hepatic cords by squeezing out between the endothelial lining cells. It is seems an ideal way, but how efficient this occurs remains unknown.

The mechanisms of MSCs based treatments of liver diseases are still unknown. Experiments in rats and humans confirmed the differentiation potential of MSCs in vitro. However, it is different to confirm in vivo especial in human bodies. It is well known that chemokines are released after tissue damage and that migratory direction follows the chemokine density gradient. In this regard, it has been recently demonstrated that MSCs express chemokine receptors and ligands that are involved in leukocyte migration during inflammation, including the stromal-derived factor-1 (SDF-1) chemokinereceptor (CXCR4) that stimulates the recruitment of progenitor cells to the site of tissue injury [10]. MSCs also express several adhesion molecules that respond to SDF-1 [11], as well as chemokines [12]. Hence, the increase of inflammatory chemokine concentration at the site of inflammation is a key mediator of MSC trafficking to the site of injury. Additional, many integrins, selectins, and chemokine receptors involved in the tethering, rolling, adhesion, and transmigration of leukocytes have also been reported to be expressed on MSCs. Several animal studies and clinical trials have demonstrated that MSCs have the potential to reverse the fibrotic process by inhibiting collagen deposition and the production of transforming growth factor-β1 [13]. Despite these encouraging results, antifibrotic effctors of MSCs is still debated, more research about this is needed in the future.

MSCs transplantation for the 55 patients with steroid-refractory acute Severe graft-versus-host disease (GVHD) was used between 2001 and 2007. Thirty patients had a complete response and nine showed improvement. Additional studies, mostly in animal models, are being conducted in solid organ transplantation, such as: heart, renal, liver and skin. MSCs have also emerged as promising candidate cells for immune- modulation therapy, especially in the setting of liver transplantation, given their ability to interact at various levels with the immune system [14]. Pan MX investigated that autologous MSCs infusion through the portal vein during allogenic living donor liver transplantation can prolong the survival of the recipient dogs [15]. Rejection of the transplanted liver is also a big problem in clinical practices, MSCs transfusion may be a new solution way.

MSCs in Liver Disease: Risks and Benefits

Liver cirrhosis is generally considered an irreversible process and represents a frequent cause of death worldwide. MSCs for liver disease means new hope. MSCs express few HLA class I and no HLA class II molecules, allowing them to evade allogeneic immune response. This is the so-called "immunoprivilege," an interesting feature in MSC biology, which makes these cells extremely suitable for both autologous and allogeneic transplantation. As previously discussed, all animal experiments and humans have showed good tolerance. No acute complication has happened. MSCs are considered a potentially relevant therapeutic tool for the treatment of liver diseases, given their high degree of plasticity andimmune-modulatory properties. The use of MSCs in the hepatologic clinical practice is hampered by the inability to monitor the transplanted cells within the patients and by the lack of standardized clinical protocols. Moreover, the antifibrotic effect of MSCs is still debated, as MSCs could also potentially differentiate into fibrogenic cells. Whether can cause the increased risk of tumor is also unknown.

Much of our knowledge of MSC is derived from in vitro experiments. Larger clinical trials have just started. However, further studies in vitro as well in vivo are needed to achieve a better understanding of the potential benefits and risks of MSCs therapeutic use in clinical settings (Table 1).

Table 1 Clinical evidence of MSCs transplantation

Type of infused cells and injection method	Number of cells	Improvement after infusion	Number and etiology of patinets	Reference
ABMMSCs Peripheral vein	31.73 × 10(6)	SF-36 questionnaire and MELD score(2/4)	4 end-stage liver disease	[13]
ABMMSCs Peripheral or the portal vein	30–50 × 10(6)	MELD score and serum Alb	4 hepatitis B, 1 hepatitis C, 1 alcoholic, and 2 cryptogenic	[8]
ABMMSCs Hepatic artery	3.4 ± 3.8 × 10(8)	MELD score, serum Alb, ALT, Tbil, and PT	53 hepatitis B/105 control	[9]
UC-MSCs Peripheral vein	2 × 10(7)	MELD score, Tbil, Alb, PT, and kindey function	30 hepatitis B/15 control	[16]

ABMSCs autologous bone marrow mesenchymal stem cells, *UC-MSCs* umbilical cord-derived mesenchymal stem cells, *Tbil* bilirubin, *Alb* album, *ALT* aminotransferase, *PT* prothrombin time

Perspective

MSCs-based therapy provides hope for treating patients with end-stage liver diseases. Though MSCs-based therapy achieved some progresses in clinic, the wide application of MSCs-based therapy in clinic should solve the following issues. First, the mechanisms of the treatment of MSCs-based therapy transplation in liver diseases and differentiation of stem cells and immune regulation should be elucidated. Specially, in the experiment in vivo, the survival, proliferation and differentiation of transplanted stem cells in hosts should be determined. Second, the transfusion route of stem cells, number of cells, and choice of treatment time have critical role in the therapeutic efficacy of liver diseases. However, these issues have no unified conclusion and further study should be taken to confirm them. Third, multi-center, double-blind and controlled clinical trial research should be designed rationally, and specified types of cells should be used to treat special liver diseases to enhance comparability. At the same time, the follow-up time should be prolonged, to further recognize the possible existed side effects and complications, such as tumor occurrence. Fourth, regulations and industry access system of clinical application of stem cells should be formulated. The formulation could guarantee that the clinical research of human stem cell transplantation could follow corresponding norms and clinical application, could ensure that the human stem cells could benefit patients, and improve medical plight of treatment of end-stage liver diseases.

References

1. Lagasse E, Connors H, Al-Dhalimy M, Reitsma M, Dohse M, Osborne L, et al. Purified hematopoietic stem cells can differentiate into hepatocytes in vivo. Nat Med. 2000;6(11):1229–34. doi:10.1038/81326.
2. Wang D, Zhang H, Liang J, Gu Z, Ma X, Huang J, et al. Effect of allogeneic bone marrow-derived mesenchymal stem cells transplantation in a polyI: C-induced primary biliary cirrhosis mouse model. Clin Exp Med. 2011;11(1):25–32. doi:10.1007/s10238-010-0105-6.
3. Hardjo M, Miyazaki M, Sakaguchi M, Masaka T, Ibrahim S, Kataoka K, et al. Suppression of carbon tetrachloride-induced liver fibrosis by transplantation of a clonal mesenchymal stem cell line derived from rat bone marrow. Cell Transplant. 2009;18(1):89–99.
4. Ju S, Teng GJ, Lu H, Jin J, Zhang Y, Zhang A, et al. In vivo differentiation of magnetically labeled mesenchymal stem cells into hepatocytes for cell therapy to repair damaged liver. Invest Radiol. 2010;45(10):625–33. doi:10.1097/RLI.0b013e3181ed55f4.
5. Petersen BE, Bowen WC, Patrene KD, Mars WM, Sullivan AK, Murase N, et al. Bone marrow as a potential source of hepatic oval cells. Science. 1999;284(5417):1168–70.
6. Kuo TK, Hung SP, Chuang CH, Chen CT, Shih YR, Fang SC, et al. Stem cell therapy for liver disease: parameters governing the success of using bone marrow mesenchymal stem cells. Gastroenterology. 2008;134(7):2111–21. doi:10.1053/j.gastro.2008.03.015.
7. Burra P, Arcidiacono D, Bizzaro D, Chioato T, Di Liddo R, Banerjee A, et al. Systemic administration of a novel human umbilical cord mesenchymal stem cells population accelerates the resolution of acute liver injury. BMC Gastroenterol. 2012;12:88. doi:10.1186/1471-230X-12-88.
8. Kharaziha P, Hellstrom PM, Noorinayer B, Farzaneh F, Aghajani K, Jafari F, et al. Improvement of liver function in liver cirrhosis patients after autologous mesenchymal stem

cell injection: a phase I-II clinical trial. Eur J Gastroenterol Hepatol. 2009;21(10):1199–205. doi:10.1097/MEG.0b013e32832a1f6c.

9. Peng L, Xie DY, Lin BL, Liu J, Zhu HP, Xie C, et al. Autologous bone marrow mesenchymal stem cell transplantation in liver failure patients caused by hepatitis B: short-term and long-term outcomes. Hepatology. 2011;54(3):820–8. doi:10.1002/hep.24434.
10. Du Z, Wei C, Yan J, Han B, Zhang M, Liu Y, et al. Mesenchymal stem cells overexpressing CXCR4 improve early liver regeneration of small-for-size liver grafts. Liver Transpl. 2013;19(2):215–25.
11. Lapidot T. Mechanism of human stem cell migration and repopulation of NOD/SCID and B2mnull NOD/SCID mice. The role of SDF-1/CXCR4 interactions. Ann N Y Acad Sci. 2001;938:83–95.
12. Ringe J, Strassburg S, Neumann K, Endres M, Notter M, Burmester GR, et al. Towards in situ tissue repair: human mesenchymal stem cells express chemokine receptors CXCR1, CXCR2 and CCR2, and migrate upon stimulation with CXCL8 but not CCL2. J Cell Biochem. 2007;101(1):135–46. doi:10.1002/jcb.21172.
13. Mohamadnejad M, Alimoghaddam K, Mohyeddin-Bonab M, Bagheri M, Bashtar M, Ghanaati H, et al. Phase 1 trial of autologous bone marrow mesenchymal stem cell transplantation in patients with decompensated liver cirrhosis. Arch Iran Med. 2007;10(4):459–66. doi:07104/AIM.008.
14. Zhang X, Jiao C, Zhao S. Role of mesenchymal stem cells in immunological rejection of organ transplantation. Stem Cell Rev. 2009;5(4):402–9. doi:10.1007/s12015-009-9076-y.
15. Pan MX, Hou WL, Zhang QJ, Gong DH, Cheng Y, Jian GD, et al. Infusion of autologous mesenchymal stem cells prolongs the survival of dogs receiving living donor liver transplantation. Nan Fang Yi Ke Da Xue Xue Bao. 2009;29(9):1783–6.
16. Zhang Z, Lin H, Shi M, Xu R, Fu J, Lv J, et al. Human umbilical cord mesenchymal stem cells improve liver function and ascites in decompensated liver cirrhosis patients. J Gastroenterol Hepatol. 2012;27 Suppl 2:112–20. doi:10.1111/j.1440-1746.2011.07024.x.

Mesenchymal Stem Cells for Bone Repair

Hongwei Ouyang, Xiaohui Zou, Boon Chin Heng, and Weiliang Shen

Abstract Some complex clinical conditions require regeneration of large bone sections, such as in the case of massive traumatic injury and atrophic nonunions. To date, regeneration of bone tissue has been an important focus for biological repair in the field of regenerative medicine. Bone tissue engineering – an interdisciplinary field at the intersection of engineering, biology and medicine, has emerged as one of the most promising approaches for developing biological bone substitutes. In the past half century, the potential use of MSCs in bone tissue engineering has been increasingly recognized because of its unique characteristics. Many studies have utilized bone marrow derived mesenchymal stem cells (BM-MSCs) as seed cells for cell sheets and biomaterial scaffolds. In the past decade, many researchers have sought to exploit their potential as seed cells for diverse applications in regen erative medicine. The availability of robust clinical and scientific data supports the use of MSC in bone repair.

Keywords MSC • Bone repair • Tissue engineering • Regenerative medicine

Mesenchymal Stem Cells (MSCs) in Bone Tissue Engineering

Bone is a highly vascularized connective tissue, which possesses intrinsic regenerative capacity in response to injury throughout adult life. However, in the case of critical sized defects, the self-regenerative capacity of bone tissue is interrupted by rapid in-growth of fibrous tissues. In addition, some complex clinical conditions

---

H. Ouyang (✉) • X. Zou • B.C. Heng • W. Shen
Center for Stem Cell and Tissue Engineering, School of Medicine,
Zhejiang University, 388 Yu Hang Tang Road,
Mailbox #39, Hangzhou 310058, China
e-mail: hwoy@zju.edu.cn

R.C. Zhao (ed.), *Essentials of Mesenchymal Stem Cell Biology and Its Clinical Translation*, DOI 10.1007/978-94-007-6716-4_13,

require regeneration of large bone sections, such as in the case of massive traumatic injury and atrophic nonunions [1]. To date, regeneration of bone tissue has been an important focus for biological repair in the field of regenerative medicine. Bone tissue engineering (BTE) – an interdisciplinary field at the intersection of medicine, biology and engineering, has emerged as one of the most promising approaches for developing biological bone substitutes.

Various different adult and embryonic sources of stem cells have been evaluated for BTE application, particularly MSCs, induced Pluripotent Stem Cells (iPSCs) and Embryonic Stem Cells (ESCs). Although iPSCs and ESCs are pluripotent [2, 3], their potential clinical applications have been hindered by several drawbacks, such as the potential immunogenicity of differentiated hESC [4, 5] and the formation of teratomas in vivo [6]. In the past half century, the potential use of MSCs in BTE has been increasingly recognized due to their unique characteristics. For example, MSCs are relatively abundant and easy to isolate [7]; the osteogenic differentiation pathway of MSCs is well defined [8]; MSCs are non-immunogenic in either the undifferentiated or differentiated states [9]; and MSCs remains stable after extensive ex vivo expansion [10]. In summary, BMSCs are readily accessible, have less ethical challenges and display lower risk of tumorigenesis after transplantation, as compared to pluripotent stem cells such as embryonic stem cells and induced pluripotent stem cells.

BM-MSCs have been utilized as seed cells together with biomaterial scaffolds in vitro. Fischer et al. have reported that the osteogenic potential of BM-MSCs was affected by the size and microporosity of hydroxyapatite (HA) microparticles. HA has been clinically utilized as an injectable bone filler. This study indicated that the BM-MSCs-HA particle composite could be utilized as an injectable formulation of tissue-engineered bone [11]. In another approach for bone tissue regeneration, BMP-2 gene was transfected into BM-MSCs by an adenoviral vector. This resulted in the endogenous secretion of BMP-2, which is known to enhance osteogenic differentiation. Ostogenesis was evaluated after these cells were cultured on 3D porous silk fibroin sponge-like scaffolds. There was substantially enhanced osteogenic differentiation as compared to BM-MSCs cultured in the presence of exogenous BMP-2 in an osteogenic culture medium. This study suggests the potential use of gene transfected MSCs together with 3D scaffolds to achieve enhanced bone tissue regeneration [12]. Recently, extensive research into the *in vitro* osteogenic differentiation of BM-MSCs for the fabrication of bone tissue-like structures with various 3D supporting matrices as well as efficient 3D culture systems such as bioreactors have been conducted. Zhang et al. [13] have established a scaffold fabrication platform and advanced biaxial rotating bioreactor system to generate an effective TE bone graft suitable for clinical application. Elcin et al. reported culturing BM-MSCs on 3D porous scaffolds within a slow turning lateral vessel rotating culture system. This system enabled the formation of a well-developed 3D bone-like tissue structural construct, suggesting the feasibility of bioreactor culture systems for *in vitro* 3D bone tissue engineering.

In vivo studies with small animal models have demonstrated that MSC transplantation enhanced bone formation, as well as the stiffness of the regenerated structures [14]. Large animal models have subsequently been used to evaluate the feasibility of a stem cell based approach for bone regeneration and repair [15, 16]. Hosseinkhani et al. reported that the combination of a perfusion culture system with BM-MSCs seeded onto a 3D collagen sponge and self-assembled peptide-amphiphile nanofiber hybrid scaffold significantly enhanced the *in vitro* and *in vivo* osteogenic differentiation of BM-MSCs, as compared to static culture on tissue culture plates [17]. Ouyang et al. [18] investigated the effect of nHAp/CTS together with BM-MSCs on the repair of bone defects *in vivo*. They found that nHAp/CTS supported the adhesion and proliferation of BMSCs and promoted their osteogenic differentiation by activating BMP signaling. In conjunction with the utilization of MSC as a cell source for BTE applications, Zhang et al. [13] have established a scaffold fabrication platform and advanced biaxial rotating bioreactor system for generating an effective TE bone graft suitable for clinical application. By applying these converging technologies of stem cells, scaffold design and manufacture, and bioreactors, the generation of tissue engineered bone with high clinical efficacy can be achieved. The tissue engineered bone derived from MSCs demonstrated its efficacy for healing in a clinically relevant load-bearing critical sized defect rat model, where the influence of stress loading, tissue injury and consequential inflammatory response all play a part in the healing process. Twelve weeks after implantation, tissue engineered bone derived from MSCs displayed 2.1 fold more new bone tissue formation, with greater compactness, and a 9.8 fold increase in stiffness compared to implanted acellular scaffolds. Furthermore, the implantation of tissue engineered bone resulted in a 3.9 fold increase in the vasculature network within the defect area, suggesting a potential role for MSCs in promoting neo-vasculogenesis [19]. Due to the encouraging experimental data obtained, a first-in-Man phase I clinical trial is being planned.

These advances in coupling MSCs to different types of porous scaffolds have achieved very exciting and promising results for bone tissue engineering [20–22]. Currently, cell-seeding techniques employ either cell-gel composites or cell suspensions to deliver cells into the scaffold [23, 24]. However, there are some disadvantages in these current techniques, such as the low efficiency of cell attachment to dense fibrous matrix or scaffolds and the weak mechanical strength of gel systems. These disadvantages make it very difficult to seed a large number of cells on dense tissue grafts. The limitations of current technology platforms prohibit the application of stem cells to improve the efficacy of large tissue grafts for tissue repair. Several researchers have attempted to fabricate cell sheets with vascular smooth muscle cells, dermal fibroblasts, and keratinocytes for tissue engineering blood vessels or skin without biomaterials and scaffolds [25]. This gave some cues on how to overcome the challenges faced with applying current cell seeding technology to large dense tissue grafts. Ouyang et al. [26] successfully incorporated about 15×10^6 MSCs into 1 cm long non-porous cortical bone graft by assembling MSC sheets with dense allografts. This overcame the inherent disadvantages of current cell seeding techniques in bone tissue engineering. It is known that the key factor in tissue repair is the availability of appropriate cells. The presence of cells is crucial because of their proliferation potential, cell-to-cell signaling, biomolecule synthesis, and deposition of extracellular matrix, all of which strongly influence the nature of skeletal tissue formation. It seems clear that a threshold quantity of cells is required at the repair sites for normal neo-tissue formation. With the MSC sheets, an excess of the threshold quantity of seed cells can be delivered to the repair site. The cartilage-like layers around de-mineralized bone graft observed *in vitro* suggested that MSC sheets could act as fresh periosteum. More importantly, they displayed abundant periosteal bone formation on MSC-allografts, which was much less than the allograft alone. The difference in bone formation between the groups was very similar to what had been reported in previous studies. For example, Guldberg et al. [27] transplanted live bone grafts harvested from Rosa 26A mice into murine segmental femoral bone defects and showed that approximately 70 % of osteogenesis on the graft was attributed to the expansion and differentiation of donor periosteal progenitor cells. Arasi et al. compared the use of fresh bone graft and frozen allograft for bone regeneration in a rabbit model and demonstrated that free autogenous periosteum wrapped around frozen allografts dramatically stimulated bone healing and repair [28]. Moreover, Gray and Elves using a free subcutaneous isograft demonstrated that marrow cells and osteocytes made little or no contribution to early osteogenesis, while Kadiyala et al. showed that live cells in periosteum and endosteum plus stromal cells are responsible for 90 % of early osteogenesis [29, 30]. Also, it is known from clinical experience that preservation of the periosteum or use of a periosteum tube graft significantly improves cortical bone graft incorporation and remodeling [31, 32]. Based on these promising findings, it appears that MSC sheets can act as fresh tissue-engineered periosteum to repopulate allografts and enhance bone formation.

From the Laboratory to Clinic

The use of MSCs for treatment of orthopedic diseases and bone defects is currently the most common clinical application of MSCs. A rise in the number of registered clinical trials and published clinical reports with MSCs is evident, with 89.4 % of registered trials and 78.0 % of published reports performed after 2005 being related to MSCs, thus reflecting growing interest in the therapeutic applications of these cells [13]. With regards to the clinical use of MSCs to promote bone formation and regeneration, Quarto et al. [33] reported the first-in-man clinical trial of MSC-based BTE therapy in 2001. In the past decade, many researchers have sought to exploit the potential of MSCs as seed cells for diverse clinical applications. For example, Quarto et al. utilized BM-MSCs cultured in vitro for 3 weeks and seeded onto macroporous HA scaffolds to treat nonunions. After 7 months the three treated patients displayed good integration of the implants. Angiographic evaluation after 7 years showed vascularization of the grafted zone, which is vital for the survival and future stability of the graft [33]. Marcacci et al. [34] investigated the clinical use of culture-expanded osteoprogenitor cells in combination with porous hydroxyapatite (HA) scaffolds for the treatment of 4 patients. Good integration of the implants with pre-existing bone was maintained during the follow-up period, and no major adverse reactions were observed. All patients showed recovery of limb function and at the last followup (6–7 years after surgery) good integration of the implants was maintained. Morishita et al. [35] used a HA scaffold to differentiate MSCs into osteoblasts ex vivo, for healing bone defect in a patient after curettage of a tumour. This illustrated that tissue-engineered osteogenic cell-scaffold composites may be an alternative to autologous bone grafting.

In addition, MSCs have also been used together with growth factors (GFs) and other biologics. For example, Warnke et al. [36] described a new bone-muscle-flap technique for the treatment of a mandibular defect. The scaffold was placed in an external titanium mesh loaded with HA blocks coated with BMP-7 and seeded with MSCs, followed by implantation into the latissimus dorsi for growth of blood vessels and bone. After 7 weeks, the constructed mandible was removed and fixed to the stumps of the original mandible. The patient regained full function of his jaw 4 weeks after the operation. Kitoh et al. [37] conducted a number of clinical trials whereby the integrated use of autologous MSC with platelet rich plasma accelerated new bone formation, which reduces the complication rate during distraction osteogenesis. These findings have several clinical implications for BTE whereby MSCs are used for the fabrication of tissue-engineered bone *in vitro*.

Conclusion

After a few decades of intensive research, the field of bone tissue engineering is finally heading towards translation into clinical therapy. The availability of robust clinical and scientific data supports the use of MSC for bone repair. Further studies are also required to address a range of pertinent technical issues including, the growth of MSCs in either 2D or 3D culture conditions, the use of serum replacement or serum free media, and the need to supplement with different cytokine combinations, with all new developments being assessed in pre-clinical studies with large animal models that mimic human disease and condition [38].

References

1. Dimitriou R, Jones E, McGonagle D, Giannoudis PV. Bone regeneration: current concepts and future directions. BMC Med. 2011;9:66.
2. Robertson NJ, Brook FA, Gardner RL, Cobbold SP, Waldmann H, Fairchild PJ. Embryonic stem cell-derived tissues are immunogenic but their inherent immune privilege promotes the induction of tolerance. Proc Natl Acad Sci USA. 2007;104:20920–5.
3. Kao CL, Tai LK, Chiou SH, Chen YJ, Lee KH, Chou SJ, Chang YL, Chang CM, Chen SJ, Ku HH, Li HY. Resveratrol promotes osteogenic differentiation and protects against dexamethasone damage in murine induced pluripotent stem cells. Stem Cells Dev. 2010;19:247–58.
4. Drukker M, Benvenisty N. The immunogenicity of human embryonic stem-derived cells. Trends Biotechnol. 2004;22:136–41.

5. Zhao T, Zhang ZN, Rong Z, Xu Y. Immunogenicity of induced pluripotent stem cells. Nature. 2011;474:212–5.
6. Fong CY, Gauthaman K, Bongso A. Teratomas from pluripotent stem cells: a clinical hurdle. J Cell Biochem. 2010;111:769–81.
7. Simmons PJ, Torok-Storb B. Identification of stromal cell precursors in human bone marrow by a novel monoclonal antibody, STRO-1. Blood. 1991;78:55–62.
8. Kahn A, Gibbons R, Perkins S, Gazit D. Age-related bone loss. A hypothesis and initial assessment in mice. Clin Orthop Relat Res. 1995;313:69–75.
9. Le BK, Tammik L, Sundberg B, Haynesworth SE, Ringden O. Mesenchymal stem cells inhibit and stimulate mixed lymphocyte cultures and mitogenic responses independently of the major histocompatibility complex. Scand J Immunol. 2003;57:11–20.
10. Sawada R, Ito T, Tsuchiya T. Changes in expression of genes related to cell proliferation in human mesenchymal stem cells during in vitro culture in comparison with cancer cells. J Artif Organs. 2006;9:179–84.
11. Fischer EM, Layrolle P, Van Blitterswijk CA, De Bruijn JD. Bone formation by mesenchymal progenitor cells cultured on dense and microporous hydroxyapatite particles. Tissue Eng. 2003;9:1179–88.
12. Meinel L, Hofmann S, Betz O, Fajardo R, Merkle HP, Langer R, Evans CH, Vunjak-Novakovic G, Kaplan DL. Osteogenesis by human mesenchymal stem cells cultured on silk biomaterials: comparison of adenovirus mediated gene transfer and protein delivery of BMP-2. Biomaterials. 2006;27:4993–5002.
13. Zhang ZY, Teoh SH, Hui JH, Fisk NM, Choolani M, Chan JK. The potential of human fetal mesenchymal stem cells for off-the-shelf bone tissue engineering application. Biomaterials. 2012;33:2656–72.
14. Tatebe M, Nakamura R, Kagami H, Okada K, Ueda M. Differentiation of transplanted mesenchymal stem cells in a large osteochondral defect in rabbit. Cytotherapy. 2005;7:520–30.
15. Brodke D, Pedrozo HA, Kapur TA, Attawia M, Kraus KH, Holy CE, Kadiyala S, Bruder SP. Bone grafts prepared with selective cell retention technology heal canine segmental defects as effectively as autograft. J Orthop Res. 2006;24:857–66.
16. Viateau V, Guillemin G, Bousson V, Oudina K, Hannouche D, Sedel L, Logeart-Avramoglou D, Petite H. Long-bone critical-size defects treated with tissue-engineered grafts: a study on sheep. J Orthop Res. 2007;25:741–9.
17. Hosseinkhani H, Hosseinkhani M, Tian F, Kobayashi H, Tabata Y. Ectopic bone formation in collagen sponge self-assembled peptide-amphiphile nanofibers hybrid scaffold in a perfusion culture bioreactor. Biomaterials. 2006;27:5089–98.
18. Peng H, Yin Z, Liu H, Chen X, Feng B, Yuan H, Su B, Ouyang H, Zhang Y. Electrospun biomimetic scaffold of hydroxyapatite/chitosan supports enhanced osteogenic differentiation of mMSCs. Nanotechnology. 2012;23:485102.
19. Zhang ZY, Teoh SH, Chong MS, Lee ES, Tan LG, Mattar CN, Fisk NM, Choolani M, Chan J. Neo-vascularization and bone formation mediated by fetal mesenchymal stem cell tissue-engineered bone grafts in critical-size femoral defects. Biomaterials. 2010;31:608–20.
20. Arinzeh TL, Tran T, Mcalary J, Daculsi G. A comparative study of biphasic calcium phosphate ceramics for human mesenchymal stem-cell-induced bone formation. Biomaterials. 2005;26:3631–8.
21. Mygind T, Stiehler M, Baatrup A, Li H, Zou X, Flyvbjerg A, Kassem M, Bunger C. Mesenchymal stem cell ingrowth and differentiation on coralline hydroxyapatite scaffolds. Biomaterials. 2007;28:1036–47.
22. Xin X, Hussain M, Mao JJ. Continuing differentiation of human mesenchymal stem cells and induced chondrogenic and osteogenic lineages in electrospun PLGA nanofiber scaffold. Biomaterials. 2007;28:316–25.
23. Colton CK. Implantable biohybrid artificial organs. Cell Transplant. 1995;4:415–36.
24. Zund G, Ye Q, Hoerstrup SP, Schoeberlein A, Schmid AC, Grunenfelder J, Vogt P, Turina M. Tissue engineering in cardiovascular surgery: MTT, a rapid and reliable quantitative method to assess the optimal human cell seeding on polymeric meshes. Eur J Cardiothorac Surg. 1999;15:519–24.

25. Pouliot R, Larouche D, Auger FA, Juhasz J, Xu W, Li H, Germain L. Reconstructed human skin produced in vitro and grafted on athymic mice. Transplantation. 2002;73:1751–7.
26. Zou XH, Cai HX, Yin Z, Chen X, Jiang YZ, Hu H, Ouyang HW. A novel strategy incorporated the power of mesenchymal stem cells to allografts for segmental bone tissue engineering. Cell Transplant. 2009;18:433–41.
27. Zhang X, Xie C, Lin AS, Ito H, Awad H, Lieberman JR, Rubery PT, Schwarz EM, O'Keefe RJ, Guldberg RE. Periosteal progenitor cell fate in segmental cortical bone graft transplantations: implications for functional tissue engineering. J Bone Miner Res. 2005;20:2124–37.
28. Karaoglu S, Baktir A, Kabak S, Arasi H. Experimental repair of segmental bone defects in rabbits by demineralized allograft covered by free autogenous periosteum. Injury. 2002;33:679–83.
29. Gray JC, Elves MW. Donor cells' contribution to osteogenesis in experimental cancellous bone grafts. Clin Orthop Relat Res. 1982;163:261–71.
30. Kadiyala S, Young RG, Thiede MA, Bruder SP. Culture expanded canine mesenchymal stem cells possess osteochondrogenic potential in vivo and in vitro. Cell Transplant. 1997; 6:125–34.
31. Emery SE, Brazinski MS, Koka A, Bensusan JS, Stevenson S. The biological and biomechanical effects of irradiation on anterior spinal bone grafts in a canine model. J Bone Joint Surg Am. 1994;76:540–8.
32. Stevenson S, Li XQ, Davy DT, Klein L, Goldberg VM. Critical biological determinants of incorporation of non-vascularized cortical bone grafts. Quantification of a complex process and structure. J Bone Joint Surg Am. 1997;79:1–16.
33. Quarto R, Mastrogiacomo M, Cancedda R, Kutepov SM, Mukhachev V, Lavroukov A, Kon E, Marcacci M. Repair of large bone defects with the use of autologous bone marrow stromal cells. N Engl J Med. 2001;344:385–6.
34. Marcacci M, Kon E, Moukhachev V, Lavroukov A, Kutepov S, Quarto R, Mastrogiacomo M, Cancedda R. Stem cells associated with macroporous bioceramics for long bone repair: 6- to 7-year outcome of a pilot clinical study. Tissue Eng. 2007;13:947–55.
35. Morishita T, Honoki K, Ohgushi H, Kotobuki N, Matsushima A, Takakura Y. Tissue engineering approach to the treatment of bone tumors: three cases of cultured bone grafts derived from patients' mesenchymal stem cells. Artif Organs. 2006;30:115–8.
36. Warnke PH, Springer IN, Wiltfang J, Acil Y, Eufinger H, Wehmoller M, Russo PA, Bolte H, Sherry E, Behrens E, Terheyden H. Growth and transplantation of a custom vascularised bone graft in a man. Lancet. 2004;364:766–70.
37. Kitoh H, Kitakoji T, Tsuchiya H, Katoh M, Ishiguro N. Transplantation of culture expanded bone marrow cells and platelet rich plasma in distraction osteogenesis of the long bones. Bone. 2007;40:522–8.
38. Arthur A, Zannettino A, Gronthos S. The therapeutic applications of multipotential mesenchymal/stromal stem cells in skeletal tissue repair. J Cell Physiol. 2009;218:237–45.

Mesenchymal Stem Cells for Diabetes and Related Complications

Vladislav Volarevic, Majlinda Lako, and Miodrag Stojkovic

Abstract Mesenchymal stem cells (MSCs) are self-renewing cells with multipotent differentiation characteristics and capability for the regulation of immune response. Accordingly, in the context of diabetes research, the ability of MSC to generate insulin-producing cells and to enhance islet engraftment and survival makes them attractive as new therapeutic agents for treatment of diabetes and related complications. In this chapter we emphasized the role of MSCs in the repair of β cell mass and function and we described the capacity of MSCs to modulate the autoimmune response during the pathogenesis of diabetes mellitus. In addition, we also presented here the mechanisms, promises and potential obstacles for MSC therapy of diabetic complications: cardiomyopathy, critical limb ischemia, nephropathy, polyneuropathy, retinopathy and diabetic wounds.

Keywords Mesenchymal stem cell • Diabetes mellitus • Cardiomyopathy • Nephropathy • Neuropathy • Wound

V. Volarevic
Centre for Molecular Medicine and Stem Cell Research, Faculty of Medical Sciences, University of Kragujevac, Kragujevac, Serbia

M. Lako
Institute of Genetic Medicine, International Centre for Life, Newcastle University, Newcastle, UK

M. Stojkovic (✉)
Centre for Molecular Medicine and Stem Cell Research, Faculty of Medical Sciences, University of Kragujevac, Kragujevac, Serbia

Department of Genetics, Spebo Medical, Leskovac, Serbia
e-mail: mstojkovic@spebo.co.rs

R.C. Zhao (ed.), *Essentials of Mesenchymal Stem Cell Biology and Its Clinical Translation*, DOI 10.1007/978-94-007-6716-4_14,

Introduction

The pathophysiology of diabetes mellitus (DM) type 1 is related to an abnormal T-cell immune response culminating in a loss of self-tolerance and destruction of the insulin producing β-cells [1–3]. Prior to disease onset, a humoral B-cell immune response has taken place, producing antibodies to β cell antigens detected in the peripheral blood [4]. Together, these processes result in reduced and insufficient β cell mass to maintain glucose homeostasis, rendering patients dependent on exogenous insulin [4]. On contrary, chronic hyperglycemia due to an imbalance between insulin production and insulin action is the key metabolic abnormality important in the pathogenesis of diabetes mellitus type 2 [1–3].

Although different initial events characterize pathogenesis of diabetes mellitus type 1 and 2, several lines of evidence suggest that there are shared mechanisms of β-cell dysfunction in both types of diabetes including the increase of pro-inflammatory cytokines and oxidative stress induced by glucotoxicity and lipotoxicity which promotes β-cells apoptosis and leads to damage of pancreatic islets [5].

Currently, there is no definitive cure for diabetes. Insulin injection does not mimic the precise regulation of β-cells on glucose homeostasis, and long term insulin therapy leading to the development of various complications [6]. In addition to insulin administration, other treatment options currently available to patients with diabetes type 1 are transplantation of whole pancreas or islets only [4]. Islet transplantation, primarily indicated in patients with unstable hypoglycemia, is advantageous compared with whole pancreas transplantation because it is relatively non-invasive. Nevertheless, revascularization of the islet cell graft, apoptosis of islets, prevention of inflammation, rejection and autoimmune destruction of the graft, requirement for lifelong immunosuppression (which can be harmful to islet β cell function) and limited supply of donor islets for widespread clinical therapies are important and significant challenges that have to be solved for successful islet transplantation [4].

The worldwide increase in the prevalence of diabetes mellitus reinforces the search for new approaches to prevent and to oppose the development and the progression of diabetes and its complications [7]. The new and optimal therapeutic approach for diabetes treatment should ideally preserve the remaining β-cells, restore β-cell function, and protect the replaced insulin-producing cells from autoimmune response [8]. The field of regenerative medicine is rapidly evolving, paving the way for cellular therapies and tissue engineering as new approach for treatment of neurodegenerative and autoimmune diseases. The remarkable ability of different stem cells obtained from human embryonic and adult tissues has sparked research endeavors evaluating therapeutic use of these cells for treatment of diabetes and its complications [9].

In this sense, mesenchymal stem cells (MSCs) can offer a promising possibility that deserves to be explored [10]. In this chapter we emphasize the role of MSCs in the repair of β cell mass and function and we describe the capacity of MSCs to modulate the autoimmune response during the pathogenesis of diabetes mellitus. In addition, we also present here the mechanisms, promises and potential obstacles for MSC therapy of diabetic complications.

MSC and Islet Pathology

MSCs, also known as multipotent mesenchymal stromal cells, are self-renewing cells that can be found in almost all postnatal organs and tissues [11]. Mesenchymal stem cells are most frequently isolated from bone marrow but can generally be derived from any organ [12]. Depending on their intended purpose, experimental or therapeutic use, MSCs can be isolated from adipose tissue, umbilical cord blood, compact bone, dental pulp and other tissues [12, 13]. Mesenchymal stem cells show variable expression levels of several molecules: CD105 (SH2), CD73 (SH3/4), stromal antigen 1, CD44, CD90, CD166 (vascular cell adhesion molecule), CD54/CD102 (intracellular adhesion molecule), and CD49 (very late antigen) [14, 15]. Conversely MSCs lack the expression of surface markers characteristic for hematopoietic cells (CD14, CD45, and CD11a/lymphocyte function-associated antigen 1 (LFA-1)), erythrocytes (glycophorin A), and platelet and endothelial cell markers (CD31) [16] (Fig. 1).

The main functional characteristics of MSCs are their immunomodulatory ability, capacity for self-renewal, and differentiation into tissues of mesodermal origin [17, 18]. Previous studies have shown that MSCs are able to differentiate into various cells derived from mesoderm (connective stroma, cartilage, fat and bone cells), making them a potentially important source for the treatment of debilitating human diseases (Fig. 2) [19]. Such multipotent differentiation characteristics coupled to their capacity for self-renewal and capability for the regulation of immune responses, described MSCs as potentially new therapeutic agents for treatment of the complications of diabetes mellitus (DM) [19].

Accordingly, in the context of diabetes research, MSCs have been used to generate insulin-producing cells and enhance islet engraftment and survival [9]. For example, mouse bone marrow derived MSCs (BM-mMSCs) have the potential to differentiate into insulin-producing cells when a combination of *PDX-1* (*pancreatic and duodenal homeobox-1*), *NeuroD1* (*neurogenic differentiation-1*), and *MafA*

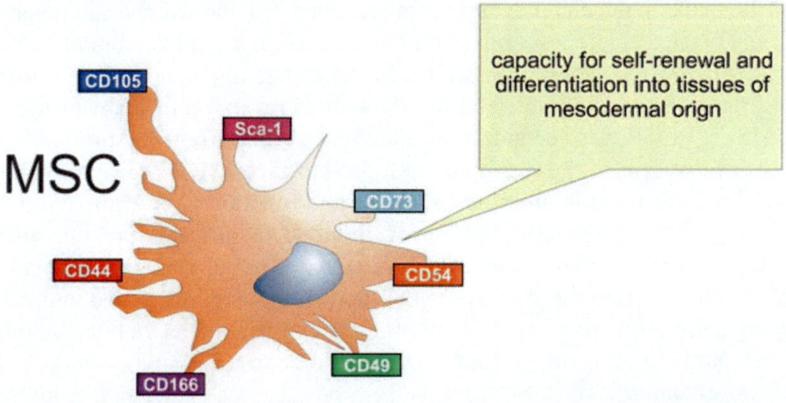

Fig. 1 Membrane markers expressed on MSC (Reproduced from Volarevic et al. [10])

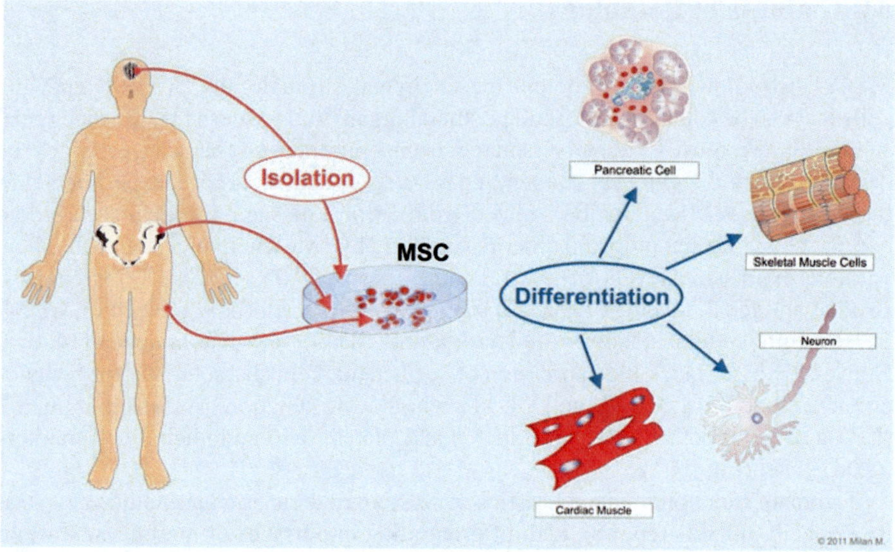

Fig. 2 Differentiation potential of MSC

(*V-maf musculoaponeurotic fibrosarcoma oncogene homolog A*) genes are overexpressed in these cells [20]. The amount of induced insulin in the BM-mMSCs transfected with all three factors was significantly higher than in those mMSCs that were only transfected with one or two of these three genes. Transplantation of the transfected cells into mice with streptozotocin-induced diabetes resulted in insulin expression and the reversal of the glucose challenge, suggesting that genetic manipulation of MSCs and their subsequent transplantation can provide a novel strategy for generation of functional β-cells [20].

Human bone marrow derived MSCs (BM-hMSCs) expressing vascular endothelial growth factor (VEGF) and *PDX1* also reversed hyperglycemia in more than half of the diabetic mice and overall improved survival and weight maintenance in all mice [21]. Detectable level of serum human insulin was noticed in mice treated with either hBMSCs-VEGF or hBMSCs-PDX1 indicating de novo β-cell differentiation from transplanted human MSCs. Sustained reversion of diabetes mediated by hBMSCs-VEGF was secondary to endogenous β-cell regeneration and correlated with activation of the insulin/insulin growth factor (IGF) receptor signaling pathway involved in maintaining β-cell mass and function [21]. Human umbilical cord derived MSC were also efficiently induced to differentiate into insulin-producing cells both *in vitro* and *in vivo* [22]. After portal vein transplantation of differentiated cells into the diabetic rats, blood sugar level decreased and insulin-producing cells containing human C-peptide and human nuclei were located in the liver [22]. However, it is important to note that some of these studies indicated that single transplantation of MSC or insulin-producing cells derived from MSCs exhibited only short-term effects [23]. Compared to single intravenous injection, multiple MSC transplantations effectively restored blood glucose homeostasis in streptozocin-induced diabetic mice for 6 months [23]. At the end of 6 months, histopathology examination revealed that MSCs engrafted preferentially into liver tissue and about 51 % of human MSCs in the recipient mice liver co-expressed human insulin, especially those surrounding the central veins. Multiple MSC transplantations efficiently restored and maintained glucose homeostasis through decreasing systemic oxidative stress in the early stage and insulin production in the late stage while liver engraftment and differentiation into insulin-producing cells were responsible for the long-term therapeutic effects of MSCs [23].

As previously discussed, transplantation of isolated islets from donor pancreata into a diabetic recipient, represents an important therapeutic approach for the treatment of insulin dependent diabetes mellitus. However, about 90 % of patients require insulin 5 years after islet transplantation. As BM-hMSCs have strong binding affinity to human islets, these cells are often used as gene delivery vehicle for successful islet transplantation in streptozotocin-induced diabetic non-obese diabetic/severe combined immunodeficient (NOD-SCID) mice [24]. Transduction of BM-hMSCs with adenovirus encoding human hepatocyte growth factor (HGF) and *interleukin 1 receptor antagonist* (*IL-1Ra*) (Adv-hHGF-hIL-1Ra) prior to co-culturing with islets further protects islets from apoptosis, helps in maintaining 3D structures and morphology, and enhances insulin secretion. Transplantation of human islets reconstituted with Adv-hHGF-hIL-1Ra transduced BM-hMSCs under the kidney capsule of diabetic mice reverses diabetes for up to 15 weeks and reduces the number of islets required to achieving normoglycemia [24]. Sakata et al. [25] confirmed improved islet function when co-transplanted with MSCs suggesting that beneficial impact of MSCs should be attributed to their immunomodulatory and angiogenic effects.

Immunomodulatory and Angiogenic Effects of MSC in Diabetes Therapy

The mechanism underlying the immunomodulatory effects of MSCs is likely to be multifactorial and results from the interaction between MSCs and immune cells: natural killer (NK) cells, dendritic cells (DCs), T and B lymphocytes to production of immunosuppressive cytokines interleukin (IL)-10 and transforming growth factor beta (TGF-β) by MSC (Fig. 3) [10, 26]. Mesenchymal stem cells are able to inhibit the proliferation, activation, cytotoxic activity and interferon-gamma (IFN-γ) production of NK cells [27]. These effects are mediated through production of *prostaglandin E2* (*PGE2*) and *indoleamine 2,3-dioxygenase* (*IDO*) [27]. Through production of PGE2, MSCs inhibit TNF-α production by DCs and upregulate IL-10 production by plasmacytoid DCs (pDCs) [28]. In addition, MSCs can alter the secretion profile of conventional DCs resulting in increased production of anti-inflammatory cytokine IL-10 and decreased production of pro-inflammatory IFN-γ and IL-12 [17, 18, 26]. MSCs are able to render T cells anergic by blocking

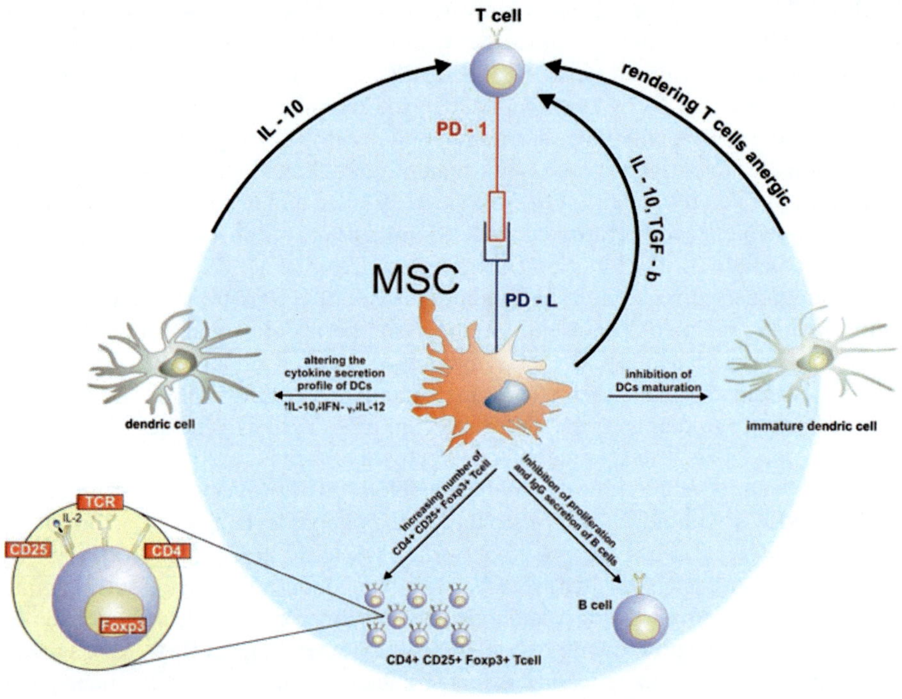

Fig. 3 Immunomodulatory characteristics of MSC (Reproduced from Volarevic et al. [10])

differentiation of monocytes to DCs or by inhibiting DC maturation [17]. Mesenchymal stem cells can inhibit T-cell proliferation by engagement of the inhibitory molecule *programmed death 1(PD-1)* to its ligands PD-L1 and PD-L2, thereby producing soluble factors that suppress T-cell proliferation (such as TGF-ß or IL-10) and through interacting with DCs [18, 26]. Inhibition of CD4+ T cells impairs B cell proliferation and antibody production. Through production of soluble factors, MSCs can inhibit proliferation and IgG secretion of B cells [26].

MSCs can also increase the number of CD4+CD25+FoxP3+ T-regulatory cells (Tregs) that suppress the immune response [17, 18, 26]. MSCs are also capable of selectively suppressing effector T(eff) cells and fostering the generation of Tregs [29]. Xu et al. showed that T(eff) cells, but not Tregs, failed to respond to IL-2 and undergo profound apoptosis in the presence of MSCs [29]. The differential regulations of these two T cell subsets by MSCs are associated with their distinct expressions of CD25, with MSCs specifically reducing the expression of CD25 on T(eff) and sparing Tregs resulting with increased protection of allogeneic islet grafts in MSC-treated animals [29].

In accordance with these immunomodulatory mechanisms, it was recently showed that human MSCs also use secreted galectins to suppress T cell mediated immune response [30]. Galectins are a family of beta-galactoside-binding proteins that has recently emerged as novel molecules with immunoregulatory functions [31, 32].

Among them, galectin-3 is the most ubiquitously expressed by immunocompetent and inflammatory cells, either constitutively or in an inducible fashion [31, 32]. Inhibition of galectin-1 and galectin-3 gene expression with small interfering RNAs abrogated the suppressive effect of MSC on allogeneic T cells [30]. Specific gene silencing of galectin-3 reduced the expression of galectin-1 in MSCs, suggesting a possible interaction between these two galectins in MSC-mediated suppression of immune response. In addition, suppression of T-cell proliferation by MSCs could be abrogated by exogenous addition of β lactose, a competitive inhibitor for galectin-3 binding to cell surface glycoproteins [30]. The restoration of T-cell proliferation in the presence of β lactose clearly indicates that the carbohydrate-recognition domain of galectin-3 is responsible for the immunosuppression of T cells and supports an extracellular mechanism of action of MSC-secreted galectin-3.

Injection of adipose-derived mesenchymal stem cells (AD-MSCs) efficiently ameliorates autoimmune diabetes in diabetic NOD mice by attenuating the Th1 immune response concomitant with the expansion/proliferation of Tregs. Injection of AD-MSCs reversed the hyperglycemia and elevated levels of insulin, amylin, and glucagon-like peptide 1 in sera [33]. This improved outcome was associated with expansion of Tregs in the pancreatic lymph nodes, reduced number of inflammatory cells and IFN-γ and increased expression of insulin, PDX-1 and TGF-β1 in pancreatic islets. In addition, *in vitro* AD-MSCs induced the expansion/proliferation of Tregs in a cell contact-dependent manner mediated by PD-L1 [33].

Recently, Ezquer et al. confirmed that the antidiabetic effect of intravenously administered MSCs is unrelated to their transdifferentiation potential but to their capability to restore the balance between Th1 and Th2 immunological responses along with the modification of the pancreatic microenvironment [34]. Mesenchymal stem cells were engrafted into secondary lymphoid organs 7 and 65 days after transplantation while no MSC-derived insulin-producing cells were noticed in pancreatic islets. This correlated with a systemic and local reduction in the abundance of auto-aggressive T cells, an increase in number of Tregs while a cytokine profile was shifted from pro-inflammatory to anti-inflammatory. Transplantation of MSCs did not reduce pancreatic cell apoptosis but recovered local expression and increased the circulating levels of epidermal growth factor (EGF), a pancreatic trophic factor [34]. Thus, it appears that immunomodulatory effects of MSCs are primarily based on their release of trophic and immunomodulatory factors which promote immunological tolerance and facilitate the survival and function of allogeneic islets [10, 26].

Through production of pro-angiogenic factors, MSCs promote vascularization and contribute to successful islet engraftment. One of the issues that can prevent successful islet transplantation of MSCs is islet ischemia [25]. The process of islet isolation destroys the vascular network between the islet and surrounding tissue [25, 35]. As a result, islets undergo prolonged ischemia during the reconstruction of the vascular network, and many islets become damaged [25, 35]. Mesenchymal stem cells express *platelet-derived growth factor* (*PDGF*) receptors and respond to PDGF production by endothelial cells during revascularization [36] promoting endothelial cell proliferation in both donor and recipient sides as well as enhance migration of endothelial cells by producing proteases that facilitate immature endothelial cell sprouting [37] and upregulating the expression of factors that promote angiogenesis and stability of the developing vasculature (angiopoietin and VEGF) in endothelial cells [38]. In addition, MSCs promote proliferation and migration of endothelial cells to the surface of islets where these cells form a "coat" [39]. "Coated" islets survived for a long time in culture and exhibit improved insulin release [39].

MSC: Potentially New Therapeutic Agents for the Treatment of Diabetes Type 2

MSC are able to improve metabolic control in experimental models of type 2 diabetes (T2D) [9]. Recently published study by Si et al. [40] showed that infusion of autologous MSCs managed to ameliorate hyperglycemia in type 2 diabetic rats. Improved metabolic control, measured by enhanced insulin secretion, amelioration of insulin sensitivity, and increased islet numbers in the pancreas, was observed in MSC treated animals particularly when MSCs were infused early (7 days) after STZ treatment. As previously discussed [23], transplanted MSC are short-lived (for a period of 4 weeks), and MSC re-injection provides an additional, comparable, and transient effect. Beneficial effects of MSC therapy were associated with improved insulin sensitivity via increased signaling (*insulin receptor substrate-1* [*IRS-1*] and Akt phosphorylation upon feeding, as well as translocation of *glucose transporter type 4* (*GLUT-4*) on cell membrane upon insulin administration) in the muscle, liver, and adipose tissue of MSC treated animals [40]. Mesenchymal stem cells mainly accumulated in pancreatic islets and liver where they have contributed to tissue repair and preservation of β-cell mass [40]. It is important to note that despite the fact that experimental model used in this study does not fully reflect the physiopathology of the progressive development of human T2D [41], data presented by Si and colleagues shed new light on the effects of autologous MSCs transplantation on insulin target tissues in T2D [9, 40].

A pilot clinical study conducted by Jiang et al. [42] showed that three intravenous infusions of human placenta-derived MSCs (PD-hMSCs) managed to significantly reduce a daily dose of insulin, improve renal and cardiac function in type 2 diabetic patients with islet cell dysfunction. However, despite of promising results, it should be emphasized that further in-depth mechanistic studies are still needed to understand how MSCs affect metabolic function in T2D.

MSC Treatment of Diabetic Cardiomyopathy and Critical Limb Ischemia

Development of ventricular dysfunction in patients with DM in the absence of coronary artery disease, valvular heart disease, or hypertension is defined as diabetic cardiomyopathy (DCM) [43]. Chronic hyperglycemia is responsible for myocardial remodeling and is a central feature in the progression of DCM, which is characterized by hypertrophy and apoptosis of cardiomyocytes and alterations in the quantity and composition of the extracellular matrix (ECM), resulting in increased collagen deposition [44]. An additional feature that contributes to the pathogenesis of DCM is the activity of matrix metalloproteinase (MMP)-2 and MMP-9 [45, 46]. The diabetic myocardium is characterized by decreased activity of MMP-2, leading to increased collagen accumulation, and increased activity of the proapoptotic factor MMP-9, which is responsible for apoptosis of endothelial cells, reduction of capillary density, and poor myocardial perfusion [45, 46]. Microcirculatory defects, necrosis and apoptosis of cardiomyocytes, and interstitial fibrosis are the main pathological characteristics of DCM [43, 45].

Mesenchymal stem cells can also induce myogenesis and angiogenesis by releasing different angiogenic, mitogenic, and antiapoptotic factors including vascular endothelial growth factor (VEGF), insulin- like growth factor-1 (IGF-1), adrenomedullin (AM), and hepatocyte growth factor (HGF) [47]. This was demonstrated using a rat model of DCM [47] wherein intravenous (i.v.) administration of bone marrow-derived rat MSCs improved cardiac function of treated animals. Transplanted MSCs differentiated into cardiomyocytes and improved myogenesis and angiogenesis [43]. In addition, MMP-2 activity significantly increased and MMP-9 activity decreased after MSCs transplantation [47]. This phenomenon increased myocardial arteriolar density and decreased collagen volume resulting in attenuation of cardiac remodeling and improved myocardial function [47] (Fig. 4).

Improvement in cardiac function following MSC therapy may also be attributed to the release of MSC-derived paracrine factors capable of cardioprotection. These factors include secreted frizzled-related protein 2, Bcl 2, heat shock protein 20, hypoxia-regulated heme oxygenase-1, hypoxic Akt-regulated stem cell factor, VEGF, HGF, AM, and stromal-derived factor [48]. A growing body of evidence strongly suggests that these factors affect remodeling, regeneration, and neovascularization leading to the improvement of myocardium contractility and viability, ameliorating consequences of infarction [48–53]. Double-blind, placebo-controlled trials showed that i.v. autologous MSCs transplantation increased left ventricular ejection fraction, reduced episodes of ventricular tachycardia, and led to reverse remodeling in postinfarction patients reducing the mortality rate in patients with ischemic stroke [10].

Mesenchymal stem cells treatment of diabetic rats results in a significant increase in heart rate, left ventricular pressure, contractility index, and notable reduction of systolic blood pressure [54]. The improvement in cardiac condition can be explained by differentiation of MSCs into insulin producing cells, cardiomyocytes and vascular endothelial cells [55] and also by the immunomodulation ability of MSCs [10, 26]. As previously described, significant increase of serum insulin levels leads to endothelial cell protection, and this is accompanied with enhanced myogenesis, angiogenesis, and attenuation of cardiac remodeling, all of which are crucial for the improvement of cardiac function in diabetic animals [10, 25, 39].

Similar effects and MSC-derived proangiogenic factors have also been implicated in the therapy of diabetic limb ischemia [51, 52]. VEGF and hypoxia-inducible

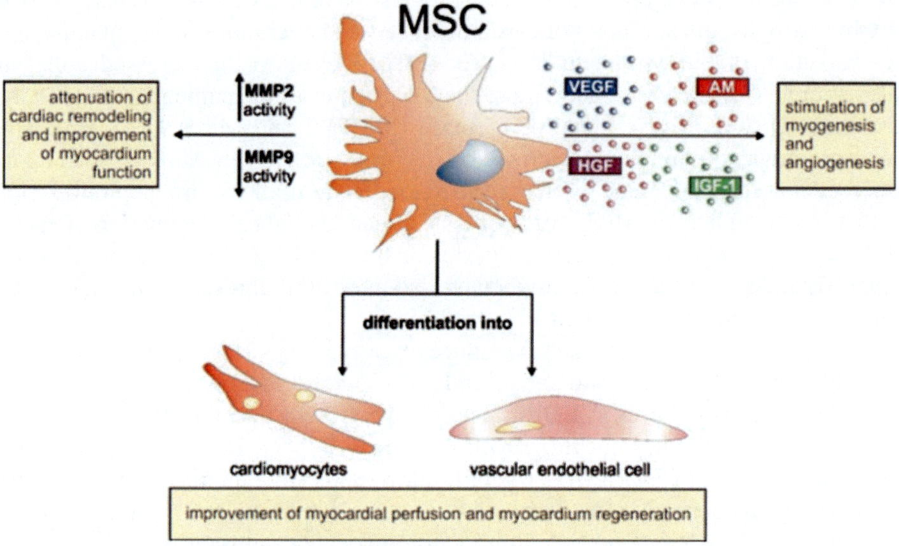

Fig. 4 Effects of transplanted MSCs on diabetic cardiomyopathy (Reproduced from Volarevic et al. [10])

factor are responsible for restoring blood flow and vasculogenesis in the ischemic hindlimb of type II diabetic (db[−]/db[−]) mice [51] or for improvement of arterial perfusion in type 1 diabetic patients with bilateral upper extremity digital gangrene [52]. *Peroxisome proliferator-activated receptor-γ coactivator-1α* (*PGC-1α*) is a key regulator important for linking angiogenesis and metabolism. Overexpression of *PGC-1α* within MSCs, enhanced the engraftment and angiogenesis of MSCs in animal model of diabetic hindlimb ischemia [56]. *In vitro*, under hypoxia condition, the overexpression of *PGC-1α* in MSCs resulted in a higher expression level of *hypoxia-inducible factor-1α* (*Hif-1α*), a greater ratio of *B-cell lymphoma leukemia-2* (*Bcl-2*)/*Bcl-2-associated X protein* (*Bax*), and a lower level of caspase 3, followed by an increased survival rate and an elevated expression level of pro-angiogenic factors. *In vivo*, the MSCs modified with PGC-1α significantly increased the blood perfusion and capillary density of ischemic hind limb of the diabetic rats, which was correlated to an improved survivability of MSCs and an increased level of several pro-angiogenic factors secreted by MSCs [56].

Recently, Lee and colleagues showed that multiple intramuscular injections of AD-MSCs might be a safe alternative to achieve therapeutic angiogenesis in patients with critical limb ischemia who are refractory to other treatment modalities [57]. Results obtained in this pilot study indicated that transplantation of AD-MSC induced formation of numerous vascular collateral networks across affected arteries resulting in significant clinical improvements noted on pain rating scales and in claudication walking distance [57].

MSC Therapy of Diabetic Nephropathy

Mesenchymal stem cells administration can prevent and treat diabetic nephropathy, which is a complication of DM and is defined as progressive kidney disease caused by angiopathy of the capillaries supplying the kidney glomeruli [10]. MSCs have been used for the treatment of diabetic nephropathy in NOD/SCID and C57 black 6 (C57/BL6) mice, which succumb to DM after application of multiple low doses of STZ [58, 59]. About 30–60 days after STZ injection, kidneys of treated mice showed the presence of abnormal glomeruli characterized by increased deposits of ECM protein in the mesangium, hyalinosis, and increased number of macrophages in the glomeruli [58, 59].

Data obtained from studies using NOD/SCID mice transplanted with human MSCs and C57Bl/6 mice that received murine MSCs indicate that injected MSCs engraft in damaged kidneys, differentiate into renal cells, and regulate the immune response resulting in an efficient treatment of diabetic nephropathy [58, 59]. Additionally, Lee at al. [59] showed that small percentage of hMSCs in the transplanted kidneys differentiated into endothelial cells as evidenced by de novo expression of CD31 [59].

The result of systemic administration of MSCs in diabetic mice was improvement of kidney function and regeneration of glomerular structure [58, 59] as MSCs are able to reconstitute necrotic segments of diabetic kidneys [60]. However, it is not clear whether MSCs can propagate after engraftment in the kidney [58, 59]. Lee et al. showed that 1 month after MSC treatment, only a few human MSCs were detected in kidneys, suggesting that they were unable to proliferate [59] so an alternative scenario for improvement of kidney function could be the ability of MSCs to scavenge cytotoxic molecules or to promote neovascularization [10]. In addition, successful MSC treatment of diabetic nephropathy could be explained by MSCs competence to differentiate into insulin-producing cells followed by decrease of glycemia and glycosuria, factors that are important for damaging renal cells [10]. Recently, Park et al. [61] demonstrated that only few umbilical cord blood-derived human mesenchymal stem cells (UCB-hMSC) managed to engraft in kidneys of STZ-induced diabetic rats. Intravenously administered UCB-hMSC effectively reduced proteinuria, renal fibronectin, and α-smooth muscle actin up-regulation, as well as renal E-cadherin down-regulation in diabetic rats without a significant effect on blood glucose. Thus, renoprotective effect of transplanted UCB-hMSC was noticed due to secretion of humoral factors suggesting UCB-hMSC as a possible treatment modality for diabetic renal injury [61].

Most recently, Fang and colleagues showed that the key mechanisms underlying the positive therapeutic impact of AD-MSC treatment in kidneys could be due to the suppression of inflammatory response and oxidative stress [62]. Autologous transplantation of AD-hMSC ameliorates STZ-induced diabetic nephropathy in rats by inhibiting oxidative stress, pro-inflammatory cytokines and the *p38:MAPK* signaling pathway [62]. Transplantation of AD-MSC minimized pathological alterations, reduced oxidative damage and suppressed the expression of pro-inflammatory cytokines, while all important molecules of the *MAPK* signaling pathway: *p-p38, p-ERK and p-JNK* decreased expression in the renal tissues of AD-MSC treated diabetic rats [62]. Taken together, data obtained from preclinical studies indicate that MSC transplantation prevents the pathological changes in the glomeruli and enhances their regeneration resulting in improved kidney function in diabetic animals.

MSC Transplantation Improves Diabetic Polyneuropathy and Retinopathy

Diabetic polyneuropathy (DPN), the most common complication of DM, is characterized by damage to nerve fibers [63]. The central features in the development and progression of DPN are neural cell degeneration and decreased nerve blood flow (NBF) [63]. Previous studies have shown that angiogenic cytokines such as basic fibroblast growth factor (bFGF) and VEGF could be useful for the treatment of DPN [64, 65]. Spontaneous pain, hyperalgesia, and diminished sensation are main symptoms of DPN [63]. Thus, the relief from symptoms of DPN is still an important issue and many novel therapeutic approaches were conducted for this purpose. It was shown in diabetic rats that MSCs, because of their ability to secrete bFGF and VEGF [10], could be used as a new and effective therapeutic agent for the treatment of DPN [64, 65]. Four weeks after intramuscular injection, MSCs settled in the gap between muscle fibers, produced bFGF and VEGF and led to increase in the ratio of capillaries to muscle fibers that was followed by improvement of hyperalgesia, and a corresponding functional improvement of neural fibers, delayed motor nerve conduction velocity, reduced sciatic NBF, and decreased axonal circularity at the site of transplantation [65] (Fig. 5).

Although several studies [11, 14, 19, 66] have suggested that MSCs have the capacity to differentiate into neural cells, such as astrocytes, oligodendrocytes, and

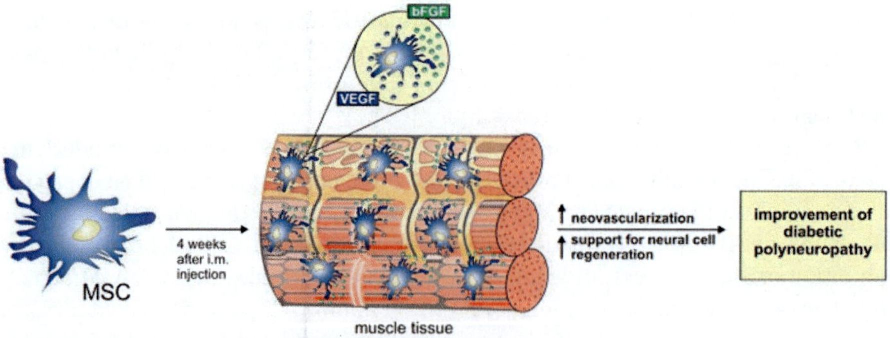

Fig. 5 MSC as potentially new therapeutic agent in polyneuropathy treatment (Reproduced from Volarevic et al. [10])

Schwann cells, this was not noted after MSCs transplantation in diabetic rats [65]. After intramuscular injection, MSCs remained at the transplant sites. They did not have any systemic effects and did not differentiate into neural cells, which suggest that systemic MSCs injection could be a better way for the improvement of all affected nerve fibers [65].

On contrary, transplanted MSCs improved the integrity of the blood-retinal barrier thus ameliorating diabetic retinopathy in STZ diabetic rats through differentiation into photoreceptor and glial-like cells in the retina and by releasing paracrine factors [67]. In addition, intravitreal injection of human MSCs has been shown to be effective in slowing the progression of diabetic retinopathy in an animal model of chemically induced diabetes mellitus [68]. Increased intravitreal and retinal concentrations of neuroprotective growth factors were identified in MSC-treated diabetic rats further confirming the neuroprotective activity of human MSCs in diabetic retinopathy [68].

MSC Treatment of Diabetic Wounds

Prolonged and incomplete wound healing, caused by reduced production of growth factors, impaired angiogenesis, and compromised formation of collagen matrixes, were observed as a complication of DM [10, 69]. The characteristics of diabetic wounds are poor neovascularization, presence of abundant inflammatory infiltrates mostly containing polymorphonuclear cells, and foci of necrotic tissue composed of neutrophils [69]. Disturbances in collagen metabolism and compromised production and functionality of growth factors such as transforming growth factor β (TGF-β), EGF, VEGF, platelet-derived growth factor (PDGF), and keratinocyte growth factor (KGF) are the main factors responsible for the pathogenesis of poor wound healing [69]. Systemic and local administration of bone marrow-derived MSCs improves healing of diabetic wounds in rats and mice [70]. After i.v. injection of MSCs, diabetic wounds showed significantly increased collagen levels followed by increased wound-breaking strength [70]. The increased production of collagen, the major component of ECM crucial for strength, integrity, and structure of normal tissues and important for repairing tissue defects created by injuries, was noticed after MSC administration [70]. MSC injection resulted in moderate (TGF-β, KGF) or significant (EGF, PDGF, and VEGF) increase in the production of growth factors involved in the repair of injured tissue that was crucial for successful diabetic wound healing [70]. These factors stimulated cell adhesion at the site of injury and induced cells to secrete more chemokines resulting in neovascularization and formation of inflammation infiltrate, containing predominantly mononuclear cells, without tissue necrosis [10, 70].

By using diabetic rats, Kuo et al. [71], confirmed that MSC therapy significantly enhanced diabetic wound healing. Significant increases in *EGF*, *VEGF*, *prolyl 4-hydroxylase*, and *Ki-67* expression were noted in the MSC-treated group as compared with the control group [71]. Beside these paracrine effects, MSCs can help to improve diabetic wounds through their differentiation ability [72] and ability

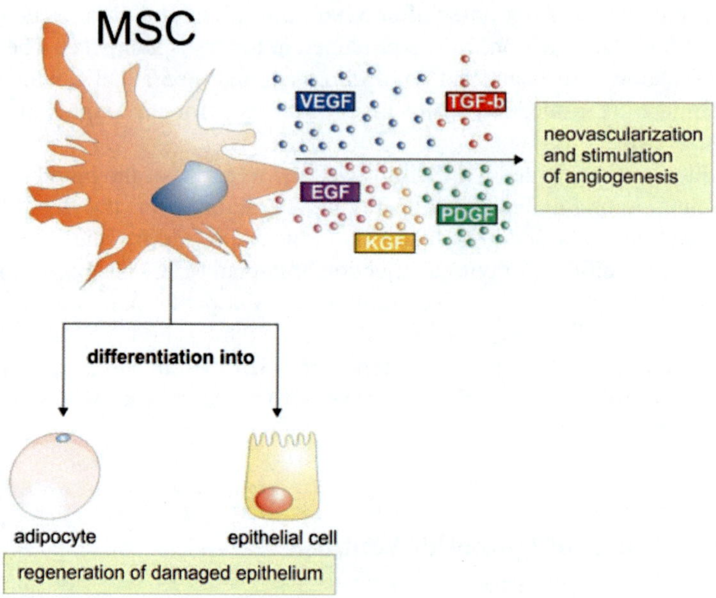

Fig. 6 MSC treatment of diabetic wounds (Reproduced from Volarevic et al. [10])

to regenerate damaged epithelium through differentiation and fusion [73]. In diabetic mice, some of the MSCs transplanted in the wound, coexpressed cytokeratin, whereas others formed sweat or sebaceous gland-like structures of the skin [72] (Fig. 6).

Most recently it was shown that amniotic MSCs enhanced wound healing in diabetic NOD/SCID mice through angiogenic and engraftment/differentiation capabilities. Transplanted amniotic MSCs exhibited high engraftment rates and expressed keratinocyte-specific proteins and cytokeratin in the wound area, indicating a direct contribution to cutaneous closure [74].

In diabetic wounds, MSCs settled predominantly in the newly formed dermis, and to a lesser extent in the epidermis, however, none were detected in the undamaged skin [10]. Although MSCs were not found in the vascular structures of diabetic wounds, it was documented that, after MSC treatment, there was enhanced capillary density in those, suggesting that MSCs promoted angiogenesis that was very important for successful healing [72, 73]. Recently, Shen et al. [75] showed that neurotrophin-3 (NT-3) significantly promotes human MSC secretion of VEGF, neural growth factor (NGF) and other vasoactive factors and that it accelerates wound healing by inducing angiogenesis through improved activation of vascular endothelial cells. NT-3 improved activation of the human MSC conditioned medium, promoted human umbilical vein endothelial cell (HUVEC) proliferation and migration and significantly improved the closure rate of HUVEC scratches [75].

MSCs have already shown efficacy in the treatment of foot ulcerations in diabetic patients [76]. Vojtassak and colleagues [76] reported complete wound resolution of a 25-year open wound within 4 weeks of MSC application. Autologous biografts composed of skin fibroblasts seeded on biodegradable collagen membranes in combination with autologous MSCs, derived from the patient's bone marrow, were successfully used for closing and healing diabetic foot ulcerations [76]. Lataillade et al. [77] presented a new approach to radiation burn treatment by dosimetry-guided surgery combined with autologous MSC therapy. However, it is important to note that both studies [76, 77] were case studies with single patient reports and larger randomized trials are needed to confirm obtained results. The largest study to date using bone marrow-derived MSCs in extremity based wounds was level 1 randomized controlled trial, conducted by Dash et al. [78] confirming that the MSC therapy is able to significantly reduce wound size and increase several clinical parameters in diabetes.

There is a difference in the efficacy between systemic and local MSCs therapy for diabetic wound healing [10, 68, 75]. For example, better effects are noticed after local administration of MSCs, possibly because of the presence of arterial-venous shunts in diabetic skin, which may complicate migration of systemically injected MSCs to the wounds [72]. In addition, it is difficult for intravenously injected MSC to home and locate to the lesion. In order to make MSCs directly act on the wound, Hou and colleagues choose a solid collagen as a carrier, planted MSCs into it and covered ischemic wounds of diabetic mice with the complex of MSCs and collagen [79]. The complex of MSCs and collagen biomaterials significantly promoted angiogenesis and wound healing [79].

Due to their ability to differentiate into osteoblasts [66], MSC represent a novel powerful orthopedic tool for diabetic bone healing and regeneration. Yu and colleagues [80] showed that bone volume ratio and trabecular thickness significantly increased while trabecular separation decreased in tibias of diabetic rats after MSC transplantation.

Limitations and Future Perspectives

There are several problems that limit the use of MSCs for diabetes therapy [10, 81]. Poor engraftment and limited differentiation under *in vivo* conditions are major obstacles for efficient therapeutic use of MSCs [81]. Large portion of transplanted MSC undergo apoptosis after transplantation in diabetic animals [82]. Hyperglycemia is reported to exert effects by different levels of reactive oxygen species overproduction which triggers the apoptosis and thereby decreases MSC viability after transplantation. Recent studies have shown that hypoxia preconditioning (HPC) may improve viability of transplanted MSCs [82]. Montaghi et al. [28] showed that apelin; the endogenous ligand for the previously orphaned G protein-coupled receptor APJ exerted anti-apoptotic effects on oxidative stress-induced apoptosis in MSCs. Thus, pretreatment of HPC-MSCs with apelin 13 could be an effective approach to modify and possibly enhance the engraftment and efficacy of MSCs during diabetes therapy.

The frequency of spontaneous differentiation of MSCs in the host tissue is extremely rare, therefore, therapeutic use of MSCs depends on the ability to control their *in vivo* differentiation into functional cells with high efficiency and purity [81]. An additional limitation is the potential of MSCs to differentiate into unwanted mesenchymal lineages [83], which could impair their therapeutic use. There are data suggesting the restriction of such unwanted differentiation by a variety of factors [84]; however, this problem is still largely unsolved because the precise roles of factors that could be responsible for the fate of MSCs after their administration are not completely understood [10, 84].

Additional limitations are possible malignant transformation and cytogenetic aberrations of MSCs. This was observed after *in vitro* expansion of murine MSCs derived from the bone marrow of Bagg albino (BALB/c) and C57BL/6 mice [81]. However, malignant transformation of transplanted hMSCs has not yet been reported [10].

Taken together, MSCs offer new opportunities for the treatment of diabetes and its complications, but they also raise many scientific questions such as potential risk of malignant transformation, unwanted mesenchymal lineages differentiation, and suboptimal targeted differentiation, which should be addressed before MSCs can be defined as a novel and efficient therapeutic agent in the treatment of diabetes and its complications.

Conclusions

Compared to other stem cells, MSCs have several advantages for therapeutic use such as ability to migrate to the sites of tissue injury, strong immunosuppressive effects, better safety after infusion of allogeneic MSCs, and lack of ethical issues, such as those related to the application of human embryonic stem cells [10]. Because of their immunomodulatory ability, self-renewal, and multi-lineage differentiation capacity, MSCs are expected to become promising therapeutic agents for treatment of diabetes type 1 and its complications [10]. Nevertheless, further in-depth mechanistic studies are needed to understand how MSCs affect metabolic function in diabetes type 2 and how diabetic microenvironment and/or comorbidities alter the quality and efficacy of MSCs isolated from patients with diabetes [9].

At the end, the increasing body of evidence obtained from preclinical studies justifies cautious optimism concerning development of effective MSC based therapy for treatment of diabetes and its complications. However, it should be emphasized that there is a paucity of validated clinical studies to prove the efficacy of MSC-based therapy. Accordingly, large, well-designed randomized clinical trials should be conducted in the future to determine the value of MSC-based therapy for the treatment of diabetes and its complications.

Key Points

- Mesenchymal stem cells have potential to treat diabetes mellitus and its complications due to their immunomodulatory and angiogenic effects and capacity for differentiation into tissues of mesodermal origin.
- Transplanted MSCs differentiated into cardiomyocytes and improved myogenesis and angiogenesis resulting in attenuation of cardiac remodeling and improvement of myocardial function.
- MSC transplantation prevents the pathological changes in the glomeruli and enhances their regeneration resulting in improved kidney function in diabetic animals.
- Injection of MSCs induces neovascularization and support regeneration of neural cells that results with improvement of diabetic polyneuropathy.
- MSC transplantation resulted in significant increase in the production of growth factors involved in the repair of injured tissue that was crucial for successful diabetic wound healing.
- The main problems that limit the use of MSCs for diabetes therapy are poor engraftment and unwanted mesenchymal lineages differentiation.
- There is a paucity of validated clinical studies to prove the efficacy of MSC-based therapy. Accordingly, large, well-designed randomized clinical trials should be conducted in the future to determine the value of MSC-based therapy for the treatment of diabetes and its complications.

References

1. Eizirik DL, Mandrup-Poulsen T. A choice of death. The signal transduction of immune-mediated b-cell apoptosis. Diabetologia. 2001;44:2115–33.
2. DeFronzo RA. Pathogenesis of type 2 diabetes: metabolic and molecular implications for identifying diabetes genes. Diabetes Rev. 1997;5:177–269.
3. Hani EH, Stoffers DA, Chevre JC, Durand E, Stanojevic V, Dina C, Habener JF, Froguel P. Defective mutations in the insulin promoter factor-1 (IPF-1) gene in late-onset type 2 diabetes mellitus. J Clin Invest. 1999;104:R41–8.
4. Davis NE, Hamilton D, Fontaine MJ. Harnessing the immunomodulatory and tissue repair properties of mesenchymal stem cells to restore β cell function. Curr Diab Rep. 2012;12:612–22.
5. Rother K. Diabetes treatment-bridging the divide. N Engl J Med. 2007;356:1499–501.
6. Mabed M, Shahin M. Mesenchymal stem cell-based therapy for the treatment of type 1 diabetes mellitus. Curr Stem Cell Res Ther. 2012;7:179–90.
7. Bernardi S, Severini GM, Zauli G, Secchiero P. Cell-based therapies for diabetic complications. Exp Diabetes Res. 2012;2012:872504.
8. Fiorina P, Voltarelli J, Zavazava N. Immunological applications of stem cells in type 1 diabetes. Endocr Rev. 2011;32:725–54.
9. Pileggi A. Mesenchymal stem cells for the treatment of diabetes. Diabetes. 2012;61:1355–6.
10. Volarevic V, Arsenijevic N, Lukic ML, Stojkovic M. Concise review: mesenchymal stem cell treatment of the complications of diabetes mellitus. Stem Cells. 2011;29:5–10.
11. Porada CD, Zanjani ED, Almeida-Porad G. Adult mesenchymal stem cells: a pluripotent population with multiple applications. Curr Stem Cell Res Ther. 2006;1:365–9.

12. da Silva Meirelles L, Chagastelles PC, Nardi NB. Mesenchymal stem cells reside in virtually all post-natal organs and tissues. J Cell Sci. 2006;119:2204–13.
13. Kanjevac T, Milovanovic M, Volarevic V, Lukic ML, Arsenijevic N, Markovic D, Zdravkovic N, Tesic Z, Lukic A. Cytotoxic effects of glass ionomer cements on human dental pulp stem cells correlate with fluoride release. Med Chem. 2012;8:40–5.
14. Chamberlain G, Fox J, Ashton B, Middleton J. Concise review: mesenchymal stem cells: their phenotype, differentiation capacity, immunological features, and potential for homing. Stem Cells. 2007;25:2739–49.
15. Dominici M, Le Blanc K, Mueller I, Slaper-Cortenbach I, Marini F, Krause D, Deans R, Keating A, Prockop D, Horwitz E. Minimal criteria for defining multipotent mesenchymal stromal cells. The International Society for Cellular Therapy position statement. Cytotherapy. 2006;8:315–7.
16. Prockop DJ. Marrow stromal cells as stem cells for nonhematopoietic tissues. Science. 1997;276:71–4.
17. Abdi R, Fiorina P, Adra CN, Atkinson M, Sayegh MH. Immunomodulation by mesenchymal stem cells: a potential therapeutic strategy for type 1 diabetes. Diabetes. 2008;57:1759–67.
18. Nauta AJ, Fibbe WE. Immunomodulatory properties of mesenchymal stromal cells. Blood. 2007;110:3499–506.
19. Pittenger MF, Mackay AM, Beck SC, Jaiswal RK, Douglas R, Mosca JD, Moorman MA, Simonetti DW, Craig S, Marshak DR. Multilineage potential of adult human mesenchymal stem cells. Science. 1999;284:143–7.
20. Guo QS, Zhu MY, Wang L, Fan XJ, Lu YH, Wang ZW, Zhu SJ, Wang Y, Huang Y. Combined transfection of the three transcriptional factors, PDX-1, NeuroD1, and MafA, causes differentiation of bone marrow mesenchymal stem cells into insulin-producing cells. Exp Diabetes Res. 2012;2012:672013.
21. Milanesi A, Lee JW, Li Z, Da Sacco S, Villani V, Cervantes V, Perin L, Yu JS. β-Cell regeneration mediated by human bone marrow mesenchymal stem cells. PLoS One. 2012;7(8):e42177.
22. Tsai PJ, Wang HS, Shyr YM, Weng ZC, Tai LC, Shyu JF, Chen TH. Transplantation of insulin-producing cells from umbilical cord mesenchymal stem cells for the treatment of streptozotocin-induced diabetic rats. J Biomed Sci. 2012;19:47.
23. Ho JH, Tseng TC, Ma WH, Ong WK, Chen YF, Chen MH, Lin MW, Hong CY, Lee OK. Multiple intravenous transplantations of mesenchymal stem cells effectively restore long-term blood glucose homeostasis by hepatic engraftment and β-cell differentiation in streptozocin-induced diabetic mice. Cell Transplant. 2012;21:997–1009.
24. Wu H, Lu W, Mahato RI. Mesenchymal stem cells as a gene delivery vehicle for successful islet transplantation. Pharm Res. 2011;28:2098–109.
25. Sakata N, Goto M, Yoshimatsu G, Egawa S, Unno M. Utility of co-transplanting mesenchymal stem cells in islet transplantation. World J Gastroenterol. 2011;17:5150–5.
26. Volarevic V, Al-Qahtani A, Arsenijevic N, Pajovic S, Lukic ML. Interleukin-1 receptor antagonist (IL-1Ra) and IL-1Ra producing mesenchymal stem cells as modulators of diabetogenesis. Autoimmunity. 2010;43:255–63.
27. Spaggiari GM, Capobianco A, Abdelrazik H, Becchetti F, Mingari MC, Moretta L. Mesenchymal stem cells inhibit natural killer-cell proliferation, cytotoxicity, and cytokine production: role of indoleamine 2,3-dioxygenase and prostaglandin E2. Blood. 2008;111:1327–33.
28. Aggarwal S, Pittenger MF. Human mesenchymal stem cells modulate allogeneic immune cell responses. Blood. 2005;105:1815–22.
29. Xu DM, Yu XF, Zhang D, Zhang MX, Zhou JF, Tan PH, Ding YC. Mesenchymal stem cells differentially mediate regulatory T cells and conventional effector T-cells to protect fully allogeneic islet grafts in mice. Diabetologia. 2012;55:1091–102.
30. Sioud M. New insights into mesenchymal stromal cell-mediated T-cell suppression through galectins. Scand J Immunol. 2011;73:79–84.
31. Volarevic V, Milovanovic M, Ljujic B, Pejnovic N, Arsenijevic N, Nilsson U, Leffler H, Lukic ML. Galectin-3 deficiency prevents concanavalin A-induced hepatitis in mice. Hepatology. 2012;55:1954–64.

32. Radosavljevic G, Volarevic V, Jovanovic I, Milovanovic M, Pejnovic N, Arsenijevic N, Hsu DK, Lukic ML. The roles of Galectin-3 in autoimmunity and tumor progression. Immunol Res. 2012;52:100–10.
33. Bassi EJ, Moraes-Vieira PM, Moreira-Sá CS, Almeida DC, Vieira LM, Cunha CS, Hiyane MI, Basso AS, Pacheco-Silva A, Câmara NO. Immune regulatory properties of allogeneic adipose-derived mesenchymal stem cells in the treatment of experimental autoimmune diabetes. Diabetes. 2012;61:2534–45.
34. Ezquer F, Ezquer M, Contador D, Ricca M, Simon V, Conget P. The antidiabetic effect of mesenchymal stem cells is unrelated to their transdifferentiation potential but to their capability to restore Th1/Th2 balance and to modify the pancreatic microenvironment. Stem Cells. 2012;30:1664–74.
35. Emamaullee JA, Rajotte RV, Liston P, Korneluk RG, Lakey JR, Shapiro AM, Elliott JF. XIAP overexpression in human islets prevents early posttransplant apoptosis and reduces the islet mass needed to treat diabetes. Diabetes. 2005;54:2541–8.
36. Ball SG, Shuttleworth CA, Kielty CM. Mesenchymal stem cells and neovascularization: role of platelet-derived growth factor receptors. J Cell Mol Med. 2007;11:1012–30.
37. Ghajar CM, Blevins KS, Hughes CC, George SC, Putnam AJ. Mesenchymal stem cells enhance angiogenesis in mechanically viable prevascularized tissues via early matrix metalloproteinase upregulation. Tissue Eng. 2006;12:2875–88.
38. Zacharek A, Chen J, Cui X, Li A, Li Y, Roberts C, Feng Y, Gao Q, Chopp M. Angiopoietin1/Tie2 and VEGF/Flk1 induced by MSC treatment amplifies angiogenesis and vascular stabilization after stroke. J Cereb Blood Flow Metab. 2007;27:1684–91.
39. Johansson U, Rasmusson I, Niclou SP, Forslund N, Gustavsson L, Nilsson B, Korsgren O, Magnusson PU. Formation of composite endothelial cell-mesenchymal stem cell islets: a novel approach to promote islet revascularization. Diabetes. 2008;57:2393–401.
40. Si Y, Zhao Y, Hao H, Liu J, Guo Y, Mu Y, Shen J, Cheng Y, Fu X, Han W. Infusion of mesenchymal stem cells ameliorates hyperglycemia in type 2 diabetic rats: identification of a novel role in improving insulin sensitivity. Diabetes. 2012 Jun;61:1616–25.
41. Pantic J, Volarevic V, Djukic A. Experimental models of diabetes mellitus. Ser J Exp Med. 2011;12:29–35.
42. Jiang R, Han Z, Zhuo G, Qu X, Li X, Wang X, Shao Y, Yang S, Han ZC. Transplantation of placenta-derived mesenchymal stem cells in type 2 diabetes: a pilot study. Front Med. 2011;5:94–100.
43. Poornima IG, Parikh P, Shannon RP. Diabetic cardiomyopathy: the search for a unifying hypothesis. Circ Res. 2006;98:596–605.
44. Jesmin S, Sakuma I, Hattori Y, Kitabatake A. Role of angiotensin II in altered expression of molecules responsible for coronary matrix remodeling in insulin-resistant diabetic rats. Arterioscler Thromb Vasc Biol. 2003;23:2021–6.
45. Camp TM, Tyagi SC, Senior RM, et al. Gelatinase B (MMP-9) an apoptotic factor in diabetic transgenic mice. Diabetologia. 2003;46:1438–45.
46. Yoon YS, Uchida S, Masuo O, Cejna M, Park JS, Gwon HC, Kirchmair R, Bahlman F, Walter D, Curry C, Hanley A, Isner JM, Losordo DW. Progressive attenuation of myocardial vascular endothelial growth factor expression is a seminal event in diabetic cardiomyopathy: restoration of microvascular homeostasis and recovery of cardiac function in diabetic cardiomyopathy after replenishment of local vascular endothelial growth factor. Circulation. 2005;111:2073–85.
47. Zhang N, Li J, Luo R, Jiang J, Wang JA. Bone marrow mesenchymal stem cells induce angiogenesis and attenuate the remodeling of diabetic cardiomyopathy. Exp Clin Endocrinol Diabetes. 2008;116:104–11.
48. Wang X, Zhao T, Huang W, Wang T, Qian J, Xu M, Kranias EG, Wang Y, Fan GC. Hsp20-engineered mesenchymal stem cells are resistant to oxidative stress via enhanced activation of Akt and increased secretion of growth factors. Stem Cells. 2009;27:3021–31.
49. Hare JM, Traverse JH, Henry TD, Dib N, Strumpf RK, Schulman SP, Gerstenblith G, DeMaria AN, Denktas AE, Gammon RS, Hermiller Jr JB, Reisman MA, Schaer GL, Sherman W.

A randomized, double-blind, placebo-controlled, dose-escalation study of intravenous adult human mesenchymal stem cells (prochymal) after acute myocardial infarction. J Am Coll Cardiol. 2009;54:2277–86.
50. Lee JS, Hong JM, Moon GJ, Lee PH, Ahn YH, Bang OY. A long-term follow-up study of intravenous autologous mesenchymal stem cell transplantation in patients with ischemic stroke. Stem Cells. 2010;28:1099–106.
51. Amin AH, Abd Elmageed ZY, Nair D, Partyka MI, Kadowitz PJ, Belmadani S, Matrougui K. Modified multipotent stromal cells with epidermal growth factor restore vasculogenesis and blood flow in ischemic hind-limb of type II diabetic mice. Lab Invest. 2010;90:985–96.
52. Comerota AJ, Link A, Douville J, Burchardt ER. Upper extremity ischemia treated with tissue repair cells from adult bone marrow. J Vasc Surg. 2010;52:723–9.
53. Fazan Jr R, Dias da Silva VJ, Ballejo G, Salgado HC. Power spectra of arterial pressure and heart rate in streptozotocin-induced diabetes in rats. J Hypertens. 1999;17:489–95.
54. Abdel Aziz MT, El-Asmar MF, Haidara M, Atta HM, Roshdy NK, Rashed LA, Sabry D, Youssef MA, Abdel Aziz AT, Moustafa M. Effect of bone marrow-derived mesenchymal stem cells on cardiovascular complications in diabetic rats. Med Sci Monit. 2008;14:249–55.
55. Volarevic V, Ljujic B, Stojkovic P, Lukic A, Arsenijevic N, Stojkovic M. Human stem cell research and regenerative medicine-present and future. Br Med Bull. 2011;99:155–68.
56. Lu D, Zhang L, Wang H, Zhang Y, Liu J, Xu J, Liang Z, Deng W, Jiang Y, Wu Q, Li S, Ai Z, Zhong Y, Ying Y, Liu H, Gao F, Zhang Z, Chen B. Peroxisome proliferator-activated receptor-γ coactivator-1α (PGC-1α) enhances engraftment and angiogenesis of mesenchymal stem cells in diabetic hindlimb ischemia. Diabetes. 2012;61:1153–9.
57. Lee HC, An SG, Lee HW, Park JS, Cha KS, Hong TJ, Park JH, Lee SY, Kim SP, Kim YD, Chung SW, Bae YC, Shin YB, Kim JI, Jung JS. Safety and effect of adipose tissue-derived stem cell implantation in patients with critical limb ischemia: a pilot study. Circ J. 2012;76:1750–60.
58. Ezquer FE, Ezquer ME, Parrau DB, Carpio D, Yañez AJ, Conget PA. Systemic administration of multipotent mesenchymal stromal cells reverts hyperglycemia and prevents nephropathy in type 1 diabetic mice. Biol Blood Marrow Transplant. 2008;14:631–40.
59. Lee RH, Seo MJ, Reger RL, Spees JL, Pulin AA, Olson SD, Prockop DJ. Multipotent stromal cells from human marrow home to and promote repair of pancreatic islets and renal glomeruli in diabetic NOD/scid mice. Proc Natl Acad Sci USA. 2006;103:17438–43.
60. Herrera MB, Bussolati B, Bruno S, Fonsato V, Romanazzi GM, Camussi G. Mesenchymal stem cells contribute to the renal repair of acute tubular epithelial injury. Int J Mol Med. 2004;14:1035–41.
61. Park JH, Park J, Hwang SH, Han H, Ha H. Delayed treatment with human umbilical cord blood-derived stem cells attenuates diabetic renal injury. Transplant Proc. 2012;44:1123–6.
62. Fang Y, Tian X, Bai S, Fan J, Hou W, Tong H, Li D. Autologous transplantation of adipose-derived mesenchymal stem cells ameliorates streptozotocin-induced diabetic nephropathy in rats by inhibiting oxidative stress, pro-inflammatory cytokines and the p38 MAPK signaling pathway. Int J Mol Med. 2012;30:85–92.
63. Vinik AI, Park TS, Stansberry KB, Pittenger GL. Diabetic neuropathies. Diabetologia. 2000;43:957–73.
64. Kinnaird T, Stabile E, Burnett MS, Shou M, Lee CW, Barr S, Fuchs S, Epstein SE. Local delivery of marrow-derived stromal cells augments collateral perfusion through paracrine mechanisms. Circulation. 2004;109:1543–9.
65. Shibata T, Naruse K, Kamiya H, Kozakae M, Kondo M, Yasuda Y, Nakamura N, Ota K, Tosaki T, Matsuki T, Nakashima E, Hamada Y, Oiso Y, Nakamura J. Transplantation of bone marrow-derived mesenchymal stem cells improves diabetic polyneuropathy in rats. Diabetes. 2008;57:3099–107.
66. Volarevic V, Erceg S, Bhattacharya S, Stojkovic P, Horner P, Stojkovic M. Stem cell based therapy for spinal cord injury. Cell Transplant. 2012 Oct 3. doi:10.3727/096368912X657260.

67. Yang Z, Li K, Yan X, Dong F, Zhao C. Amelioration of diabetic retinopathy by engrafted human adipose-derived mesenchymal stem cells in streptozotocin diabetic rats. Graefes Arch Clin Exp Ophthalmol. 2010;248:1415–22.
68. Scalinci SZ, Scorolli L, Corradetti G, Domanico D, Vingolo EM, Meduri A, Bifani M, Siravo D. Potential role of intravitreal human placental stem cell implants in inhibiting progression of diabetic retinopathy in type 2 diabetes: neuroprotective growth factors in the vitreous. Clin Ophthalmol. 2011;5:691–6.
69. Medina A, Scott PG, Ghahary A, Tredget EE. Pathophysiology of chronic nonhealing wounds. J Burn Care Rehabil. 2005;26:306–19.
70. Kwon DS, Gao X, Liu YB, Dulchavsky DS, Danyluk AL, Bansal M, Chopp M, McIntosh K, Arbab AS, Dulchavsky SA, Gautam SC. Treatment with bone marrow-derived stromal cells accelerates wound healing in diabetic rats. Int Wound J. 2008;5:453–63.
71. Kuo YR, Wang CT, Cheng JT, Wang FS, Chiang YC, Wang CJ. Bone marrow-derived mesenchymal stem cells enhanced diabetic wound healing through recruitment of tissue regeneration in a rat model of streptozotocin-induced diabetes. Plast Reconstr Surg. 2011;128:872–80.
72. Wu Y, Chen L, Scott PG, Tredget EE. Mesenchymal stem cells enhance wound healing through differentiation and angiogenesis. Stem Cells. 2007;25:2648–59.
73. Falanga V, Iwamoto S, Chartier M, Yufit T, Butmarc J, Kouttab N, Shrayer D, Carson P. Autologous bone marrow-derived cultured mesenchymal stem cells delivered in a fibrin spray accelerate healing in murine and human cutaneous wounds. Tissue Eng. 2007;13:1299–312.
74. Kim SW, Zhang HZ, Guo L, Kim JM, Kim MH. Amniotic mesenchymal stem cells enhance wound healing in diabetic NOD/SCID mice through high angiogenic and engraftment capabilities. PLoS One. 2012;7:e41105.
75. Shen L, Zeng W, Wu YX, Hou CL, Chen W, Yang MC, Li L, Zhang YF, Zhu CH. Neurotrophin-3 accelerates wound healing in diabetic mice by promoting a paracrine response in mesenchymal stem cells. Cell Transplant. 2012 Oct 3. doi: 10.3727/096368912X657495.
76. Vojtassák J, Danisovic L, Kubes M, Bakos D, Jarábek L, Ulicná M, Blasko M. Autologous biograft and mesenchymal stem cells in treatment of the diabetic foot. Neuro Endocrinol Lett. 2006;27:134–7.
77. Lataillade JJ, Doucet C, Bey E, Carsin H, Huet C, Clairand I, Bottollier-Depois JF, Chapel A, Ernou I, Gourven M, Boutin L, Hayden A, Carcamo C, Buglova E, Joussemet M, de Revel T, Gourmelon P. New approach to radiation burn treatment by dosimetry-guided surgery combined with autologous mesenchymal stem cell therapy. Regen Med. 2007;2:785–94.
78. Dash NR, Dash SN, Routray P, Mohapatra S, Mohapatra PC. Targeting nonhealing ulcers of lower extremity in human through autologous bone marrow-derived mesenchymal stem cells. Rejuvenation Res. 2009;12:359–66.
79. Hou C, Shen L, Huang Q, Mi J, Wu Y, Yang M, Zeng W, Li L, Chen W, Zhu C. The effect of heme oxygenase-1 complexed with collagen on MSC performance in the treatment of diabetic ischemic ulcer. Biomaterials. 2013;34:112–20.
80. Yu M, Zhou W, Song Y, Yu F, Li D, Na S, Zou G, Zhai M, Xie C. Development of mesenchymal stem cell-implant complexes by cultured cells sheet enhances oseointegration in type 2 diabetic rat model. Bone. 2011;49:387–94.
81. Tolar J, Nauta AJ, Osborn MJ, Panoskaltsis Mortari A, McElmurry RT, Bell S, Xia L, Zhou N, Riddle M, Schroeder TM, Westendorf JJ, McIvor RS, Hogendoorn PC, Szuhai K, Oseth L, Hirsch B, Yant SR, Kay MA, Peister A, Prockop DJ, Fibbe WE, Blazar BR. Sarcoma derived from cultured mesenchymal stem cells. Stem Cells. 2007;25:371–9.
82. Mottaghi S, Larijani B, Sharifi AM. Apelin 13: a novel approach to enhance efficacy of hypoxic preconditioned mesenchymal stem cells for cell therapy of diabetes. Med Hypotheses. 2012;79(6):717–8. doi:10.1016/j.mehy.2012.08.007.
83. Le Blanc K, Pittenger M. Mesenchymal stem cells: progress toward promise. Cytotherapy. 2005;7:36–45.
84. Atsma DE, Fibbe WE, Rabelink TJ. Opportunities and challenges for mesenchymal stem cell-mediated heart repair. Curr Opin Lipidol. 2007;18:645–9.

Mesenchymal Stromal Cell (MSC) Therapy for Crohn's Disease

Jignesh Dalal

Abstract Promising results of MSCs infusion for GI graft versus host disease and fistulizing Crohn's Disease (CD) have been published. Treatment of Inflammatory Bowel Disease (IBD) requires a comprehensive treatment approach to maintain symptomatic control, improve health-related quality of life measures, and minimize complications from the disease. In this chapter, we will discuss the role of cellular therapies in Crohn's disease. Success of these phase I, II and III trials have set the stage for usage of this novel treatment in combination with other therapies for CD

Keywords MSC • Crohn's disease • Inflammatory bowel disease

Introduction

Crohn's Disease (CD) and Ulcerative Colitis (UC) are the major forms of Inflammatory Bowel Disease (IBD) causing significant morbidity, economic burden and even mortality [1, 2]. Exact etiology remains unknown but a variety of factors have been thought to be contributors, including environment, the individual adaptive-innate immune responses, epithelial barrier function. Even though use of anti TNF therapies are becoming more wide spread only one third of patients have long lasting remission [3]. Despite systemic approach to therapy and addition of new biologics, the need for intestinal resection in CD has remained stable [4]. Primary and secondary failure to respond to approved therapies, inability to provide a surgical solution to fistulising manifestations and the recurrent need of surgeries are still challenges requiring novel therapies in these disorders [5].

J. Dalal, M.D. (✉)
Department of Pediatrics, Children's Mercy Hospital,
2401 Gillham Rd, Kansas City, MO 64108, USA
e-mail: jddalal@cmh.edu

R.C. Zhao (ed.), *Essentials of Mesenchymal Stem Cell Biology and Its Clinical Translation*, DOI 10.1007/978-94-007-6716-4_15,

Incidence

Although incidence and prevalence of UC and CD are beginning to stabilize in high-incidence areas such as northern Europe and North America, they continue to rise in low-incidence areas such as southern Europe, Asia, and much of the developing world. As many as 1.4 million people in the United States (US) and 2.2 million people in Europe suffer from these diseases [6]. Approximately 20,000 new cases are diagnosed each year in US [7]. The incidence in children is increasing and the overall prevalence in children may be 10–25 % of all patients [7]. The prevalence of CD in North America ranges from 26.0 to 198.5 cases per 100,000 people. The incidence rates range from 3.1 to 14.6 cases per 100,000 people/year [8]. Most patients have a chronic intermittent disease course, 13 % have an unremitting disease course, and 10 % have a prolonged remission. Less than half require corticosteroids at any point and all requires surgery at least once.

MSC

When cells from a bone marrow aspirate are cultured in plastic flasks, haematopoietic cells and Hematopoietic stem cells (HSC)s do not adhere to the plastic and are removed with change of media. The remaining plastic-adherent cells were originally termed colony-forming unit fibroblasts because they formed fibroblast-like colonies ex vivo. Subsequently, these adherent cells have been termed MSCs, an abbreviation for both mesenchymal stromal cells and mesenchymal stem cells [9, 10]. Similar to HSCs, MSCs are rare in the bone marrow, representing 1 in 10,000 nucleated cells. MSCs are multipotent bone-marrow cells able to differentiate in vitro and in vivo into tissues of mesenchymal origin. Moreover, these cells provide support for the growth and differentiation of hematopoietic progenitor cells in bone-marrow microenvironments, and in animal models, promote engraftment of hemopoietic cells [11]. The unexpected observation that MSCs inhibited T-cell proliferation *in vitro* in 2002–2003 by three different investigators [10, 12, 13] opened up the door for use of MSCs for autoimmune disorders first in animal models and then in humans. In co-culture experiments with allogeneic lymphocytes, MSCs do not induce lymphocyte proliferation, interferon-γ production, or upregulation of activation markers. MSCs suppress proliferation of activated lymphocytes in vitro in a dose-dependent, non-HLA-restricted, manner [12]. In a baboon skin-graft model, Bartholomew and coworkers showed that infusion of ex-vivo expanded donor-derived or third-party cells prolonged the time to rejection of histoincompatible skin grafts [12]. Furthermore, infused cells improve the outcome of acute renal, neural, and lung injury, possibly by promoting a shift from production of proinflammatory cytokines to anti-inflammatory cytokines at the site of injury [11]. In 2005, Zappia and coworkers demonstrated that the intravenous (IV) injection of syngeneic MSCs ameliorated the clinical course of myelin oligodendrocyte glycoprotein (MOG)-induced experimental autoimmune encephalomyelitis (EAE) with reduction of demyelination and leukocytes infiltration of the CNS [14]. MSCs have proven to be safe and not result in toxicity or ectopic tissue growth in an increasing number of human trials. However, they do have this potential which is something that needs to be considered when designing trials [15].

Pathophysiology of CD

Role of Genetics and CD

Twin and family studies confirmed a strong genetic influence on the acquisition of CD. For example, approximately 50 % of monozygotic twins and 30 % of offspring of two affected parents developed the disease [16]. Linkage analysis and positional cloning strategies, together with subsequent genome-wide association (GWA) studies, have identified over 30 distinct genetic loci that confer susceptibility. Some of the most strongly associated genes included CARD15, the IBD5 locus, the autophagy genes ATG16L1 and IRGM, and the IL-23 receptor [17]. In CD, it is estimated that all the genes identified so far account for less than 20 % of the total genetic risk [18]. The unifying findings in all CD patients are therefore phenotypic abnormalities, rather than defects in single genes. Polymorphisms in the identified genes, and many more that are as of yet undiscovered, may confer susceptibility by contributing to mucosal barrier dysfunction, the innate immunodeficiency state, or by influencing the propagation of chronic inflammation in the tertiary phase of the disease [17].

Inflammatory Cytokine Involvement in CD

Cytokine is a collective term for a group of low-molecular-weight peptides that are active at very low concentrations and bind to specific receptors to produce autocrine, paracrine, and other endocrine effects. In CD, the major cytokines arise from T-helper cell (Th) 1 and Th17, CD4 T-cell differentiation and consist of interferon and interleukin (IL)-17/IL-22 generated by these types of differentiation [19]. In contrast, in UC, a Th2-like differentiation process is paramount, which results in expansion of natural killer T-cells producing IL-13 and IL-5. IFN-gamma production remains normal and IL-4 is also not increased. Human trials with anti-cytokine therapy have been successful. This clearly establishes a major role for cytokines in this disease and stresses the importance of understanding the pathophysiology when devising therapeutic approaches [20, 21].

Role of Microbiota

A lot of literature suggesting an association between the microbiota and the immune system has recently emerged [21, 22]. Shifts in the intestinal microenvironment may lead to changes in the microbiota known as dysbiosis. Dysbiosis may increase susceptibility to intestinal inflammation [23]. In support of this hypothesis, T-bet$^{-/-}$ Rag2$^{-/-}$ (TRUC) mice spontaneously develop dysbiosis and colitis, which can eventually progress into colorectal cancer [24]. Remarkably, microbiota transfer from these donors into wild type mice can confer disease [25]. Subsequent studies identified two proteobacteria over-represented in TRUC mice, Proteus mirabilis and Klebsiella pneumoniae, as the colitogenic microbes [26]. Recently, a new clinical syndrome of culture-negative, antibiotic-responsive diarrhea is described in 10 % of patients undergoing a cord blood transplant. This is distinct from Intestinal GVH and responds to Metronidazole with or without fluoroquinolones. On histologic examination, all patients with the cord colitis syndrome had chronic active colitis, with granulomatous inflammation present in 7 out of 11 patients (64 %). Five of the eleven patients (45 %) had recurrent diarrhea shortly after discontinuation of antibiotics, and all patients who had a relapse had a response to reinitiation of antibiotic therapy. Again, this clinical entity supports an association between dysbiosis and inflammation. Another argument in support of this hypothesis is association of chronic granulomatous disease (CGD), an immunodeficiency disorder caused by mutations in NADPH oxidase, with bowel inflammation that is indistinguishable from CD [27]. The cellular defect in this condition is not limited to bacterial killing; digestion is also severely impaired as a consequence of abnormal pH and charge compensation [28]. This may lead to inability of neutrophils to degrade and remove the bacteria and other bowel contents leading to inflammation mimicking CD.

Immunologic Effects of MSCs in IBD

The study of human MSCs derived from subcutaneous adipose tissue (hASC) in 5 % dextran sulfate sodium (DSS) mouse model has shown that hASCs inhibit T-cell activation with the superantigen staphylococcal enterotoxin E (SEB), as measured by cytokine secretion and T-cell proliferation. The inhibitory effect was partially reversed when peripheral blood mononuclear cells (PBMC)s and hASCs were separated by a semi-permeable transwell membrane suggesting a partial cell–cell contact dependence [29]. Moreover, the co-culture of allogeneic PBMCs and hASCs in the same chamber of the transwell system fully restored their inhibitory activity on SEB-activated PBMCs situated in the other chamber, suggesting that PBMC–hASC contact induces the secretion of an immunosuppressive factor(s) for T-cells. IL-10 production increased in a cell–cell contact-dependent manner in co-cultures of hASCs with PBMCs or monocytes, but not with T-cells. IL-10 blockade partially reversed the inhibitory activity of hASCs on T-cells.

Colons of hASC-treated mice contained reduced levels of inflammatory cytokines (TNFa, IFNc, IL6, IL1b and IL12), chemokines (RANTES), macrophage inflammatory protein-2 (MIP-2), and increased levels of the anti-inflammatory/regulatory cytokine IL-10, in comparison with untreated DSS colitic mice. This effect was not only a consequence of a diminished inflammatory infiltration in the mucosa because mononuclear cells isolated from the lamina propria of hASC-treated mice produced less TNFa, IL12 and IFNc on ex vivo culture, suggesting that hASCs deactivated the colonic inflammatory response [29]. hASC-treated colitic mice had significantly higher numbers of CD4+CD25+Foxp3+ Treg in mesenteric lymph nodes than untreated colitic mice that persisted for a long period of time. In vivo depletion of IL10 or CD25+ T-cells partially reversed the beneficial action of hASCs on colitis, demonstrating the involvement of IL-10 and Treg in their therapeutic effect. Both syngeneic and allogeneic murine ASCs (mASCs) were as efficient as hASCs in ameliorating the colitis suggesting that the immunosuppressive action of ASCs is non-major histocompatibility complex (MHC)-restricted and that the infused ASCs are immune-tolerated by the host, which is very convenient for a future clinical application of these cells in IBD [29].

Mesenchymal Stem Cells and Regeneration of Colonic Epithelium

MSCs can differentiate to different cell types like myofibroblast, epithelial- and endothelial cells in the gastrointestinal tract [30]. After differentiation ISEMFs (intestinal subepithelial myofibroblasts) form pericryptal fibroblast sheet adjacent to of the basal lamina of cryptal epithelium in the lamina propria [31]. They provide important microenvironment by secreting secrete several factors, like cytokines (IL-1, -6, -10, TNF-α), growth factors (GM-CSF, PDGF, bFGF, KGF, HGF), chemokines (IL-8, MCP1, MIP-1α,2) and inflammatory mediators (PGE2, prostacyclin, PAF) [32]. These cytokines are important for regeneration of epithelium. They also play key role in epithelial differentiation, migration and formation of new basement membrane [31]. The number of myofibroblast originating from the bone marrow significantly increased in lamina propria in colitis patients as compared to healthy control [33, 34].

Second possible mechanism is by ability of MSCs to differentiate into functional epithelial cells. If there is strong regenerative stimulus MSCs become polarized to epithelium by process called mesenchymal to epithelial transition (MET) [35]. Eventually, MSCs may perform lineage specific functions, such as nutrition, absorption, production of mucin, cytokeratin and chromogranin.

Cellular Therapy Clinical Trials in IBD

Hematopoietic Stem Cell (HSC) Therapy

CD is heterogenenous disease as evidence by GWAS study. Cellular therapy similarly as treatment, is heterogeneous as evidence from literature review. Autologous and allogenic BMT, local and systemic infusion of MSC from different sources, and regulatory T cell therapy has been employed successfully for treatment of CD. Since the report in 1993 of the first CD patient who underwent autologous HSCT for lymphoma resulting in improvement of CD, at least 25 patients have been reported who underwent HSCT for cancer, achieving remission in 22 cases, over a median follow up of 20 months [4]. In phase I study from Chicago, 12 patients with active moderate to severe CD refractory to conventional therapies including anti-tumour necrosis factor (TNF) treatment underwent auto transplant with high dose cyclophosphamide ± ATG (Equine or rabbit) as preparative regimen. Eleven of twelve patients entered a sustained remission after a median follow-up of 18.5 months. Only one patient had developed recurrence of active CD [36]. Second study from Milan published in 2008 included four patients who had failed immunosuppressant treatment had undergone multiple surgical resections. Three months after transplant, all patients had achieved clinical remission and complete endoscopic remission was achieved in two out of three patients. Three of the four patients had sustained remission after a median follow-up of 16.5 months. No mortality was observed in the two series [37]. A Phase III study from Nottingham, UK is testing the role of early versus late autologous HSCT in treatment of refractory CD, has completed accrual of 45 patients recruiting patients of CD. Initial analysis will be performed in March 2013 [38]. Inspite of good initial response of autologous BMT relapse free survival at 5 years is 19 % for 18 patients who are more than 5 years from initial transplantation [39].

After first report of IL10 receptor mutation leading to severe early onset CD in 2009 allogenic BMT is now becoming established treatment for that subset of patients who usually remains resistant to standard medical and surgical treatment. In recently published article from Germany, 5 out of 16 patients with infantile onset CD, underwent allogenic HSCT, achieving sustained remission in CD. There was no transplant related mortality [40].

After reported success of Treg therapy in SCID mice in 1997 [41], data from Crohn's And Treg Cells Study (CATS1) study is published in 2012 in humans. In phase 1/2 clinical study in 20 patients with refractory CD, infusion of Ovalbumin-specific Treg cells (ova-Tregs) resulted in significant improvement in 40 % of patients at weeks 5 and 8 [42].

Systemic MSC Infusion Therapy

Allogenic MSC

In a first human trial of MSc in refractory CD, Onken et al. from Duke University treated ten patients with CD. Randomization scheme was to receive either low (two million cells/kg) or high (eight million cells/kg) dose IV infusions; two doses 7 days apart. All nine patients responded with decrease in CDAI score with significant clinical response (defined as a ≥100-point reduction in CDAI) achieved in three patients (33 %) by day 14. All clinical responders had previously failed infliximab therapy. Although not statistically significant, the mean reduction in the CDAI score at day 28 was greater in the high dose than the low dose group (−137 vs. −65, $p=0.39$, WRS). All infusions were well tolerated and there were no treatment-related serious adverse events [43].

Very recently Jiang et al. reported experience of seven patients with IBD, who received single IV infusion of 1×10^6 allogenic MSC from related healthy donor. All seven patients experienced improvement in symptoms and five of seven patients had complete remission [44].

The largest, randomized, placebo controlled, double-blind Phase III study of Prochymal (allogenic bone marrow derived MSC) in CD was initiated in 2007 by Osiris [45]. The plan is to enroll 270 patients with active CD (CDAI 250–450); who had a history of treatment failure with or intolerance to steroids, immunosuppressants and biologics. Patients are randomized to receive four infusions over 2 weeks of either 600 million cells (low dose: two infusions of 200×10^6 hMSC in week 1, then two infusions of 100×10^6 hMSC in week 2), 1,200 million cells (high dose: two infusions of 400×10^6 hMSC in week 1, then two infusions of 200×10^6 hMSC in week 2) or placebo. The primary endpoint of the study was remission at day 28 with secondary endpoints being clinical response, improved quality of life (increased IBDQ score) and decreased number of draining fistulae. The trial was put on hold after enrollment of 207 patients as an interim analysis showed an unexpected high rate of response in the placebo arm. After further analysis of the blinded data, last year, it was determined that the endpoint of remission had been significantly met in the treatment-per-protocol group and was approaching significance in the intent-to-treat group [46]. The trial was reopened after FDA permission to complete enrollment of 270 patients with a 1:1:1 distribution in low dose, high dose and placebo arm and complete accrual by December 2014. Patients who reported a response in induction became eligible for an open-label extension. Retrospectively, this aspect of the trial design was thought to be responsible for the high reported response in the placebo arm. Therefore, the open label protocol for response maintenance has been discontinued, with approval by the FDA, to remove this potential source of bias [47].

Autologous MSC

Another phase I study of autologous MSCs for luminal refractory CD was published from the Netherlands. Nine patients received two doses of $1–2\times10^6$ cells/kg body weight, intravenously 7 days apart. All patients had previously failed corticosteroids, at least two anti-tumour necrosis factor (TNF) drugs, and the majority (9/10) also had failed two immunosuppressants (a thiopurine and methotrexate). In this study, no clear signal of efficacy was observed; remission was not achieved in any patient, three patients had a reduction of at least 70 points in CDAI, but the disease worsened significantly in four patients requiring surgery (three cases) or rescue medication (one case) within 14 weeks after cell treatment. Endoscopy improved in two cases but no significant changes in C-reactive protein levels were seen [48].

Local Autologous MSC Therapy for Fistula

An internal pocket in the fistula tract remaining unhealed after infliximab treatment is presumed to be the predominant reason for recurrence. Standard management of Crohn's fistulae involves combination of surgical intervention and infliximab. The first trial of cell therapy using autologous MSCs (ASC) obtained from a lipoaspirate for local treatment of fistulae for CD in five patients was published in 2005 [49]. The same group published a phase 2 multicenter randomized controlled trial describing effectiveness and safety of ASCs in the treatment of complex perianal fistulas in 2009. Patients with complex perianal fistulas (cryptoglandular origin, $n = 35$; associated with Crohn's disease, $n = 14$) were randomly assigned to intralesional treatment with fibrin glue or fibrin glue plus 20 million ASCs. Fistula healing and quality of life (SF-12 questionnaire) were evaluated at 8 weeks and 1 year. If healing was not seen at 8 weeks, a second dose of fibrin glue or fibrin glue plus 40 million ASCs was administered. Fistula healing was observed in 17 (71 %) of 24 patients who received ASCs in addition to fibrin glue, compared with 4 (16 %) of 25 patients who received fibrin glue alone (relative risk for healing, 4.43; confidence interval, 1.74–11.27; $P < 0.001$). The proportion of patients with healing was similar in Crohn's and non-Crohn's subgroups. ASCs were also more effective than fibrin glue alone in patients with a suprasphincteric fistulous tract ($P = 0.001$). Quality of life scores were higher in patients who received ASCs than in those who received fibrin glue alone. At 1 year follow-up, the recurrence rate in patients treated with ASCs was 17.6 %. Both treatments were well tolerated [50]. In a second study published recently, a local injection of MSCs was given to nine patients with perianal and one patient with enterocutaneous fistulas. Injections of a median of 20×10^6 cells (range 15–30) were given every 4 weeks until a response was obtained or 'no more cells were available.' Complete fistula closure sustained for 1 year was obtained in seven and a response (reduction of at least 50 % of fistula tracts) in three. Furthermore, all nine patients with perianal fistulas had active disease in the rectum at baseline, and healing of rectal lesions was observed in the seven patients who underwent endoscopy at month 12 of follow-up. Thus, the latter study suggests a considerable therapeutic benefit of local injection of MSCs in fistulising lesions [5]. After the first publication of success of MSC in steroid refractory gut and liver GVHD by Le Blanc et al. in 2004, many trials have published its role in treatment of acute and chronic GVHD from single or multiple donors from the US and Europe in multicenter trials [11, 51–54]. More than 300 patients including adults and children have been treated. Reported response rates range from 60 to 80 % with more responses for pediatric patients for Gut and Liver GVH [11, 53]. Infusion of MSCs appeared to be safe and no major toxicities, including ectopic tissue growth, were observed. There is a significant difference in survival between complete responders, partial and non-responding patients [11]. At this stage, reasons for the apparent discrepancies between efficacy of local injection of MSCs for treatment of fistulas compared to systemic administration for treatment of luminal Crohn's disease are not completely clear. MSCs have been reported to home to sites of injury and disease following intravenous infusion and contribute to the repair process. The expression of adhesion molecules and chemokine receptors on MSCs may be responsible for their ability to migrate selectively to sites of inflammation through a ICAM1- and VCAM1- dependent interaction with endothelial cells [55]. In an experimental model of colitis, it has been demonstrated that systemically injected MSCs are detected in the mesenteric lymph nodes and spleen of the recipient colitic mice 1–3 days post-injection [29]. Interestingly, labeled MSCs were recruited by the inflamed colon but not by the non-inflamed intestine. However, the proportion of cells recruited to inflamed or damaged organs and the survival of cells at sites of inflammatory lesions remains to be clarified, a necessary prerequisite for optimizing a potential systemic treatment. In studies showing efficacy of local injections, $30–60 \times 10^6$ MSCs are injected in a single fistulous tract and these injections are generally repeated [5, 50]. In the study using systemic injection for treatment of luminal disease [48], a total amount of $100–400 \times 10^6$ MSCs were injected (depending on the patient's weight). Considering the extension of the inflamed intestine, and the fact that only a proportion of MSCs will reach the inflamed organ, cell density at sites of luminal inflammation would be considerably lower than that achieved in fistula tracts by local injection. To circumvent this issue Ghosh et al. from UK injected haploidentical MSCs after catheterization of the mesenteric artery via the femoral route into a 35 year old patient with severe refractory fistulizing CD failing all conventional therapies, biological therapies and surgical defunctioning ileostomy. The patient received 10^5/kg MSCs and 4 weeks later a second dose of 10^6/kg [56]. CDAI pre treatment was 384 and dropped to 258 two weeks after the first infusion and remained as such at the time of the second infusion administered after 4 weeks. All of above studies point toward a difference in cell density achieved at inflammatory sites, with systemic and local injections.

Conclusion and Future Direction

Meta-analysis of MSC for human use has confirmed the safety in large numbers of patients. Paradigm for refractory CD treatment is gradually shifting towards cellular therapies. There is an apparent discrepancy in responses to MSCs for CD based on route of administration (IV vs. intralesional vs. intra-arterial).

Many challenges remain ahead, including determining the best source of MSCs, the best administration route, and the density of cells needed at the site of lesion to guarantee effective therapy. In addition, it will be especially important to determine which combination of this modality with other approaches, including biologics, are effective in treating in IBD. Better understanding of MSC priming and the molecular mechanisms is needed. Clinicians need to establish algorithm based approach depending on genetic mutations, response to standard treatment, for better treatment of CD patients.

References

1. Kappelman MD et al. Direct health care costs of Crohn's disease and ulcerative colitis in US children and adults. Gastroenterology. 2008;135(6):1907–13.
2. Hanauer S. Inflammatory bowel disease: epidemiology, pathogenesis, and therapeutic opportunities. Inflamm Bowel Dis. 2006;12(supplement 1):S3–9.
3. Peyrin-Biroulet L, Lemann M. Review article: remission rates achievable by current therapies for inflammatory bowel disease. Aliment Pharmacol Ther. 2011;33(8):870–9.
4. Garcia-Bosch O, Ricart E, Panes J. Review article: stem cell therapies for inflammatory bowel disease – efficacy and safety. Aliment Pharmacol Ther. 2010;32(8):939–52.
5. Ciccocioppo R et al. Autologous bone marrow-derived mesenchymal stromal cells in the treatment of fistulising Crohn's disease. Gut. 2011;60(6):788–98.
6. Loftus Jr EV. Clinical epidemiology of inflammatory bowel disease: incidence, prevalence, and environmental influences. Gastroenterology. 2004;126(6):1504–17.
7. Parashette KR, Makam RC, Cuffari C. Infliximab therapy in pediatric Crohn's disease: a review. Clin Exp Gastroenterol. 2010;3:57–63.
8. Andres PG, Friedman LS. Epidemiology and the natural course of inflammatory bowel disease. Gastroenterol Clin North Am. 1999;28(2):255–81. vii.
9. Lazarus HM et al. Ex vivo expansion and subsequent infusion of human bone marrow-derived stromal progenitor cells (mesenchymal progenitor cells): implications for therapeutic use. Bone Marrow Transplant. 1995;16(4):557–64.
10. Le Blanc K et al. Mesenchymal stem cells inhibit and stimulate mixed lymphocyte cultures and mitogenic responses independently of the major histocompatibility complex. Scand J Immunol. 2003;57(1):11–20.
11. Le Blanc K et al. Mesenchymal stem cells for treatment of steroid-resistant, severe, acute graft-versus-host disease: a phase II study. Lancet. 2008;371(9624):1579–86.
12. Bartholomew A et al. Mesenchymal stem cells suppress lymphocyte proliferation in vitro and prolong skin graft survival in vivo. Exp Hematol. 2002;30(1):42–8.
13. Di Nicola M et al. Human bone marrow stromal cells suppress T-lymphocyte proliferation induced by cellular or nonspecific mitogenic stimuli. Blood. 2002;99(10):3838–43.
14. Zappia E et al. Mesenchymal stem cells ameliorate experimental autoimmune encephalomyelitis inducing T-cell anergy. Blood. 2005;106(5):1755–61.
15. Mohseny AB, Hogendoorn PCW. Mesenchymal tumors: when stem cells go mad. Stem Cells. 2011;29:397–403.
16. Halme L et al. Family and twin studies in inflammatory bowel disease. World J Gastroenterol. 2006;12(23):3668–72.
17. Sewell GW, Marks DJ, Segal AW. The immunopathogenesis of Crohn's disease: a three-stage model. Curr Opin Immunol. 2009;21(5):506–13.
18. Barrett JC et al. Genome-wide association defines more than 30 distinct susceptibility loci for Crohn's disease. Nat Genet. 2008;40(8):955–62.
19. Strober W, Fuss IJ. Proinflammatory cytokines in the pathogenesis of inflammatory bowel diseases. Gastroenterology. 2011;140(6):1756–67.
20. Mannon PJ et al. Suppression of inflammation in ulcerative colitis by interferon-beta-1a is accompanied by inhibition of IL-13 production. Gut. 2011;60(4):449–55.
21. Neish AS. Microbes in gastrointestinal health and disease. Gastroenterology. 2009;136(1): 65–80.
22. Porter CK et al. Infectious gastroenteritis and risk of developing inflammatory bowel disease. Gastroenterology. 2008;135(3):781–6.
23. Chow J, Tang H, Mazmanian SK. Pathobionts of the gastrointestinal microbiota and inflammatory disease. Curr Opin Immunol. 2011;23(4):473–80.
24. Garrett WS et al. Colitis-associated colorectal cancer driven by T-bet deficiency in dendritic cells. Cancer Cell. 2009;16(3):208–19.

25. Garrett WS et al. Communicable ulcerative colitis induced by T-bet deficiency in the innate immune system. Cell. 2007;131(1):33–45.
26. Garrett WS et al. Enterobacteriaceae act in concert with the gut microbiota to induce spontaneous and maternally transmitted colitis. Cell Host Microbe. 2010;8(3):292–300.
27. Marks DJ et al. Inflammatory bowel disease in CGD reproduces the clinicopathological features of Crohn's disease. Am J Gastroenterol. 2009;104(1):117–24.
28. Reeves EP et al. Killing activity of neutrophils is mediated through activation of proteases by K+ flux. Nature. 2002;416(6878):291–7.
29. Gonzalez-Rey E et al. Human adult stem cells derived from adipose tissue protect against experimental colitis and sepsis. Gut. 2009;58(7):929–39.
30. Yabana T et al. Enhancing epithelial engraftment of rat mesenchymal stem cells restores epithelial barrier integrity. J Pathol. 2009;218(3):350–9.
31. Valcz G et al. The role of the bone marrow derived mesenchymal stem cells in colonic epithelial regeneration. Pathol Oncol Res. 2011;17(1):11–6.
32. Powell DW et al. Myofibroblasts. I. Paracrine cells important in health and disease. Am J Physiol. 1999;277(1 Pt 1):C1–9.
33. Brittan M et al. Bone marrow derivation of pericryptal myofibroblasts in the mouse and human small intestine and colon. Gut. 2002;50(6):752–7.
34. Brittan M, Wright NA. Stem cell in gastrointestinal structure and neoplastic development. Gut. 2004;53(6):899–910.
35. Baum B, Settleman J, Quinlan MP. Transitions between epithelial and mesenchymal states in development and disease. Semin Cell Dev Biol. 2008;19(3):294–308.
36. Oyama Y et al. Autologous hematopoietic stem cell transplantation in patients with refractory Crohn's disease. Gastroenterology. 2005;128(3):552–63.
37. Cassinotti A et al. Autologous haematopoietic stem cell transplantation without CD34+ cell selection in refractory Crohn's disease. Gut. 2008;57(2):211–7.
38. Group BaMTfE. Autologous stem cell transplantation for Crohn's disease: ASTIC Autologous Stem Cell Transplantation for Crohn's Disease. ClinicalTrials.gov, 2006(NCT00297193). 2006.
39. Burt RK et al. Autologous nonmyeloablative hematopoietic stem cell transplantation in patients with severe anti-TNF refractory Crohn disease: long-term follow-up. Blood. 2010;116(26):6123–32.
40. Kotlarz D et al. Loss of interleukin-10 signaling and infantile inflammatory bowel disease: implications for diagnosis and therapy. Gastroenterology. 2012;143(2):347–55.
41. Groux H et al. A CD4+ T-cell subset inhibits antigen-specific T-cell responses and prevents colitis. Nature. 1997;389(6652):737–42.
42. Desreumaux P et al. Safety and efficacy of antigen-specific regulatory T-cell therapy for patients with refractory Crohn's disease. Gastroenterology. 2012;143(5):1207–1217.e2.
43. Onken JE, Norata GD, Hanson J, Pandak M, Custer L. Successful outpatient treatment of refractory Crohn's disease using adult mesenchymal stem cells. In: American College of Gastroenterology annual meeting, Las Vegas; 2006.
44. Liang J et al. Allogeneic mesenchymal stem cell transplantation in seven patients with refractory inflammatory bowel disease. Gut. 2012;61(3):468–9.
45. Therapeutics O. A phase III, multicenter, placebo-controlled, randomized, double-blind study to evaluate the safety and efficacy of Prochymal[tm] (ex vivo cultured adult human mesenchymal stem cells)intravenous infusion for the induction of remission in subjects experiencing treatment-refractory moderate-to-severe Crohn's disease. ClinicalTrials.gov, 2010(NCT00482092). 2010.
46. Theraputics O. Evaluation of PROCHYMAL for treatment-refractory moderate-to-severe Crohn's disease. ClinicalTrials.gov, 2010(NCT01233960). 2010.
47. Osiris Therapeutics I. Osiris resumes enrollment in stem cell trial for crohn's disease following positive interim analysis. 2010.
48. Duijvestein M et al. Autologous bone marrow-derived mesenchymal stromal cell treatment for refractory luminal Crohn's disease: results of a phase I study. Gut. 2010;59(12):1662–9.

49. Garcia-Olmo D et al. A phase I clinical trial of the treatment of Crohn's fistula by adipose mesenchymal stem cell transplantation. Dis Colon Rectum. 2005;48(7):1416–23.
50. Garcia-Olmo D et al. Expanded adipose-derived stem cells for the treatment of complex perianal fistula: a phase II clinical trial. Dis Colon Rectum. 2009;52(1):79–86.
51. Le Blanc K et al. Treatment of severe acute graft-versus-host disease with third party haploidentical mesenchymal stem cells. Lancet. 2004;363(9419):1439–41.
52. Martin PJU, Uberti JP, Soiffer RJ, Klingemann H, Waller EK, Daly AS, et al. Prochymal improves response rates in patients with steroid-refractory acute graft versus host disease (SR-GVHD) involving the liver and gut: results of a randomized, placebo-controlled, multicenter phase iii trial in GVHD. Biol Blood Marrow Transplant. 2010;16:S169–70.
53. Prasad VK et al. Efficacy and safety of ex vivo cultured adult human mesenchymal stem cells (Prochymal) in pediatric patients with severe refractory acute graft-versus-host disease in a compassionate use study. Biol Blood Marrow Transplant. 2011;17(4):534–41.
54. Ringden O, Le Blanc K. Mesenchymal stem cells for treatment of acute and chronic graft-versus-host disease, tissue toxicity and hemorrhages. Best Pract Res Clin Haematol. 2011;24(1):65–72.
55. Chamberlain G et al. Concise review: mesenchymal stem cells: their phenotype, differentiation capacity, immunological features, and potential for homing. Stem Cells. 2007;25(11): 2739–49.
56. Dinesen L, Wang A, Vianello F. Mesenchymal stem cells administered via novel selective mesenteric artery cannulation for the treatment of severe refractory Crohn's disease. J Crohns Colitis. 2010;3:S51e2.

The Summary of Stroke and Its Stem Cell Therapy

Renzhi Wang, Ming Feng, Xinjie Bao, Jian Guan, Yang liu, and Jin Zhang

Abstract Stroke is one of the most common diseases affecting human health. It has a high mortality rate and many clinically difficult sequelae such as hemiplegia, numbness, dysphasia and cognitive deficits. Studies have confirmed that stem cells have therapeutic effects following stroke. However, there are still a number of uncertainties for the success of this treatment. What is the mechanism of action? The production of trophic and angiogenic factors or the replacement of the apoptotic or necrotic neurons? Which kind of cell is most suitable for transplantation?

Which is the most appropriate route for transplantation? In this article, we summarize the exiting studies about stem cells for stroke, including animal experiments and clinical trials, to explore the best scheme of stem cell therapy for stroke.

Keywords MSC • Stroke • Transplantation

Introduction

Stroke is a sudden onset of cerebral blood circulation disorders. It is also called the cerebral vascular accident. Because of various predisposing factors causing cerebral artery stenos is, occlusion or rupture, some patients with cerebrovascular disease, encounter acute disturbances of cerebral blood circulation, whose clinical performance is reflected as a temporary or permanent brain dysfunction with particular symptoms and signs. Stroke can be divided into ischemic stroke and hemorrhagic stroke [1].

R. Wang (✉) • M. Feng • X. Bao • J. Guan • Y. liu
Department of Neurosurgery, Peking Union Medical College Hospital, Chinese Academy of Medical Sciences and Peking Union Medical College, Beijing, 100730, P. R. China
e-mail: wangrz@126.com

J. Zhang
Department of Biochemistry, Dalian Medical University, Dalian, 116044, P. R. China

R.C. Zhao (ed.), *Essentials of Mesenchymal Stem Cell Biology and Its Clinical Translation*, DOI 10.1007/978-94-007-6716-4_16,

Ischemic stroke accounts for approximately 80 % of all strokes. It means that the local brain tissue encounters ischemia, hypoxia and softened necrosis due to disturbance of blood circulation. Because atherosclerosis and thrombosis happen on the arteries that supply blood to the brain, stenosis or occlusion occurs, causing acute cerebral blood insufficiency in focal area. In some other cases, certain abnormal objects (solid, liquid, gas) enter the brain arteries or the arteries of the neck that supply blood to the brain along the blood circulation, causing vascular occlusion or blood flow loss, then the brain tissue softening necrosis occurs. The former is called Arteriosclerotic Thrombotic Cerebral infarction, the latter is called the cerebral embolism.

Hemorrhagic stroke was divided into two subtypes: intracranial hemorrhage (ICH) and subarachnoid hemorrhage (SAH). The volume of the blood determines the severity of stroke. The mortality of Hemorrhagic stroke is significantly higher than that of ischemic stroke [2].

Why Is Stroke Treatment Urgent?

Stroke is one of the leading causes of death in the world, second only to ischemic heart disease. About more than five million people die from stroke each year. And this is also the most common cause of making people encounter disabilities. In Europe, about 250,000 people are disabled because of a stroke every year. Though the current health conditions are gradually improved, emergency care conditions are getting better and better, stroke is still a huge burden worldwide, because it would consume too much health care resources and property [3]. Thrombolysis with tissue plasminogen activator is a recognized effective method for the treatment of acute ischemic stroke, but its use is limited because it can only be effective within the first 4.5 h after the stroke happened. In the UK, only 5 % of stroke patients are undergoing this kind of treatment.

Stem Cell Therapy in Animal Experiments

Stem cell therapy is an emerging therapeutic modality in the treatment of stroke. Cell-based therapies have the potential to open up new avenues of treatment in this arena. Targets for stem cell therapy include neuroprotective approaches aimed at protecting at-risk tissue during the acute phase of stroke, as well as neuroreparative approaches which may involve direct replacement of damaged brain tissue, or alternatively promotion of the brain's endogenous repair processes.

Broadly speaking, stem cell therapy in clinic can be divided into "endogenous" and "exogenous" method. The endogenous approach aims to mobilize the stem cell within the body. This kind of method includes the use of granulocyte colony-stimulating factor (G-CSF) to mobilize hematopoietic stem cells to enter the peripheral blood. The endogenous approach is application of stem cell from extraneous. Neural stem cells (NSCs) have the ability to differentiate into neurons, astrocytes and oligodendrocytes [4–7]. Exogenous neural stem cells have many kinds of sources. One of the most common sources is embryonic and fetal neural stem cells. However, because of a series of ethical problems, embryonic or fetal neural stem cells use is limited. Further more, we are unable to clarify the possibility if embryonic or fetal neural stem cells in the future would have the problem of abnormal development. Seeing that, its use is more restricted. The emergence of induced pluripotency, whereby skin fibroblasts from patients can be transformed into ESC-like cells, may overcome the ethical issues of using fetal or embryonic tissue in future studies, while also providing a source of autologous cells, thereby removing any questions regarding immune rejection. Some studies also considered extracting the neural stem cells from the adult central nervous system to replace embryonic or fetal neural stem cells [6]. There have been reports that recorded the whole process of the extraction and isolation of neural stem cells from adult mouse brain. One study showed that, in adult rat subventricular zone, the status of neural stem cells' survival and migration have a tight relationship with whether the rats encount ischemic stroke and its severity. Although possible in principle, brain biopsies for isolation of adult human NSCs and autologous transplantation poststroke is technically difficult, and no clinical trials utilizing adult NSCs have been undertaken [6, 8, 10].

What Is the Mechanism of Action?

Explanation of the potential mechanisms is very important for further development of the stem cell treatment techniques and finally pushing the kind of treatment method to clinic in large scale. There are a number of proposed mechanisms that have been investigated in preclinical stroke models. We did many animal models experiments for the mechanisms all that can explain the effect of treatment of stem cell.

MSCs Promotes Function Recovery in Small Animal with MACO

The capacity of MSCs to release growth and trophic factors, or to stimulate their release from resident bran cells, has been suggested to contribute to the beneficial effect in cerecral ischemia. Indeed, delivery of MSCs in stroke models leads to reduced apoptosis of cells at the lesion boundary [8, 9] and promotes endogenous cells proliferation [11, 12]. Low-level basal secretion of multiple neurotrophic factors by MSCs has been observed in culture, and ischemic rat brain extracts can induce production of neuroptrophins and neurotrophic factor (BDNF) is constitutively and is increased in ischemic lesions following MSCs treatment in the rat middle cerebral artery occlusion (MCAO) model. Transplantation of BDNF gene-modified human MACs results in increased BDNF levels in ischemic lesions and stronger therapeutic effects than MACs alone.

Cultured bone marrow-derived MSCs secrete angiogenic cytokines including vascular endothelial growth factor (VEGF) and angiopoietin-1 (Ang-1). Angiogenesis, the growth of new blood vessels, may help to restore oxygen and nutrient supply to the ischemic tissue. Significantly active angiogenesis appears at 3–4 days after cerebral ischemia, especially in the penumbra, and greater microvessel density in the ischemic border correlates with better improvement and longer survival in stroke patients [13]. In the present study, transplantation of hBMSCs significantly promoted endothelial cell proliferation and induced angiogenesis in the IBZ. These findings indicate that hBMSCs induced angiogenesis may largely depend on the proliferation of endogenous endothelial cells or recruitment of endothelial progenitor cells toward the ischemic brain, rather than the transdifferentiation of grafted hBMSCs.

The proliferation of nestin + or musashi + NSPCs was accelerated in the SVZ and SGZ, and there was enhanced migration of DCX + progenitor cells toward the ischemic lesion, higher differentiation of NSPCs to mature neurons or glial cells, and less apoptosis of newly formed cells in the IBZ. Ultimately, these effects resulted in improved neurological functional recovery. Thus, in the hBMSCs treated rats, NSPCs could survive at a higher rate in the process of neurogenesis and participate in the repair of ischemic injury. This is supported by our observation that migrating NSPCs were present more frequently near blood vessels. Similar to other studies, we reported that BMSCs could differentiate into neural cells in vivo, and replacement of neurons was often considered the main goal of cell therapy for cerebral ischemia. Most surviving hBMSCs were located close to the ischemic lesion. The number of hBMSCs that survived at 12 weeks after transplantation was relatively small, consistent with other reports [14]. The death of hBMSCs was primarily attributed to the deleterious microenvironment after cerebral ischemia, whereas immunological rejection can not be fully excluded.

MSCs Promotes Function Recovery in Large Animal with MACO

Human mesenchymal stem cell transplantation protects against cerebral ischemic injury and in Macaca fascicularis. MSC is a promising candidate for cell therapy. The therapeutic ability of MSCs has been documented in animal models of focal cerebral ischemia [15]. In the present study, a non-human primate ischemia model was used to assess the therapeutic effects of hBMSCs transplantation and the potential mechanisms underlying these effects. One week after ischemia. the results demonstrated that hBMSC treatment exerted neuroprotective and anti-apoptotic effects on cerebral ischemia. In addition, the results suggested that hBMSC transplantation inhibited astroglial reactivity and provided a favorable microenvironment for the proliferation of progenitor cells. We also found that treatment with hBMSC upregulated expression of IL-10 peri-ischemia. Decrease of neuronal apoptosis and the proliferation of progenitor cells in the SVZ are interesting and potential mechanisms for therapeutic purposes. The cells within the boundary zone of ischemic lesions are highly vulnerable to apoptosis. So, neuroprotection of cells peri-ischemia against death is very useful for functional recovery. The subependymal layer of the ventricular zone is an important source of neuronal and glial progenitors during development. Increase of progenitor cells exhibiting migratory and neural differentiation potentials is present after stroke [16]. The reduction of cell apoptosis within the peri-ischemic tissue and increased induction of SVZ cell proliferation in the MSC-treated group could enhance neuronal functional recovery. In early period of ischemic injury, increase of astroglial activity can produce more neurotrophic factors for support of functional recovery.

We transplanted hBMSCs 1 week after ischemic injury and observed attenuation of astroglial reactivity 2 weeks after hBMSC treatment. Inhibition of astroglial reactivity may aid in plasticity of axonal regeneration, neural circuitry, and recovery of neurologic function. At present study, we also demonstrated that administered hBMSCs increased IL-10 expression. The upregulation of endogenous IL-10 plays a potential neuroprotective role against cerebral ischemia and provides a favorable microenvironment for neurogenesis after ischemic stroke.

MSCs Promotes Function Recovery in Small Animal with ICH

The potential use of BMSCs for brain repair and regeneration has been reported in different animal models of injury, such as ischemic and ICH stroke, trauma, and neonatal anoxia. There is a discrepancy about the effects of neural stem cell (NSC) or MSC transplantation on infarct volume. Some authors have reported that NSC or BMSC transplantation reduced infarct volume. The present results reported that better protection appeared in the BMSCs transfected with treatment in rats with ICH.

The present study has clearly demonstrated that the BMSCs group showed a stronger antiapoptotic effect compared with the PBS group. During 12 h to 5 days of reoxygenation, BMSCs exhibited stable antiapoptotic action on damaged neurons, NSCs or BMSCs can significantly decrease the infarct volume [17–19]. BMSCs infusion after ICH significantly improved functional recovery evaluated by mNSS and reduced the damaged volume compared with ICH group, and better behavioral recovery and a greater reduction in lesion volume in the BMSCs group were observed compared with PBS group.

Intravenously administered MSCs enter the rat brain, survive, and migrate to the perihematomal area. Most of the transplanted cells differentiate into glial cells (_75 %), and _10 % of them differentiate into neurons. These findings reinforce that not only ischemic infarction but also ICH are good candidates for cell transplantation therapy. Selective neuronal loss such as in the case of Parkinson's disease has already been considered a good candidate for neural replacement therapy [20]. ICH is associated with considerable mechanical disruption of tissue in a large portion of the brain, including the neural and glial cells, and it was generally accepted that this disorder would be less likely to benefit from neural transplantation. Recently, human bone marrow stromal cell transplantation in the rat focal ischemia model induced an increase in the amount of brain-derived neurotrophic factor and nerve growth factor 7 days after ischemia. Measurements of the time course of neuronal differentiation and measurements of trophic factors likely would allow the separation between functional neural integration and trophic effects of MSCs. Moreover, functional improvement was observed and the grafts survived when the cells were transplanted. This brief window of opportunity might arise because the microenvironment in the host changes rapidly after injury. Locally injected cells died within 1 day, but intravenously injected MSCs were successfully recruited into the injured brain. Even though many inflammatory cytokines might be neurotoxic or have an astrocyte-inducing effect, intravenously infused MSCs in our experiment gave rise to functional improvement. These differences might be related to both the administration route and the characteristics of MSCs.

MSCs Promotes Function Recovery in Large Animal with ICH

Stem cell transplantation holds the promise of a cure for many degenerative diseases, including cerebrovascular disease. Human mesenchymal stem cells (hMSCs), usually obtained from bone marrow, are multipotent stem cells. hMSCs have many advantages over other stem cells, such as embryonic stem cells. Human mesenchymal cells have low immunogenicity, arouse fewer ethical disputes, allow for autografting, and show positive immunomodulation effects. Stem cell treatment has shown benefits in a few studies using experimental ICH models in rodents [21–24], but no comparative study has been reported in primates. In this study, an experimentally induced ICH model was established in the Macaca fascicularis monkey. In this study, an experimentally induced ICH model was established in the Macaca fascicularis monkey.

The pathologic processes after an ICH are complicated. An ICH can induce neurologic damage by local tissue deformation and subsequent development of excitotoxicity, apoptosis, and inflammation [24]. Transplantation of mesenchymal stem cells (MSCs) might promote neuroprotection and functional recovery by several mechanisms as demonstrated by previous studies. First and to a large extent, the therapeutic potential of MSCs may rely on their differentiation ability. MSCs can also regulate angiogenesis by a process dependent on fibroblast growth factor receptor and vascular endothelial growth factor receptor signaling cascades. Therefore, MSCs hold great promise for the treatment of ICH. To move the translational process forward, this study used a primate ICH model of Macaca fascicularis monkey; the hMSCs were injected using a stereotactic method near the hematoma; the efficacy of treatment was evaluated by observing the recovery of glucose metabolism of the surrounding cortex and basal ganglia through serial 18F-FDG PET; and 2 other complimentary approaches were also combined, including a neurologic deficit scoring to monitor the functional recovery of the monkey and a pathologic analysis to reveal changes in the brains at the end of the study [25]. The neurologic function scores were also significantly better in MSCs treatment groups. We adopted a modified neurologic deficit score scale that was more detailed and easier to operate than the Kito score scale. The scale included an evaluation of consciousness, skeletal muscle coordination, sensory system function, and motor system function. To be more objective, the assessment was conducted by 2 researchers who were unaware of the treatment paradigm. However, we still found that the scores varied widely within each group. The largest variation was at the fourth week after hMSC transplantation. In addition, better results were also found in the early treatment group than in the late treatment group, especially at the later period of the stable phase.

This study indicated the benefit of stereotactic transplantation of hMSCs in a primate model of ICH, neurologic function scoring, and pathologic analysis. The efficacy reported in this article and the safety of hMSC transplantation may warrant further translational research on promoting the hMSC treatment for ICH in humans.

Clinical Trials with Stem Cells

Whether stem cells can induce substantial symptomatic relief in patients has not been demonstrated. Clinical trial to some extend can provide evidence that MSC transplantation is safe and beneficial. It is reported that 30 patients in 2005 [26] and 85 patients in 2010 [27] after MSC transplantation get better recovery, but the mechanism of MSC action in those patients remains uncertain.

Now, several clinical trials using intravenous or intra-arterial infusion of non-MSCs or MSCs in patients with stroke are undergoing or planned, and some have been registered in www.clinicaltrials.gov. UK-based company ReNeuron started first clinical trials with transplantation of NSCs in patients with stoke [28]. It implanted NSCs into putamen of 12 patients between 6 and 24 months after stoke, and these cells differentiated into neuron-like cells and oligodendroglial and endothelial phenotypes without signs of tumorigenicity, and improvements were observed between 6 and 12 weeks. However, the cells died beyond 6 months and the mechanisms for the improvements were still confused.

Trails in animals are commonly seen and the improved clinical behaviors are have been observed, but whether it occurs in patients is still unknown.

Treatment for Stroke

It is improved that neuronal replacement is possible for neurological disorders such as Parkinson's disease, but it's still uncertain for the stroke. However, although there is insufficient clinical evidence for stem cell therapy, the therapy has been used worldwide.

Vivek Misra [29] from Texas Stroke Institute tells that stem cell therapy for stroke is promising. The stem cell therapy contains lots of methods of implantation, such as direct intracerebral injection, intracisternal/cerebroventricular or intravascular routes like intravenous or intra-aterial infusion. The basic principle of intracerebral transplantation is directly transplanting neural stem cells into lesions at particular area, but the damage on the brain tissue caused by it is a problem. Using a general anesthetic for intracerebral delivery, particularly for the acute stroke patient, is problematic, though alternatively a local anesthetic can be used. Considering safety and efficiency. The followings are some common ways in clinical work.

Intracisternal/Cerebroventricular Route

Although it is a invasive way, compared with direct intracerebral implantation, it is a precious method for less lesion to the target point. However, there are still some side effects. They treated 7 ischemic and 3 hemorrhagic patients, and some of them suffered fever and meningeal signs 48 h after cellular delivery [30].

Intravenous Route

Infusion is the least invasive method. With the potential guide, the stem cells will distribute in the target lesion, but commonly will spread the whole brain even other organs. Since patients with ischemic stroke usually have associated with cardiac or renal impairment, the intravenous infusion way will benefit these organs. Thirty patients treated with this way shows that this method was safe and feasible in the short term as well as on the long-term follow-up [26], and results in improved neurological recovery. Honmou reported 12 patients demonstrated intravenous infusion of autologous mesenchymal cells with 36–133 days of follow-ups [31] and proved that it was feasible. Other doctors who chose these way also proved this results. But still similar to animal experiments, this study also found cell sequestered in the spleen, lung, and kidney.

Intra-arterial Route

Actually cells delivery to the ischemic target directly in the artery. Compared with the intravenous way, intra-arterial route results in high concentration in the target ischemic tissues, and this can also be improved in the animal experimental [32]. One preclinical study comparing intravenous and intra-arterial infusion autologous bone marrow mononuclear cell delivery found significant reduction in infarct volume, higher cell engraftment and improved motor function with intra-arterial delivery [33]. Despite of the advantages of intra-arterial infusion way, some animal experiments showed that it would worsen the ischemia and raised the mortality [34].

Recently, a study showed that after the stem cell infusion through the artery, the blood flow would reduce and it would cause the microvascular occlusion [35].

Challenges for Stem Cell Therapy in Clinical Trials

Although all the animal experiments appeared promising and the less invasive route such as the intravenous and intra-arterial way are improved in clinical experiments, several key question should be solved like when to give the stem cell therapy, and the dose of cells delivered or implanted or how to choose the therapy. Despite these problems, stem cell therapy for strokes have already worldwide.

Choice of Cells

There are several donor cell types for stroke, especially for ischemic stroke. The curative effect is limited because usually it takes long time to get the stem cells from the autologous sources. It is effective if we give the stem cell therapy thorough intravenous or intra-arterial way in the first few hours after a stroke [36].

Timing of Therapy

Preclinical studies have shown that stem cell therapy in the treatment of acute and chronic stroke models have many advantages and great potential. The ideal treatment timing and possible mechanisms are closely related. In accordance with the requirements of the nerve protective mechanism, neural stem cell transplantation should be as soon as possible, but it will be subjected to the harmful injury of ischemia reperfusion caused by various free radicals, neurotransmitters and other inflammatory cytokines [21]. A neuroreparative approach aiming for direct replacement of damaged tissue with new neuronal circuitry (using intracerebral transplantation), would benefit from a later timescale, once neuroinflammation has subsided. Such a strategy would be looking at treatment delivered weeks to months after the stroke, in keeping with ongoing host plasticity during that time [37–40]. So, unlikely the usual intravenous or intra-arterial therapy, it is not clear about the exact time windows for stem cell therapy. Since most of the ongoing human studies get the cells from autologous sources, the therapy can last from few days to several weeks.

Dose of the Therapy

In the preclinical studies, intravenous and intra-arterial route are the two common ways. In the intravenous way, it is demonstrated that the higher the dose is, the smaller infarct volumes is [41], and also the stem cells can be detected in the peripheral organs. And compared between the two ways, it shows that the intra-arterial delivery could produce similar results with lesser cell doses. However, the optimal dose for the therapy is still uncertain, the next clinical studies should focus on the feasibility of autologous cell procurement and the maximum viable cell dose that could be safely obtained.

Other Effects

Several clinical studies shows that patients in stroke combined with post-myocardial infarction treated with stem cell therapy can demonstrate a trend toward improved clinical outcomes as well as physiologic parameters [42]. And these trails did not report any significant adverse events.

Cases

One case reported is a patient with left middle cerebral artery distribution infarct, 3 days after symptom onset [43]. Before given the intra-arterial infusion, the intracranial artery patency had been demonstrated by the transcranial Doppler. The authors also reported a decrease in hypoperfusion on SPECT as well as increased metabolism in the ischemic tissue, 7 days after intra-arterial cell delivery. And they also reported that stem cell therapy was safe in humans and could last good clinical recovery.

Perspectives

More and more clinical trials have begun to confirm the safety and feasibility of different stem cell transplantation methods. But there are still many unanswered questions. The great variation in the stem cell trials completed to date means that it is difficult to make any meaningful comparisons between them.

In conclusion, there have been significant advances made in the field of stem cell research over the last two decades, with evidence of significant benefits in both acute and chronic animal models of stroke. Stem cell infusion therapy is a potential adjunct therapeutic modality to strokes and it shows improved clinical neurological recovery. Despite the challenges in clinical translation from animal trails to clinical work, the initial pilot studies have demonstrated the safe and feasible way like the intravenous or intra-arterial infusion. The ongoing and next trails using various delivery routes, choices of cells, timing of therapy and doses of the stem cells are likely to bridge the gaps that exist in clinical work, and the results should be promising.

References

1. Intercollegiate Stroke Working Party. National Sentinel Stroke Clinical Audit 2010: Public Report for England, Wales and Northern Ireland. London: Royal College of Physicians; 2011.
2. Eriksson PS, Perilieva E, Bjork-Eriksson T, et al. Neurogenesis in the adult human hippocampus. Nat Med. 1998;4:1313–7.
3. Curtis MA, Kam M, Nannmark U, et al. Human neuroblasts migrate to the olfactory bulb via a lateral ventricular extension. Science. 2007;315:1243–9.
4. Jin K, Wang X, Xie L, et al. Evidence for stroke-induced neurogenesis in the human brain. Proc Natl Acad Sci USA. 2006;103:13198–202.
5. Shyu WC, Lin SZ, Yang HI, et al. Functional recovery of stroke rats induced by granulocyte colony-stimulating factor–stimulated stem cells. Circulation. 2004;110:1847–54.
6. Schabitz WR, Kollmar R, Schwaninger M, et al. Neuroprotective effect of granulocyte colony stimulating factor after focal cerebral ischemia. Stroke. 2003;34:745–51.
7. Weissman IL, Anderson DJ, Gage F. Stem and progenitor cells: origins, phenotypes, lineage commitments, and transdifferentiations. Annu Rev Cell Dev Biol. 2001;17:387–403.
8. Thomson JA, Itskovitz-Eldor J, Shapiro SS, et al. Embryonic stem cell lines derived from human blastocysts. Science. 1998;282:1145–7.
9. Chu K, Kim M, Jeong SW, Kim SU, Yoon BW. Human neural stem cells can migrate, differentiate, and integrate after intravenous transplantation in adult rats with transient forebrain ischemia. Neurosci Lett. 2003;343:129–33.
10. Chu K, Kim M, Park KI, et al. Human neural stem cells improve senso-rimotor deicits in the adult rat brain with experimental focal ischemia. Brain Res. 2004;1016:145–53.
11. Chen J, et al. Intravenous bone marrow stromal cell therapy reduces apoptosis and promotes endogenous cell proliferation after stroke in female rat. J Neurosci Res. 2003;73:778–86.
12. Nomura T, et al. I.V. infusion of brain-derived neurotrophic factor gene-modified human mesenchymal stem cells protects against injury in a cerebral ischemia model in adult rat. Neuroscience. 2005;136:161–9.
13. Krupinski J, Kaluza J, Kumar P, Kumar S, Wang JM. Role of angiogenesis in patients with cerebral ischemic stroke. Stroke. 1994;25:1794–8.
14. Shen LH, Li Y, Chen J, Zacharek A, Gao Q, Kapke A, Lu M, Raginski K, Vanguri P, Smith A, Chopp M. Therapeutic benefit of bone marrow stromal cells administered 1 month after stroke. J Cereb Blood Flow Metab. 2007;27:6–13.
15. Borlongan CV, Lind JG, Dillon-Carter O, Yu G, Hadman M, Cheng C, Carroll J, Hess DC. Bone marrow grafts restore cerebral blood flow and blood brain barrier in stroke rats. Brain Res. 2004;1010:108–16.
16. Arvidsson A, Kokaia Z, Lindvall O. N-methyl-D-aspartate receptor-mediated increase of neurogenesis in adult rat dentate gyrus following stroke. Eur J Neurosci. 2001;14:10–8.
17. Kurozumi K, Nakamura K, Tamiya T, et al. BDNF gene-modified mesenchymal stem cells promote functional recovery and reduce infarct size in the rat middle cerebral artery occlusion model. Mol Ther. 2004;9(2):189–97.
18. Chu K, Park K, Lee ST, et al. Combined treatment of vascular endothelial growth factor and human neural stem cells in experimental focal cerebral ischemia. Neurosci Res. 2005;53(4):384–90.
19. Mike V, Nonoguchi N, Ikeda N, Coffin R, Kuroiwa T, Miyatake S. Vascular endothelial growth factor gene-transferred bone marrow stromal cells engineered with a herpes simplex virus type 1 vector can improve neurological deficits and reduce infarction volume in rat brain ischemia. Neurosurgery. 2007;61(3):586–95.
20. Bjorklund A, Lindvall O. Cell replacement therapies for central nervous system disorders. Nat Neurosci. 2000;3:537–44.
21. Jeong SW, Chu K, Jung KH, Kim SU, Kim M, Roh JK. Human neural stem cell transplantation promotes functional recovery in rats with experimental intracerebral hemorrhage. Stroke. 2003;34:2258–63.

22. Lee HJ, Kim KS, Kim EJ, et al. Brain transplantation of immortalized human neural stem cells promotes functional recovery in mouse intracerebral hemorrhage stroke model. Stem Cells. 2007;25:1204–12.
23. Lee ST, Chu K, Jung KH, et al. Anti-inflammatory mechanism of intravascular neural stem cell transplantation in haemorrhagic stroke. Brain. 2008;131:616–29.
24. Zhang H, Huang Z, Xu Y, Zhang S. Differentiation and neurological benefit of the mesenchymal stem cells transplanted into the rat brain following intracerebral hemorrhage. Neurol Res. 2006;28:104–12.
25. Feng M, et al. Serial 18F-FDG PET demonstrates benefit of human mesenchymal stem cells in treatment of intracerebral hematoma: a translational study in a primate model. J Nucl Med. 2011;52(1):90–7.
26. Bang OY, Lee JS, Lee PH, Lee G. Autologous mesenchymal stem cell transplantation in stroke patients. Ann Neurol. 2005;57:874–82.
27. Lee JS, Hong JM, Moon GJ, Lee PH, Ahn YH, Bang OY. A long-term follow-up study of intravenous autologous mesenchymal stem cell transplantation in patients with ischemic stroke. Stem Cells. 2010;28:1099–106.
28. Mack GS. Reneuron and stem cells get green light for neural stem cell trials. Nat Biotechnol. 2011;29:95–7.
29. Misra V, Ritchie MM. Stem cell therapy in ischemic stroke. Neurology. 2012;79:207–12.
30. Rabinovich SS, Seledtsov VI, Banul N, et al. Cell therapy of brain stroke. Bull Exp Biol Med. 2005;139:126–8.
31. Honmou O, Houkin K, Matsunaga T, et al. Intravenous administration of auto serum-expanded autologous mesenchymal stem cells in stroke. Brain. 2011;134:1790–807.
32. Guzman R, De Los Angeles A, Cheshier S, et al. Intracarotid injection of fluorescence activated cell-sorted CD49d-postive neural stem cells improves targeted cell delivery and behavior after stroke in a mouse stroke model. Stroke. 2008;39:1300–6.
33. Kamiya N, Ueda M, Igarashi H, et al. Intra-arterial transplantation of bone marrow mononuclear cells immediately after reperfusion decreases brain injury after focal ischemia in rats. Life Sci. 2008;83:433–7.
34. Li L, Jiang Q, Ding G, et al. Effects of administration route on migration and distribution of neural progenitor cells transplanted into rats with focal cerebral ischemia, an MRI study. J Cereb Blood Flow Metab. 2010;30:653–62.
35. Chu JY, Pendharkar A, Wang N, et al. Intra-arterial injection of neural stem cells using a microneedle technique does not cause microembolic strokes. J Cereb Blood Flow Metab. 2011;31:1263–71.
36. Kranz A, Wagner DC, Kampral M, et al. Transplantation of placenta-derived mesenchymal stromal cells upon experimental stroke in rats. Brain Res. 2010;1315:128–36.
37. Kelly S, Bliss TM, Shah AK, et al. Transplanted human fetal neural stem cells survive, migrate, and differentiate in ischemic rat cerebral cortex. Proc Natl Acad Sci USA. 2004;101:11839–44.
38. Hicks AU, Lappalainen RS, Narkilahti S, et al. Transplantation of human embryonic stem cell-derived neural precursor cells and enriched environment after cortical stroke in rats: cell survival and functional recovery. Eur J Neurosci. 2009;29:562–74.
39. Takahashi K, Yamanaka S. Induction of pluripotent stem cells from mouse embryonic and adult fibroblast cultures by defined factors. Cell. 2006;126:663–76.
40. Takahashi K, Tanabe K, Ohruki M, et al. Induction of pluripotent stem cells from adult human fibroblasts by defined factors. Cell. 2007;131:861–72.
41. Vendrame M, Cassady J, Newcomb J, et al. Infusion of human umbilical cord blood cells in a rat model of stroke dose-dependently rescues behavioral deficits and reduces infarct volume. Stroke. 2004;35:2390–5.
42. Dimmeler S, Zeiher AM. Cell therapy of acute myocardial infarction: open questions. Cardiology. 2009;113:115–60.
43. Mendonca ML, Freitas GR, Silva SA, et al. Safety of intra-arterial autologous bone marrow mononuclear cell transplantation for acute ischemic stroke. Arq Bras Cardiol. 2006;86:52–5.

Mesenchymal Stem Cell Transplantation for Systemic Lupus Erythematosus

Lingyun Sun

Abstract Recently, a substantial progress has been made in the field of mesenchymal stem cell transplantation (MSCT). Experimental and clinical data suggest that MSCT has been a promising therapy strategy for severe and refractory systemic lupus erythematosus (SLE). From March, 2007 to now, more than 210 SLE patients in the world have received the MSCT. Over 90 % of the transplantations were carried out in the Affiliated Drum Tower Hospital of Nanjing University Medical School in China. The reported main types of cell source are allogenic bone marrow (BM, 21 %) and umbilical cord (UC, 78 %) and autologous BM (1 %). The infused MSCs dose was 1×10^6 per kg body weight. The outcome measures, such as disease activity, 24-h proteinuria, serologic features and even glomerular filtration rate have improved significantly. In addition, no serious adverse events related to MSCT were reported so far. Double UC-MSCT could not enhance therapeutic effect compared with single transplantation in refractory SLE. This chapter will review the rationale, progress and perspectives of MSCT in treatment of SLE.

Keywords SLE • Transplantation • MSCT

Introduction

Systemic lupus erythematosus (SLE) is a multifactorial autoimmune inflammatory disease with diverse clinical symptoms due to multiple organ involvement, leading to a high mortality and morbidity. The reported prevalence ranges from 15 to 100 per 100,000 individuals among the different racial groups. Conventional

---

L. Sun (✉)
Department of Immunology and Rheumatology, The Affiliated Drum Tower Hospital of Nanjing University Medical School, 321 Zhongshan Road, Nanjing, Jiangsu 210008, P. R. China
e-mail: lingyunsun2001@yahoo.com.cn

R.C. Zhao (ed.), *Essentials of Mesenchymal Stem Cell Biology and Its Clinical Translation*, DOI 10.1007/978-94-007-6716-4_17,

immunosuppressive or immunomodulatory therapy, notably corticosteroids and cyclophosphamide (CTX), can control most cases with SLE, but not all. About 20 % of refractory patients experience progressive disease that results in tissue damage, physical and psychosocial disability, or even death. In addition, the long-term use of immunosuppression carries significant risk for opportunistic infections and secondary malignancy. Fortunately, many new drugs and therapeutic strategies have emerged, as well as mesenchymal stem cell transplantation (MSCT), the most important progress in the therapy of SLE.

Mesenchymal stem cells (MSCs) are multipotential progenitor cells capable of differentiating into many different cell types of mesodermal origin, as well as ectodermal and endodermal origin [1–3]. A large quantity of evidence has demonstrated that human MSCs can inhibit, in a dose-dependent manner, T and B cell proliferation, suppress NK cell activation and modulate the cytokine secretion profile of dendritic cells and macrophages [4–6]. Other studies have suggested that human MSCs, not just as third-party cells, can also act as nonprofessional antigen-presenting cells and suppress the cytotoxic effects of Ag-primed T effector cells in a short time [7]. MSCs have also been found to release a number of soluble immunosuppressive factors involving in MSC-mediated immunoregulation, such as indoleamine 2, 3-dioxygenase (IDO), IL-6 and soluble human leukocyte antigen-G molecules (HLA-G5) [8–10]. In addition, their low immunogenicity due to their lacking expression of class II major histocompatibility complex (MHC) co-stimulatory molecules make MSCs able to escape alloantigen recognition and then unable to activate alloreactive T cells. These properties make MSCs therapeutic potential cells in many diseases. To date, MSCs have been applied successfully in patients with severe dilated cardiomyopathy, cartilage disorders, autoimmune diseases (AD), stroke, and refractory severe graft versus host disease (GVHD) with very encouraging results [11–15].

No doubt, SLE patients are the major group with AD who received MSCT. To date, above 210 SLE patients in the world have received the transplantation, showing significant safety and efficacy profiles. This chapter will focus on the rationale, current status and perspectives of MSCT in treatment of SLE.

The Rationale for MSCT in SLE

The exact etiopathogenesis of SLE has challenged investigators for many years, but researchers never stop their steps to find the truth. Only we know more about the physiology of SLE, then we will be on our way toward developing rational treatments: more effective but less toxic.

More than two decades ago, there is significant evidence that autoimmune diseases originate from defects that reside within the hematopoietic stem cells (HSCs). Isolated HSCs are sufficient to transfer autoimmune disease from susceptible mice to normal mice [16]. Murine autoimmune disorders can be prevented and treated after transplantation with allogeneic T cell-depleted or whole bone marrow [17].

Other stimulating finding is that serendipitous remissions of autoimmune disease were observed in patients receiving an HSCs transplantation (HSCT) for coexisting hematological disorders [18]. Following these observations, Ikehara et al. have suggested autoimmune diseases as HSCs disorders [19].

The story of MSCs deficiency in SLE initially began with some experimental findings. Transplantation of HSCs with bones to recruit MSCs was indeed found to have a curative effect in MRL/*lpr* mice, a murine model of lupus. When HSCs alone were infused, the mice suffer a relapse 5 months after transplantation [19]. Soon after, Kushida et al. also confirmed the crucial and necessary role of MSCs in successful bone marrow transplantation in MRL/lpr mice [20]. Another interesting finding was that bone marrow stromal cells in SLE patients failed to support allogeneic CD34+ cells growth compared with the control group [21]. These findings result in lots of researchers trying to clarify if there are MSCs defects existing in SLE patients. Lupus maybe is not only hematopoietic but also mesenchymal stem cell disease,

Up to now, many researchers have showed that bone marrow derived mesenchymal stem cells (BM-MSCs), from SLE patients and lupus mouse models, are structurally and functionally abnormal compared with healthy control. MSCs from SLE patients grew slower than those of normal controls, aged more quickly and lost vitality sooner during passage. The cells from lupus patients, compared to controls, were defective in secreting TGF-β, IL-6 and IL-7 [22]. Nie and El-Badri also confirmed structural and functional defects in MSCs population from SLE patients and lupus-prone BXSB mouse with bigger appearance, slow grow rate, increased telomerase activity, reduce level of proliferation and gap junction protein [23, 24]. Moreover, Sun and his colleagues have showed that BM-MSCs from MRL/lpr mice displayed impairment of osteogenic differentiation verified by decreased mineralization and osteogenic gene expression, and impairment of adipogenic differentiation proved by reduced lipid-specific Oil red O-positive cells and adipocyte-specific gene expression [25]. Other results also have demonstrated that the capacity of osteogenic differentiation of bone marrow derived MSCs (BM-MSCs) from SLE patients was reduced compared with that from healthy controls. The activated NF-κB signaling in SLE BM-MSCs inhibits the BMP-2 induced osteoblastic differentiation through BMP/Smad signaling pathway [26]. High oxidation status in BM-MSCs from SLE patients induced the rearrangement of F-actin cytoskeleton via downregulation of RhoA signaling pathway and then impaired their osteogenic differentiation capacity [27].

Recently, researchers have found much evidence about deficiencies of MSCs from SLE patients. There were increased frequencies of apoptotic and aging of SLE BM-MSCs, with markedly decreased expression of anti-apoptotic Bcl-2 both at mRNA and protein levels, whereas increased expression of pro-apoptotic factors such as Bax, caspase 8 and Fas [28]. Gu and coworker also reported their consistent data, showing that MSCs from SLE patients were more senescent and p16 (INK4A) maybe plays an essential role in the process by inhibiting ERK1/2 activation [29]. Moreover, levels of intracellular reactive oxygen species (ROS), which may promote cellular senescence, from SLE BM-MSCs were higher than those from normal controls, with the activation of PI3K/AKT/FoxO3 signaling pathway [28].

While all these results have stressed the deficiency in lupus MSCs population and give a strong impetus for MSCT in lupus treatment, it is difficult to claim such abnormality is from genetic or acquired factors. This issue is equally important as that in HSCT, because it will ultimately determine the most suitable source for MSCT, autologous or allogeneic. So, recently some researchers began to focus on the abnormal genetic determinants which could underly some deficient characteristics of lupus BM-MSCs. They found that there were a total of 1,905 genes which were differentially expressed by BMMSCs derived from SLE patients, of which, 652 genes were upregulated and 1,253 were downregulated. Gene ontology analysis showed that the majority of these genes were related to cell cycle and protein binding. The abnormal gene profile altered protein expression, then many intracellular signal pathways regulating actin cytoskeleton, focal adhesion and tight junction were abnormal in SLE BM-MSCs compared with normal controls. In the MSCs from lupus patients, the BMP/TGF-β signaling pathway was downregulated, while the MAPK signaling pathway was activated via phosphorylation of ERK1/2 and SAPK/JNK [30].

All these evidence has directly or indirectly illustrated both acquired and genetic deficiencies of lupus MSCs, which favored allogeneic rather than autologous MSCT as an effective treatment for patients with lupus. The allogeneic MSCs could work as substitutes for the abnormal auto-MSCs and try to reset the immune system. Some results of animal experiments are more convincing and lead a good beginning to the field of allogeneic MSCT for SLE treatment. Zhou et al. and Sun et al. show that infusion of human BM-MSCs at both early and matured stages benefited a significantly reduction in serum levels of anti-double-stranded DNA antibodies (anti-ds DNA) IgG and IgM, anti-nuclear antibodies (ANA), immunoglobulins IgG1, IgG2a, IgG2b, IgM and 24-h proteinuria in MRL/lpr mice, as well as complement C3 in renal tissue [25, 31]. Increased serum albumin levels in MRL/lpr mice were observed after MSCT. Other findings from Medical University of South Carolina show that allogeneic sources of MSCs infusion, but not autologous lupus-derived sources of MSCs, improved survival, stabilized proteinuria, and decreased glomerular IgG deposition, in both MRL/*lpr* and (NZB × NZW)F1 mice [32].

On the other hand, Carrion and coworkers have showed that autologous BM-MSCs treatment did not improve initial disease activity in two SLE patients during 14 weeks of follow-up despite of increasing $CD4^+CD25^+FoxP3^+$cell counts [33], which further prove the rationale of allogeneic MSCT in the treatment of SLE.

The Current Status of MSCT in SLE

To date, more than 210 SLE patients in the world have received MSC the transplantation. Over 90 % of the transplantations were carried out in the Affiliated Drum Tower Hospital of Nanjing University Medical School in China. The enrolled patients should have progressive and active disease with SLE Disease Activity Index (SLEDAI) score ≥8, despite continuous treatment with intravenous (IV) pulse

CTX with a total dosage of 400–800 mg every month for at least 6 months or oral mycophenolate mofetil (MMF 1,000–2,000 mg/day) for at least 3 months, and continued daily dosage of more than 20 mg of prednisone or its equivalent.

The early used MSCs in the application and clinical trials are BM-MSCs. A total of 39 refractory SLE patients have received BM-MSCT in the Drum Tower Hospital. Reports have started to emerge of small numbers of patients, in which 4 SLE patients received IV infusions of allogeneic BM-MSC cells with some promising benefit [25]. In 2010, Liang et al. reported their preliminary results of a pilot study about MSCT in 15 CTX-refractory lupus patients [34]. The median age of patients (14 female and 1 male) was 28.3 years, with a range from 12 to 44 years. The average disease duration was 91.1 months. All subjects had been previously treated with CTX and high dose of prednisone (more than 20 mg/day). Patient eligibility criteria also included lupus glomerulonephritis (class III, IV, V) with severe elevation of increment of 24-h urine protein levels and/or serum creatinine ≥1.5 mg/dl. Bone marrow was collected from patients' healthy family member and *ex vivo* expanded in culture under GLP/GMP protocols. BM-MSCT was intravenously infused at 1×10^6 cells/kg body weight. Primary outcomes were overall survival and disease remission defined as requiring no further high dose of immunosuppressive medications except the low maintenance doses of corticosteroids and CTX. Post MSCT maintenance therapy includes a tapering dose of steroid and CTX, with 2 patients on prednisone at 5–10 mg/day and completely off CTX at 6 months, and 13 patients on prednisone at 5–10 mg/day and CTX at 0.6 mg/every 2 months. Secondary outcomes included SLEDAI, anti-ds DNA antibodies and renal function monitored by 24-h urine protein and serum creatinine levels. The short-term clinical outcome in 12–18 months follow up post-MSCT showed no allogenic MSCT-related complications including cardiovascular, pulmonary insufficiencies, infection, malignancy, and metabolic disturbances. Assessment of SLEDAI indicated the improvement of disease activity in all allogenic MSCT-treated patients at each follow-up period. All recipients were followed up for 12–18 months and showed recovery of kidney function. Reduced 24-h proteinuria was commonly seen in all patients after MSCT, especially at 1 month (1,129 ± 145.6 mg), 6 months (511.8 ± 127.8 mg) and 15 months (418.5 ± 207.5 mg) *vs.* the values present at baseline (2,202 ± 243.3 mg). Four subjects experienced significant improvement in their serum creatinine levels: from 1.73 to 0.78 at 1 month follow-up, from 5.36 to 2.41 at 3 months, from 4.40 to 2.08 at 1 month follow-up and from 2.55 to 0.90 mg/dl at 3 months follow-up, respectively. Anti-ds DNA titers improved at 1-month post MSCT in all patients. Improvement in glomerular filtration rate (GFR) was noted in two patients in which formal testing was done. Other non-lupus nephritis related manifestations such as fatigue, loss of weight, low-grade fever and skin rashes also improved gradually. Moreover, increased levels of $CD4^{+}Foxp3^{+}$ cells followed allogeneic MSCT in 10 SLE patients with statistical significance at 3-month post MSCT. These early clinical data demonstrate safety and efficacy of MSCT in SLE patients and improvement of disease activities at post allogeneic MSCT.

These years, umbilical cord has been selected as one of available sources of MSCs and seems to be the most frequently employed for the transplantation in

China. The umbilical cord-derived mesenchymal stem cells (UC-MSCs) share most of the characteristics with BM-MSCs and have other distinct advantages. They are showed to have higher proliferation, improved accessibility, lower risk of viral contamination, as well as lower levels of expression of CD106 and HLA-ABC, which represent a specific and different feature compared with BM-MSCs and may favour the use of UC-MSCs for allogeneic cell therapy [35]. So far, a total of 148 refractory patients received UC-MSCT in the Drum Tower Hospital. Sun and his colleagues have reported their data about UC-MSCT in 16 patients [36]. The median followup time after MSCT was 8.25 months (range 3–28 months). Significant improvements in the SLEDAI score, levels of serum ANA, anti-ds DNA antibody, serum albumin, and complement C3, and renal function were observed. Significant reduction in disease activity was achieved in all patients, and there has been no recurrence to date and no treatment-related deaths. Clinical remission was accompanied by an increase in peripheral Treg cells and a re-established balance between Th1- and Th2-related cytokines.

The infused MSCs dose (1×10^6 per kg body weight) is same as that always used in the treatment of GVHD [37, 38]. It is unclear what the optimal cell dose and the optimal infusion numbers of MSCs are in clinical transplants, which will rely to a large extent on clinical experience. Wang and his colleagues have undertaken a study to observe whether double transplantations of MSCs are superior to single transplantation [39]. Fifty-eight refractory SLE patients were enrolled in this study, in which 30 were randomly given single MSCT, and the other 28 were given double MSCT. Patients were followed up for rates of survival, disease remission, and relapse, as well as transplantation-related adverse events. Their results showed that no remarkable differences between single and double allogeneic MSCT were found in terms of disease remission and relapse, amelioration of disease activity, and serum indexes within more than 1 year followup. This study demonstrated that single MSCs transplantation at the dose of one million MSCs per kilogram of body weight was sufficient to induce disease remission for refractory SLE patients.

Despite so many exciting results of MSCT emerging in the treatment of SLE, safety is still the most significant issue that both physicians and patients focus on. The primary concern is the possibilities of tumor formation, which is blame to the innate ability of continuing self-renewal of stem cells. Jeong discovered that after transplantation of allogeneic short-term cultured BM-MSCs, growing tumors were observed in 30 % of hearts in the experimental acute myocardial infarction model, and in 46 % of hindlimbs in the diabetic neuropathy model during the follow-up at 4–8 weeks [40]. Other researchers found allogeneic MSCs favored tumor growth in animals, maybe due to the potential side effects of immunosuppression induced by MSCs [41]. Subcutaneous injection of B16 melanoma cells led to tumor growth in allogeneic recipients only when MSCs were coinjected. However, in the reported and unreported data from the Drum Tower Hospital (the longest followup is 5 years), no serious adverse events including tumors and infections were found so far. None of the lupus patients developed acute and chronic GVHD during followup after MSCT. The common adverse events included insomnia, facial flushing, short-term low fever. Some symptoms quickly restored without any intervention.

Conclusions and Prospects

SLE patients exhibit MSCs defects, genetic and acquired factors may both contribute to the cell deficiency. Allogeneic MSCT represents an exciting approach and is more attractive than autologous MSCT in lupus treatment, with its efficacy and safety in the preliminary experience. It will give a new platform in the treatment of refractory and severe SLE patients.

However, it is still a long way to go before applying MSCs to the clinics. Many concerns needed to be addressed at this stage, such as patient entry criteria, cell dose, numbers of infusion, monitoring protocol, mechanistic studies, outcoming measures, post-transplant immunosuppression schedule, and therapy of relapses. More researchers who are interested in pursuing MSCT as a potential treatment for SLE should be able to collaborate closely. More large random clinical studies are in need to establish the safety and efficacy of MSCT in lupus treatment.

References

1. Jiang Y, Jahagirdar BN, Reinhardt RL, Schwartz RE, Keene CD, Ortiz-Gonzalez XR, Reyes M, Lenvik T, Lund T, Blackstad M, Du J, Aldrich S, Lisberg A, Low WC, Largaespada DA, Verfaillie CM. Pluripotency of mesenchymal stem cells derived from adult marrow. Nature. 2002;418:41–9.
2. Debayon P, Samuel SM, Maulik N. Mesenchymal stem cell: present challenges and prospective cellular cardiomyoplasty approaches for myocardial regeneration. Antioxid Redox Signal. 2009;11:1841–55.
3. Ju S, Teng GJ, Lu H, Jin J, Zhang Y, Zhang A, Ni Y. In vivo differentiation of magnetically labeled mesenchymal stem cells into hepatocytes for cell therapy to repair damaged liver. Invest Radiol. 2010;45(10):625–33.
4. Nauta AJ, Fibbe WE. Immunomodulatory properties of mesenchymal stromal cells. Blood. 2007;110:3499–506.
5. Jiang XX, Zhang Y, Liu B, Zhang SX, Wu Y, Yu XD, Mao N. Human mesenchymal stem cells inhibit differentiation and function of monocyte-derived dendritic cells. Blood. 2005;105:4120–6.
6. Sotiropoulou PA, Perez SA, Gritzapis AD, Baxevanis CN, Papamichail M. Interactions between human mesenchymal stem cells and natural killer cells. Stem Cells. 2006;24:74–85.
7. Potian JA, Aviv H, Ponzio NM, Harrison JS, Rameshwar P. Veto-like activity of mesenchymal stem cells: functional discrimination between cellular responses to alloantigens and recall antigens. J Immunol. 2003;171:3426–34.
8. Uccelli A, Moretta L, Pistoia V. Mesenchymal stem cells in health and disease. Nat Rev Immunol. 2008;8(9):726–36.
9. Selmani Z, Naji A, Gaiffe E, Obert L, Tiberghien P, Rouas-Freiss N, Carosella ED, Deschaseaux F. HLA-G is a crucial immunosuppressive molecule secreted by adult human mesenchymal stem cells. Transplantation. 2009;87(9 Suppl):S62–6.
10. Soleymaninejadian E, Pramanik K, Samadian E. Immunomodulatory properties of mesenchymal stem cells: cytokines and factors. Am J Reprod Immunol. 2012;67(1):1–8.
11. Chin SP, Poey AC, Wong CY, Chang SK, Teh W, Mohr TJ, Cheong SK. Cryopreserved mesenchymal stromal cell treatment is safe and feasible for severe dilated ischemic cardiomyopathy. Cytotherapy. 2010;12(1):31–7.
12. Weng JY, Du X, Geng SX, Peng YW, Wang Z, Lu ZS, Wu SJ, Luo CW, Guo R, Ling W, Deng CX, Liao PJ, Xiang AP. Mesenchymal stem cell as salvage treatment for refractory chronic GVHD. Bone Marrow Transplant. 2010;45(12):1732–40.

13. Sasaki Y, Honmou O. Bone marrow stem cell therapy for stroke. Nihon Rinsho. 2011;69(12): 2203–8.
14. Spagnoli A, Longobardi L, O'Rear L. Cartilage disorders: potential therapeutic use of mesenchymal stem cells. Endocr Dev. 2005;9:17–30.
15. Cucchiarini M, Venkatesan JK, Ekici M, Schmitt G, Madry H. Human mesenchymal stem cells overexpressing therapeutic genes: from basic science to clinical applications for articular cartilage repair. Biomed Mater Eng. 2012;22(4):197–208.
16. Ikehara S, Kawamura M, Takao F, Inaba M, Yasumizu R, Than S, Hisha H, Sugiura K, Koide Y, Yoshida TO, et al. Organ-specific and systemic autoimmune diseases originate from defects in hematopoietic stem cells. Proc Natl Acad Sci USA. 1990;87:8341–4.
17. Ikehara S, Yasumizu R, Inaba M, Izui S, Hayakawa K, Sekita K, Toki J, Sugiura K, Iwai H, Nakamura T, et al. Long-term observations of autoimmune-prone mice treated for autoimmune disease by allogeneic bone marrow transplantation. Proc Natl Acad Sci USA. 1989;86:3306–10.
18. Yin JA, Jowitt SN. Resolution of immune-mediated diseases following allogeneic bone marrow transplantation for leukemia. Bone Marrow Transplant. 1992;9(1):31–3.
19. Ikehara S, Inaba M, Yasumizu R, Nagata N, Toki J, Hisha H, Sugiura K, Oyaizu N, Kawamura M, Than S, et al. Autoimmune diseases as stem cell disorders. Tohoku J Exp Med. 1994;173(1):141–55.
20. Kushida T, Inaba M, Hisha H, Ichioka N, Esumi T, Ogawa R, Iida H, Ikehara S. Crucial role of donor-derived stromal cells in successful treatment for intractable autoimmune diseases in mrl/lpr mice by BMT via portal vein. Stem Cells. 2001;19(3):226–35.
21. Papadaki HA, Boumpas DT, Gibson FM, Jayne DR, Axford JS, Gordon-Smith EC, Marsh JC, Eliopoulos GD. Increased apoptosis of bone marrow CD34(+) cells and impaired function of bone marrow stromal cells in patients with systemic lupus erythematosus. Br J Haematol. 2001;115:167–74.
22. Sun LY, Zhang HY, Feng XB, Hou YY, Lu LW, Fan LM. Abnormality of bone marrow-derived stem cell in patients with systemic lupus erythematosus. Lupus. 2007;16(2):121–8.
23. Nie Y, Lau C, Lie A, Chan G, Mok M. Defective phenotype of mesenchymal stem cells in patients with systemic lupus erythematosus. Lupus. 2010;19:850–9.
24. El-Badri NS, Hakki A, Ferrari A, Shamekh R, Good RA. Autoimmune disease: is it a disorder of the microenvironment? Immunol Res. 2008;41(1):79–86.
25. Sun L, Akiyama K, Zhang H, Yamaza T, Hou Y, Zhao S, Xu T, Le A, Shi S. Mesenchymal stem cell transplantation reverses multi-organ dysfunction in systemic lupus erythematosus mice and humans. Stem Cells. 2009;27:1421–32.
26. Tang Y, Xie H, Chen J, Geng L, Chen H, Li X, Hou Y, Lu L, Shi S, Zeng X, Sun L. Activated NF-kB in bone marrow mesenchymal stem cells from SLE patients inhibits osteogenic differentiation through down-regulating Smad signaling. Stem Cells Dev. 2013;22(4):668–78.
27. Shi D, Li X, Sun L. High oxidation status induced the rearrangement of F-actin cytoskeleton of bone marrow-derived mesenchymal stem cells in patients with systemic lupus erythematosus via downregulation of RhoA signaling pathway. Int J Rheum Dis. 2012;15 Suppl 1:26–8.
28. Li X, Liu L, Meng D, Wang D, Zhang J, Shi D, Liu H, Xu H, Lu L, Sun L. Enhanced apoptosis and senescence of bone-marrow-derived mesenchymal stem cells in patients with systemic lupus erythematosus. Stem Cells Dev. 2012;21(13):2387–94.
29. Gu Z, Cao X, Jiang J, Li L, Da Z, Liu H, Cheng C. Upregulation of p16(INK4A) promotes cellular senescence of bone marrow-derived mesenchymal stem cells from systemic lupus erythematosus patients. Cell Signal. 2012;24(12):2307–14.
30. Tang Y, Ma X, Zhang H, Gu Z, Hou Y, Gilkeson GS, Lu L, Zeng X, Sun L. Gene expression profile reveals abnormalities of multiple signaling pathways in mesenchymal stem cell derived from patients with systemic lupus erythematosus. Clin Dev Immunol. 2012;2012:826182.
31. Zhou K, Zhang H, Jin O, Feng X, Yao G, Hou Y, Sun L. Transplantation of human bone marrow mesenchymal stem cell ameliorates the autoimmune pathogenesis in MRL/lpr mice. Cell Mol Immunol. 2008;5(6):417–24.

32. Gu F, Molano I, Ruiz P, Sun L, Gilkeson GS. Differential effect of allogeneic versus syngeneic mesenchymal stem cell transplantation in MRL/lpr and (NZB/NZW)F1 mice. Clin Immunol. 2012;145(2):142–52.
33. Carrion F, Nova E, Ruiz C, Diaz F, Inostroza C, Rojo D, Mönckeberg G, Figueroa FE. Autologous mesenchymal stem cell treatment increased T regulatory cells with no effect on disease activity in two systemic lupus erythematosus patients. Lupus. 2010;19(3):317–22.
34. Liang J, Zhang H, Hua B, Wang H, Lu L, Shi S, Hou Y, Zeng X, Gilkeson GS, Sun L. Allogenic mesenchymal stem cells transplantation in refractory systemic lupus erythematosus: a pilot clinical study. Ann Rheum Dis. 2010;69(8):1423–9.
35. Lu LL, Liu YJ, Yang SG, Zhao QJ, Wang X, Gong W, Han ZB, Xu ZS, Lu YX, Liu D, Chen ZZ, Han ZC. Isolation and characterization of human umbilical cord mesenchymal stem cells with hematopoiesis-supportive function and other potentials. Haematologica. 2006;91:1017–26.
36. Sun L, Wang D, Liang J, Zhang H, Feng X, Wang H, Hua B, Liu B, Ye S, Hu X, Xu W, Zeng X, Hou Y, Gilkeson GS, Silver RM, Lu L, Shi S. Umbilical cord mesenchymal stem cell transplantation in severe and refractory systemic lupus erythematosus. Arthritis Rheum. 2010;62(8): 2467–75.
37. Sato K, Ozaki K, Mori M, Muroi K, Ozawa K. Mesenchymal stromal cells for graft-versus-host disease: basic aspects and clinical outcomes. J Clin Exp Hematop. 2010;50(2):79–89.
38. Toubai T, Paczesny S, Shono Y, Tanaka J, Lowler KP, Malter CT, Kasai M, Imamura M. Mesenchymal stem cells for treatment and prevention of graft-versus-host disease after allogeneic hematopoietic cell transplantation. Curr Stem Cell Res Ther. 2009;4(4):252–9.
39. Wang D, Akiyama K, Zhang H, Yamaza T, Li X, Feng X, Wang H, Hua B, Liu B, Xu H, Chen W, Shi S, Sun L. Double allogenic mesenchymal stem cells transplantations could not enhance therapeutic effect compared with single transplantation in systemic lupus erythematosus. Clin Dev Immunol. 2012;2012:273291.
40. Jeong JO, Han JW, Kim JM, Cho HJ, Park C, Lee N, Kim DW, Yoon YS. Malignant tumor formation after transplantation of short-term cultured bone marrow mesenchymal stem cells in experimental myocardial infarction and diabetic neuropathy. Circ Res. 2011;108(11):1340–7.
41. Djouad F, Plence P, Bony C, Tropel P, Apparailly F, Sany J, Noël D, Jorgensen C. Immunosuppressive effect of mesenchymal stem cells favors tumor growth in allogeneic animals. Blood. 2003;102(10):3837–44.

Mesenchymal Stem Cell Transplantation for Systemic Lupus Erythematosus

Lingyun Sun

Abstract Recently, a substantial progress has been made in the field of mesenchymal stem cell transplantation (MSCT). Experimental and clinical data suggest that MSCT has been a promising therapy strategy for severe and refractory systemic lupus erythematosus (SLE). From March, 2007 to now, more than 210 SLE patients in the world have received the MSCT. Over 90 % of the transplantations were carried out in the Affiliated Drum Tower Hospital of Nanjing University Medical School in China. The reported main types of cell source are allogenic bone marrow (BM, 21 %) and umbilical cord (UC, 78 %) and autologous BM (1 %). The infused MSCs dose was 1×10^6 per kg body weight. The outcome measures, such as disease activity, 24-h proteinuria, serologic features and even glomerular filtration rate have improved significantly. In addition, no serious adverse events related to MSCT were reported so far. Double UC-MSCT could not enhance therapeutic effect compared with single transplantation in refractory SLE. This chapter will review the rationale, progress and perspectives of MSCT in treatment of SLE.

Keywords SLE • Transplantation • MSCT

Introduction

Systemic lupus erythematosus (SLE) is a multifactorial autoimmune inflammatory disease with diverse clinical symptoms due to multiple organ involvement, leading to a high mortality and morbidity. The reported prevalence ranges from 15 to 100 per 100,000 individuals among the different racial groups. Conventional

---

L. Sun (✉)
Department of Immunology and Rheumatology, The Affiliated Drum Tower Hospital of Nanjing University Medical School, 321 Zhongshan Road, Nanjing, Jiangsu 210008, P. R. China
e-mail: lingyunsun2001@yahoo.com.cn

R.C. Zhao (ed.), *Essentials of Mesenchymal Stem Cell Biology and Its Clinical Translation*, DOI 10.1007/978-94-007-6716-4_17,

Considerations of Quality Control Issues for the Mesenchymal Stem Cells-Based Medicinal Products

Bao-Zhu Yuan, Debanjan Sarkar, Simone Pacini, Mahmood Khan, Miodrag Stojkovic, Martin Zenke, Richard Boyd, Armand Keating, Eric Raymond, and Robert Chunhua Zhao

Abstract Mesenchymal stem cells (MSCs) represent a potentially attractive product for achieving successful stem-cell-based therapy. Clinical application studies of the MSCs for a variety of diseases such as graft-versus-host disease, liver fibrosis, heart disease, diabetes, osteoarthritis, and spinal cord injury have been proposed worldwide. While clinical applications of MSC-based therapy (MSCT) increasingly gain popularity among clinical practitioners and researchers, concerns have been raised on quality issues of MSCs among all stem cell-based medicinal products (SCMP). Quality issues are associated with donor's qualification and *in vitro* cell processing as well as some still unknown biological characteristics of the MSCs, especially in the context of potential tumorigenicity. Quality issues could

B.-Z. Yuan (✉)
National Institutes for Food and Drug Control, Beijing 100038, China
e-mail: yuanbaozhu@nicpbp.org.cn

D. Sarkar
Department of Chemical and Biological Engineering, Department of Biomedical Engineering, University at Buffalo, The State University of New York, Buffalo, NY 14260, USA

S. Pacini
Department of Clinical and Experimental Medicine, University of Pisa, Via Roma 56, 56124 Pisa, Italy
e-mail: simone.pacini@do.unipi.it

M. Khan
Department of Emergency Medicine, Davis Heart and Lung Research Institute, The Ohio State University, Columbus, OH, USA
e-mail: mahmood.khan@osumc.edu

M. Stojkovic
Centre for Molecular Medicine and Stem Cell Research, Faculty of Medical Sciences, University of Kragujevac, Kragujevac, Serbia

Department of Genetics, Spebo Medical, Leskovac, Serbia
e-mail: mstojkovic@spebo.co.rs

R.C. Zhao (ed.), *Essentials of Mesenchymal Stem Cell Biology and Its Clinical Translation*, DOI 10.1007/978-94-007-6716-4_18,

affect safety and efficacy of the MSCT. To ensure optimal quality assurance for both safety and efficacy of the stem cell-based therapy in clinical trials, processes for SCMP production, quality control tests are necessary at each step. More specifically, donor identity, stem cell tissue origins, purity, safety, potency, and stability, must be strictly controlled during the entire product development cycle.

Keywords MSC • Quality control • Medical product

Introduction

Stem cells are defined as cells possessing various differentiation potentials with capacity of cell renewal. Stem cell-based therapy (SCT) is an emerging therapeutic modality in which stem cells isolated from human donors are processed through *in vitro* proliferation, differentiation, and then administered to autologous or allogeneic patients for treating a variety of diseases [1].

Mesenchymal stem cells (MSCs) belong to a group of heterogeneous stem cells with multipotent potential for cell differentiation. The discovery of MSCs can be dated back to the 1960s [2] It was later found that the MSCs exist in almost all tissues throughout the body, even though, the vast majority of the MSCs used in clinical studies derived mainly from bone marrows, the Wharton's Jelly of the umbilical cords, and adipose tissues. Among all stem cell-based medicinal products (SCMP), the MSC products have increasingly gained more popularity due to their ability to self-renew, differentiate into lineages of mesodermal origin, such as osteoblasts, chondrocytes, and adipocytes in culture and transdifferentiate into

M. Zenke
Department of Cell Biology, Institute for Biomedical Engineering, Rhenish-Westphalian Technical University, Aachen University Medical School, Pauwelsstrasse 30, 52074 Aachen, Germany

R. Boyd
Monash Immunology and Stem Cell Laboratories (MISCL) Level 3, Monash University, Building 75, Wellington Road Clayton, Melbourne, VIC 3800, Australia

A. Keating
Cell Therapy Program, Princess Margaret Hospital, Institute of Biomaterials and Biomedical Engineering, Toronto, ON, Canada

Department of Medicine, University of Toronto, Toronto, ON, Canada

E. Raymond
Department of Medical Oncology (INSERM U728-Paris 7 Diderot University), Beaujon University Hospital, 100 boulevard du Général Leclerc, 92110 Clichy, France

R.C. Zhao (Co-edited)
Center of Excellence in Tissue Engineering, Institute of Basic Medical Sciences and School of Basic Medicine, Chinese Academy of Medical Sciences and Peking Union Medical College, 5# Dongdansantiao, 100005 Beijing, China, People's Republic

lineages of ectodermal and endodermal origins [2]. In addition, MSCs show very unique immune regulatory effects through some still unknown secretion functions with abilities to inhibit inflammatory and autoimmune responses in different pathological contexts [3].

Worldwide, MSCs have been applied clinically in patients of severe ischemic cardiomyopathy, osteoarthritis, liver fibrosis, autoimmune diseases, and diabetes with very encouraging results. New MSCs-based clinical studies have shown a steady increase with over 250 new studies having been registered in NIH's website clinicaltrails.org according to a recent search. Encouragingly, the first MSC-based stem cell product, Prochymal, developed by Osiris Therapeutics in Columbia, Maryland, has been approved by the Canadian regulatory authority for treating acute child steroid-resistant graft versus host diseases (GVHD) [4].

Preparation processes and clinical indications of the MSC products as well as other SCMPs display significant diversity and complexity [5]. However, as a novel biological product, the development of all SCMPs must undergo common research and development pathways, from cell sourcing, product preparation, to preclinical studies through both *in vitro* and in vivo tests, clinical trials, and applications. At each step of the entire product development cycle, quality, safety, and biological effectiveness of MSC products should be clearly addressed through implementation of effective quality control methods that are necessary to ensure both safety and efficacy in eventual SCT.

However, there are still gaps in both stem cell sciences and technologies, limiting the achievement of effective assurance processes for MSC products as well as other SCMPs. For MSC products, the understanding of critical biological characteristics of the MSCs remains still very limited such as the tumorigenic or tumor modulatory effects as well as immune regulatory effects. In addition, only limited numbers of the high quality clinical data for the MSC products are available, making extremely difficult to establish appropriate technical guidelines for specific MSC products. Thus, it is very difficult for both product developers and regulatory authorities to implement effective quality control for both development and clinical applications of MSC products. In addition, many investigators of the MSC-based clinical studies that have strong background or experiences in basic research or routine clinical practices do not have sufficient understanding of requirements in quality control and clinical trial issues that are required for the development of SCMPs. Moreover, there are still lacks of standard materials and quality standards, which can be employed in both product development and quality control [1].

From regulatory perspectives, MSC products as well as other SCMPs can be at least categorized as biological products, cell products, and/or therapeutic cell products. Therefore, at current stage, without establishing specific technical guidelines, the quality control for the MSC products should be achieved at least in part based on the best understanding of the updated stem cell sciences and technologies and through adopting relevant principles from the existing guidelines or recommendations. National or international regulatory authorities and research organizations for the quality control of biological products established guidelines for the use of cell substrates and for manufacturing biological and therapeutic cell products. These guidelines could be considered as an initial effort toward development of high quality control frameworks, from which more detailed technical recommendations or guidelines will be established along the continuous progresses in MSC studies and clinical applications.

Quality Issues

The Relevant Principles from the Existing Guidelines

According to risk analysis, all therapeutic cell products must meet requirements for cell identity, purity, safety, and biological activity [6–8]. Some basic principles addressing these quality issues already exist in various technical guidelines or recommendations for the use of cell substrates and for the production of biological and therapeutic cell products. These already existing general guidelines should be useful for initiating specific quality control for the SCMPs, including the MSC products.

Principles from the guideline for quality control of the cell substrates, especially the cell bank characterization, should be adopted from Chinese Pharmacopeia (Part III, 2010 Version) [9], European Pharmacopeia [10], WHO recommendations for evaluating animal cell cultures as substrates for the manufacture of biological medicinal products and for the characterization of cell banks (2010) [11], FDA Guidance for Industry-Characterization and Qualification of Cell Substrates (2010) [12], and International Conference for Harmonization (ICH) guidelines for evaluating cell lines of human or animal origin (Q5A-R1, 1999; Q5D, 1997) [13–15]. The principles for quality control of therapeutic cell products, especially for product's safety and potency, should be adopted from the Guidelines for development of human therapeutic somatic cell products in China (2003) [15], EMA guideline on human cell-based medicinal products (2007) [6], FDA Guidance for human somatic cell therapy and gene therapy (1998) [7], FDA Guidance-human somatic cell therapy IND application (2008) [8], FDA Guidance-Potency Tests for Cellular and Gene Therapy (2011) [16], and US Pharmacopeia <1046> Cellular and tissue-based products [17]. In addition, the principles for the quality issues of ancillary materials could be adopted from the US Pharmacopeia <1043> Ancillary Materials for Cell, Gene and Tissue-Engineered Products, especially in selection of tissue culture medium, serum, and various reagents and supplements to be used in preparation of the SCMPs [18].

In addition to these guidelines from national and international regulatory authorities, guidelines or recommendations from various international stem cell research societies should also be taken as important references for taking the initial quality control activities. These guidelines include International Society for Stem Cell Research (ISSCR) Guidelines for clinical translation of stem cells (2008) [1], and International Society for Cell Therapy (ISCT) for minimum criteria for defining multipotent stem cells (2006) [19].

The Major Quality Issues of MSC Products

Like all new biological products for medical use, MSC products must meet the requirements for identity, purity, safety, potency, and stability. The cell identity should be clearly demonstrated by the information collected from analysis in cell biology, cytogenetics, biochemistry, cell surface biomarkers, and genetic polymorphism.

Safety issues for MSC products are associated with the risks of either endogenous or exogenous origins. The risk factors, that could introduce microbial contamination, and product or process-related impurities, or possess potential of inducing tumorigenicity, modulating tumor growth, or stimulating abnormal immunological responses, could result in severe adverse consequences. In addition, some still unidentified risk factors could lead to uncontrolled, unwanted cell differentiation, or de-differentiation inside the body of the recipients with unpredictable adverse effects.

Risks associated with microbial contaminations may derive from donors during cell isolation and selection or occur during *in vitro* cell processing, thus requiring strict donor screening, processing cells in conformance with the cGMP requirements, establishing standard operating procedures (SOP) for both product preparation and management, and implementing microbial safety tests to ensure freedom from microbial contamination.

For the risk factors related to tumor formation or growth, accumulated data suggest that MSCs possess low or no tumorigenicity but may modulate tumor growth through enhancing growth of the existing tumors [20]. The tumorigenic activity of MSCs is associated with the reduced genomic stability occurring during long *in vitro* cell culturing and/or complex cell processing, whereas the tumor modulating activities might come from their intrinsic immunosuppressive effects, thus encouraging investigators or manufacturers to conduct relevant studies with appropriate designs to carefully characterize tumorigenic or tumor modulating activities of the products [2].

Information of the potency of MSC products should provide proof of principle and mechanisms of action contributing the treatment efficacy [1]. It could be evaluated by relevant assays designed from the understanding of biological characteristics of MSCs, such as the pattern of cell differentiation, the ability to regulate immune responses, especially the proliferation of different lymphocyte subpopulations and secretion of the associated cytokines. Investigators or manufacturers are always encouraged to develop product-specific or clinical application-specific potency assays. For this purpose, specific *in vitro* cell models or *in vivo* animal models should be developed during the stages of product development and preclinical studies for characterizing potency of the products as well as providing information for appropriate administration mode and dosage of the product to be used in clinical studies [1].

Process-related impurities include residual production-medium components (e.g., serum, antibodies, or exogenous cytokines), ancillary materials used in downstream processing, and possible leachables (e.g., plasticizers from tubing or culture plastic). The process-related impurities could be introduced during cell isolation or *in vitro* processing from cell debris or unwanted cells of autologous or allogenic origins when the requirements of cGMP have not been strictly respected. Unwanted cells include non-stem cells from donors, un-controlled differentiated cells, or unrelated cells, which could contaminate stem cells when they are all cultured and processed in the same cell-processing environment. Impurities may be bioactive, immunogenic, or may have other deleterious effects, depending on their concentrations in the final product [17, 21].

The stability regarding cell viability and potency of the products might be compromised during cell storage, cryopreservation, and transportation [6]. Evaluation of purity and stability should rely on careful cell characterization for identity, safety, and potency that may underwent various alterations during manufacturing and storage processes.

Quality Assurances for MSC Products

Standard operating procedures (SOPs) for managing entire preparation process or for guiding individual cell processing should be established and revised regularly according to new achievements in stem cell sciences and technologies. The individual cell processing includes cell isolation or collection, separation, purification, proliferation, modification, induction of cell differentiation, cell cryopreservation and recovery, residual removal, and packaging, among many others. SCMPs remained from each study or application should be strictly managed in a legal and ethical manor guided by well-established SOPs. All data in association with product preparation should be archived properly.

The quality of the entire preparation process should be thoroughly evaluated and validated through data and document reviewing, facility inspection and quality control tests. Appropriate acceptance criteria need to be established for effectively assessing and validating each individual process [5].

Requirements from Donors

For allogeneic MCS treatments, donor's identity and health information should be collected and carefully reviewed before stem-cell isolation. In addition, the information for techniques and routes used in cell isolation, and health history of both the individual and his/her family should be provided. The information of genetic diseases of both monogenic and polygenic conditions and of infectious diseases are particularly important in donor's qualification. Donors of the allogenic MSCs must be screened for possible infection with specific human viruses, including HIV1/2, HBV, HCV, HTLV1/2, EBV, CMV, etc., and with treponema pallidum. Donors with serious infectious disease and well-characterized genetic disorders are prohibited from being the source of allogeneic stem cells [11, 22].

For autologous MSC donors, the adjustment could be made in terms of quality requirements, screening tests and the acceptance criteria according to the characteristics, tissue source and clinical indication of the product.

Requirements for Cell Isolation and Processing

SOPs for cell processing, including cell isolation, selection, purification, proliferation, and modification, cryopreservation and transportation, and quality assuring procedures, such as personnel training, material and utility management, facility maintenance, environment monitoring, etc, should be established and strictly implemented.

To minimize variability associated with different batches of the same products, a multi-level cell banking system, such as Master Cell Bank (MCB) and Working Cell Bank (WCB), should be established for preserving the cells, especially for the cells with rich amount in each batch. The banked cells must have clear cell identity and freedom from microbial contamination [11, 13, 14].

During isolation and cultivation of cells in order to further prevent external agent contaminations, it is possible and preferable to use closed systems. Moreover, closed systems are usually equipped with automatic or semiautomatic procedures which help to reduce batch-to-batch variability due to the manual manipulation (i.e. changing media or harvesting). Quantitative monitoring and recording of several parameters as pH, CO2 tension, etc during cell preparation could help to produce a datalog that should be validated according to defined limits.

Cell Culture Medium

Cell culture medium used in preparation of all SCMPs must meet purity requirements established for cell substrates with microbial and endotoxin freedom-free media. In addition, product contaminants that may remain in the final product must induce no adverse effects in recipients. While supporting normal cell growth, the medium should not affect biological characteristics of the stem cells, i.e. their "stemness" and differentiation potentials. In addition, except for a temporary use in cell isolation from the source tissue, the use of antibiotics is not recommended as supplements in preparation of the MSC products [11, 17, 18].

If commercial medium is needed, qualified manufacturers should be selected as suppliers, ensuring that they could provide relevant quality certificates and medium component information. It is always suggested to first choose medium and other cell culturing materials that have been previously approved by regulatory authorities for the use in clinical applications. For materials that have not underwent formal approval, it is recommended to validate the quality of each batch of the material and the validation report should then be provided [17, 18].

In the consideration of serum use, it is recommended to avoid using either human or animal serum in preparation of SCMP. The use of allogeneic human serum or plasma must be strictly prohibited. If the use of animal serum is inevitable, the serum should be tested to exclude any contamination with origin-specific viruses and prions. Bovine serum from the endemic area of spongiform encephalitis must be prohibited [11, 18].

If human blood components, such as albumin, transferrin and various cytokines, are needed as supplements in the medium, the quality information of these ancillary materials including batch numbers and quality standards should be well documented and validated if necessary. It is highly recommended to choose the blood components approved by the regulatory authority for clinic applications [11, 17, 18].

Quality Control Tests

Major Principles in Quality Control Tests

To ensure both safety and efficacy of the SCT, each MSC product must be fully evaluated by well-designed quality control tests with appropriate acceptance criteria or specifications. The test design should be based on understanding of the updated achievements in science and technology for MSCs as well as the existing technical guidelines for cell substrates and therapeutic cell products.

In general, designed tests should be able to clearly demonstrate product's characteristics, safety, potency, and stability. The tests for characteristics should be able to show cell identity, viability, growth activity, and purity. Tests of safety should effectively identify possible contaminations with endogenous and exogenous microorganisms or toxins, abnormal immune responses and tumorigenicity or tumor modulatory effects of the products, whereas the biological potency assay(s) should reveal cell differentiation potential, structure and function of the differentiated cells, as well as the immune regulatory effects of the products. Selection of different tests for each product should be guided by the established principles to reflect the differences in cell derivation, characteristics and clinical indications of the product. In addition, the tests should be constantly improved according to the fast growth in both knowledge and techniques from the MSC sciences.

Two major types of the quality control tests, i.e. product quality test and release test, should be required for each product. The product quality test should be employed as a comprehensive test to identify quality issues regarding cell characteristics, safety and potency introduced possibly in different stages of product preparation process, thus also serving as a process assurance test. The release test should be used as a faster and simplified test prior to clinical application for each batch of the product on the condition that the same batch of the product has been previously tested by the product quality test. The release tests should put more emphasis on safety issues of the product [6].

To prevent variability issues arising among different batches of the product, the minimum number of batches randomly selected for testing should be required in product quality tests. In addition, the product quality testing should be repeated whenever changes happen in the preparation protocol, ancillary materials, and location of the facility. Each batch of the final product should be defined as the product prepared and processed from the same tissue of the same donor at the same time and same facility by the same technical protocols.

The test selection reflecting the level of stringency in quality control could be adjusted according to the characteristics and clinical applications of each product. For autologous MSC products without complex *in vitro* cell processing, the testing for cell identity, survival, growth activity, adventitious agents, as well as for some basic biological features of the product may prove to be appropriate. For allogeneic MSCs or autologous MSCs with complex *in vitro* cell processing and the banked MSCs, the more intense testing should be employed including the tests for cell characteristics, endogenous and exogenous microbial contaminations, purity, tumorigenicity or tumor modulation, immunological responses and regulatory effects, cell differentiation and other more specific potency assays if possible.

Product Quality Testing

1. **Cell identity**
   The tests used for revealing genotypes and phenotypes of the product, including cell morphology, cytogenetics, genetic polymorphism (e.g. STR profile), biochemical or biological features (i.e. isoenzyme profiling, cell surface markers, and specific gene or protein expressions) should be employed in concert to clearly demonstrate cell identity.

2. **Viability and growth activity**
   Various cell biology tests, such as the tests for cell counting, cell doubling time, cell cycle progression, colony formation in soft agarose gel, and telomerase activity and telomere length should be used in combination to measure cell viability and growth activity.

3. **Purity and homogeneity**
   Analyzing cell surface markers, genetic polymorphism, isoenzyme profiling, or species-specific mitochondrial rRNAs to identify possible contamination with unwanted cells during cell isolation and processing should be examine to ensure cell purity.

   The homogeneity of mixed products should be analyzed by examining the variability in cell surface markers, viability, purity and certain biological activities, including cell differentiation and immunoregulatory activity, of each individual batch of the product. The criteria for the acceptable variation for each test item should be developed through a significant amount of correlation studies before consideration of mixing different batches of the product.

4. **Sterility and mycoplasma tests**
   Possible contaminations with bacteria, fungi and mycoplasma should be tested by following the procedures described in the Chinese Pharmacopoeia, Part III of the 2010 version.

5. **Endogenous and adventitious viruses**
   In vitro and *in vivo* methods should be used in combination to test human or animal specific viruses for each batch of the product. All allogeneic products must be tested for human specific viruses, which include HIV, HBV, HCV, EBV, CMV etc. The bovine specific viruses should be tested if bovine serum was used during product preparation. If the materials of porcine origin such as trypsin are used, at least porcine parvovirus should be tested. In addition, retroviruses should be tested for all MSC products.

6. **Endotoxin**
   Endotoxin should be tested on the basis of a LAL assay described in Chinese Pharmacopeia, Part III of the 2010 version.

7. **Immune responses**
   The immune responses provoked possibly by allogeneic MSC products should be assessed by testing their effects on *in vitro* proliferation of total peripheral lymphocytes or certain lymphocyte subpopulations, or secretion of the related cytokines during co-incubation of MSCs with the peripheral blood mononuclear cells (PBMC).

8. **Tumorigenicity**
   Tumorigenicity should be evaluated for allogeneic MSC products or autologous MSC products with complex cell processing through examining their ability to form tumor in immune-compromised animals.

9. **Potency**
   To determine potency of MSC products, a variety of potency tests should be employed to reveal cell differentiation potential, structures and functions of the differentiated cells, the ability to regulate abnormal immunological or inflammatory responses, or secretion of specific cytokines, expression of specific genes and/or proteins in appropriate *in vitro* or *in vivo* models.

   To determine multipotency for cell differentiation, all MSC products, regardless of their tissue origins, should be evaluated for their capability to differentiate *in vitro* into a variety of cell types, such as adipocytes, chondrocytes and osteocytes. In addition, it is encouraged for the investigators/manufacturers to develop new potency assay(s) to evaluate specific attributes of the products to the efficacy in specific clinical applications.

10. **Removal of contaminants**
    Residual contaminants may remain in the final product from cell culture medium, which may induce adverse effects and impair safety, thus requiring removal prior administration. Contaminants from bovine serum albumin, antibodies, cytokines or other supplements in the products should be fully characterized to establish an acceptance criteria.

Release Tests

The release tests should be performed prior to clinical studies or immediately prior treatment administrations for each batch of the product. The tests for identity, purity, potency, microbial safety, viability, and packaging volume should be selected. The test selection could be adapted if the full product characterizations and/or comprehensive product quality tests have been conducted previously for the required amount of batches of each product.

Validation of the Quality Control Tests

In China, the quality control tests for MSC products and resulting quality standards should be reviewed and validated by a third independent party, who should be fully competent in conducting quality control of the cell products. The National Institutes for Food and Drug Control (NIFDC) serves as a competent independent party to review and validate both quality control tests and the resultant standards for each MSC product, and issues a validation report at the end. To facilitate the validation, especially for the product quality tests, the project report summarizing information for all aspects of the product, such as technical procedures for product preparation, characterization and clinical applications, and ancillary materials and excipients used during product preparation or in the final product, should be provided to the NIFDC. To facilitate validation for the release tests, previous reports from validation of the product quality tests should be presented as the important supporting documents.

Quality Studies for the MSC Products

On top of quality control tests, investigators should be encouraged to develop more effective new *in vitro* and *in vivo* models, upon which the relevant studies should be conducted to improve the capability for product quality control around entire product development cycle.

Studies for New Cell Characterizations

Monitoring Spontaneous Cellular Transformation

Status of cell growth, especially the independent growth from growth factors, should be continuously monitored to prevent spontaneous transformation occurring during long *in vitro* cell culture. Surrogate biomarkers, especially gene or protein expressions in association with spontaneous transformation, should be developed from new studies.

Tumorigenicity and Tumor Modulation Studies

Investigators/manufacturers should determine that sufficient amount of cells and appropriate tumorigenicity assay are provided for each specific product because there are significant variations in the feature of tumorigenicity among different products.

It is also very important to develop surrogate biomarkers through characterizing critical biological characteristics in association with tumorigenicity using genomics or proteomics techniques. Surrogate biomarkers could be used when tumorigenicity can not be directly evaluated in animal models. Critical biological characteristics include dependence on growth factor(s), sensitivity to apoptosis induction, genome stability, expression of the genes or proteins directly involved in transformation or tumorigenicity of MSCs.

In addition, investigators and/or manufacturers should design appropriate tests to determine tumor modulating activity of the MSC product according to their origin and clinical applications.

Stability Studies and Shelf Time Determination

Stability of each MSC product during cryopreservation, transportation or temporary storage prior to clinical applications should be studied. Cell viability, density, purity, sterility, and potency should serve as critical parameters in evaluating product stability. Based on stability studies, the optimum conditions for cell preservation, storage and transportation, product formulation, as well as the 'shelf time' of the product should then be determined.

Developing Quick Testing Methodology

Based on the updated achievements in MSC science and technology, investigators should develop alternative new quick testing technologies to be used in the release tests, especially for quickly determining the quality, safety and potency of the MSC products when the well-established long testing is not appropriate for achieving the same purpose.

Developing Standard Materials and Quality Standards

Development of standard materials and quality standards is extremely important in implementing quality control for SCMPs. However, since the SCT is still in its primary stage, it makes such task particularly difficult. Even though, investigators still should carry out the relevant quality studies of their own products, especially the correlation studies between the result from the selected test and the most relevant biological activities of the product with a goal of developing acceptance criteria, or standard, as well as developing standard materials to used in future quality control tests for MSC products.

In summary, the exciting achievements from clinical studies endorse the MSC-based SCT moving forward using the qualified MSC products. Although no guideline specifically for the MSC products, the most prevalent SCMPs in China, has been established, many principles from the existing guidelines for quality control of the cell substrates and therapeutic cell products, could be adopted for building up a preliminary quality control framework, upon which the detailed guidelines will be constantly added for directing the well guided and well-regulated development of MSC products as well as their clinical translations.

References

1. ISSCR guidelines for clinical translation of stem cells (2008).
2. Wong R. Mesenchymal stem cells: angels or demons? J Biomed Biotech. doi:10.1155/2011/459510:1-8.
3. Miguel MP, Julian SF, Martinez AB, Pascual CY, Aller MA, Arias J, Motiel FA. Immunosuppressive properties of mesenchymal stem cells: advances and applications. Curr Mol Med. 2012;12:574–91.
4. Prasad VK, Lucas KG, Kleiner GI, Talano JA, Jacobsohn D, Broadwater G, Monroy R, Kurtzberg J. Efficacy and safety of ex vivo cultured adult human mesenchymal stem cells (Procymal™) in pediatric patients with severe refractory acute graft-versus-host disease in a compassionate use study. Biol Blood Marrow Transplant. 2011;17(4):534–41.
5. Rayment EA, Williams DJ. Concise review: mind the gap: challenges in characterizing and quantifying cell- and tissue-based therapies for clinical translation. Stem Cells. 2010;28:996–1004.
6. EMA guideline on human cell-based medicinal products (2007).
7. FDA guidance for human somatic cell therapy and gene therapy (1998).
8. FDA guidance-content and review of CMC information for human somatic cell therapy IND application (2008).
9. Chinese Pharmacopoeia, Part III, 2010 Version.
10. European Pharmacopeia-Method 5.2.3-Cell substrates for the production of vaccines for human use.
11. WHO – Recommendations for the evaluation of animal cell cultures as substrates for the manufacture of biological medicinal products and for the characterization of cell banks (2010).
12. FDA guidance for industry-characterization and qualification of cell substrates and other biological materials used in the production of viral vaccines for infectious disease indications (2010).
13. ICH guidelines-viral safety evaluation of biotechnology products derived from cell lines of human or animal origin-Q5A(R1)-1999.
14. ICH guidelines-derivation and characterization of cell substrates used for production of biotechnological/biological products-Q5D-1997.
15. Guidelines for clinical studies and quality control of human somatic cells (in Chinese) (2003).
16. FDA guidance-potency tests for cellular and gene therapy (2011).
17. US Pharmacopeia <1046>Cellular and tissue-based products.
18. US Pharmacopeia <1043>Ancillary materials for cell, gene and tissue-engineered products.

19. Dominici M, et al. Minimum criteria for defining multipotent stem cells-The ISCT position statement. Cytotherapy. 2006;8(4):315–7.
20. Lazennec G, Jorgensen C. Concise review: adult multipotent stromal cells and cancer: risk or benefit? Stem Cells. 2008;26(6):1387–94.
21. Herberts CA, Kwa MS, Hermsen HP. Risk factors in the development of stem cell therapy [J]. J Transl Med. 2011;9:29.
22. FDA guidance for industry-current good tissue practice (CGTP) and additional requirements for manufactures of human cells, tissues, and cellular and tissue-based products (HCT/Ps).

Regulations/Ethical Guidelines on Human Adult/Mesenchymal Stem Cell Clinical Trial and Clinical Translation

Xiaomei Zhai and Renzong Qiu

Abstract There are two parts in this article. In the first part the authors describe the regulatory and ethical challenges caused by stem cell research and its clinical application, especially by the widespread use of unproven and unregulated "stem cell therapy" and the guidelines and regulations for attempting to deal with challenges in China. At the end of this part ethical guidelines for human adult (including mesenchymal) stem cell clinical trials and application is appended. In the second part the authors try to provide an overview of regulations or ethical guidelines on human adult/mesenchymal stem cell clinical translation in USA, UK, the European countries and India. The article concludes with the harmonization of regulating human adult/mesenchymal stem cell clinical translation in the world and the Guidelines for the Clinical Translation of Stem Cells drafted by the International Society for Stem Cell Research provide a basis for the harmonization.

Keywords Stem cell research • Clinical translation • Mesenchymal stem cell • Stem cell therapy • Clinical trials • Regulation • Ethical guideline • Harmonization

X. Zhai (✉)
Professor and Director, Center for Bioethics, Chinese Academy of Medical Sciences/Peking Union Medical College, Beijing, China
e-mail: xmzhai@gmail.com

R. Qiu
Professor and Chair of the Academic Committee, Center for Bioethics, Chinese Academy of Medical Sciences/Peking Union Medical College, Beijing, China
e-mail: renzong@gmail.com

R.C. Zhao (ed.), *Essentials of Mesenchymal Stem Cell Biology and Its Clinical Translation*, DOI 10.1007/978-94-007-6716-4_19,

Chinese Perspective

Background

Stem cell research in China has been developing in a socio-cultural context different from western countries. First, science and scientists still enjoy the highest respect from the public and obtain great support from the governments at all levels. Secondly, there is no ideological barrier to human therapeutic cloning and stem cell research in Chinese culture, for dominant is a gradualist and relational view of personhood that has been widely accepted by scientific community, community of the humanities and social sciences, the public and decision-makers. A person begins with birth, human fertilized egg, embryo and fetus are not person yet, though they have human biological life (but not the personal life).

In 1997 after Dolly was published in *Nature*, at the meeting convened by Chen Minzhang, the ex-Minister of Health, participants including scientists, bioethicists and law scientists unanimously suggested Ministry of Health (MOH) to concern ethical, legal and social issues (ELSI) in biomedicine and biotechnology. After the meeting MOH set up an ethical review committee to review the protocols supported by MOH, and later it became Ethics Committee, MOH China.

In 2001 Chinese government announced to not approve, not support, not permit and not accept human reproductive cloning, but support human therapeutic cloning. In March 2005 at the 59th UN Congress Chinese government rejected the UN Announcement of Prohibiting All forms of Human Cloning, and reconfirmed its basic position, that is, disapproving human reproductive cloning, but supporting human therapeutic cloning.

There are two fundamental values in stem cell research. The one is to promote such scientific frontiers of biomedicine and biotechnology as stem cell research in order to solve health problems of millions people; and the two is to appropriately address ethical issues emerging in stem cell research and its clinical application in order to protect interests/rights of patients, subjects and the public. It is scientists and bioethicists who took initiatives to draft recommendations on ethical principles and regulations for stem cell research and submitted to MOH in Beijing 1999 and Shanghai 2000 respectively. Based on these recommendations MOST and MOH drafted and finally promulgated *Ethical Guiding Principles for Human Embryonic Stem Cell Research* (2003). *Ethical Guiding Principles* (2003) is to try to maintain a delicate balance between scientific freedom and ethical constraints, while maximizing scientific freedom and minimizing ethical/regulatory constraints.

Ethical Guiding Principles stipulates which procedures are allowed and which are prohibited as follows:

Human reproductive cloning	P
hESC derived from spared gamete or embryos after IVF	A
hESC from fetal cells from accidental, spontaneous or voluntarily selected abortions	A

(continued)

(continued)

hESC from embryos obtained by somatic cell nuclear transfer technology or parthenogenetic split embryos	A
hESC from germ cells voluntarily donated	A
Embryos thus obtained, its *in vitro* culture period within 14 days	A
Hybridize human germ cells with germ cells of any other species	P
Trade of human gametes, fertilized eggs, embryos and fetal tissues	P

The experiences of 9 years implementation of the *Ethical Guiding Principles* show that the policy of maximising scientific freedom and minimising ethical constraints makes oversight impossible. For example, it is not known that how many institutions in China are doing human embryonic stem cell research, and how many institutes among them are qualified, and how many not etc. *Ethical Guiding Principles* do not specify the scientific and ethical qualifications an institute must meet, and there is no legal imperative to be licensed from any regulatory bodies, or even to register or be put on file. The *Ethical Guiding Principles* specify which research conduct is permitted and which is prohibited, but how many institutes are observing these requirements? The *Ethical Guiding Principles* requires that any protocol proposing to carry out human embryo stem cell research must be reviewed and approved by an IRB. But how is the quality of ethical review? What guidelines do IRBs use to review the protocol? How is the IRB established? How many IRB are qualified in their composition and functioning, and in their capacity for scientific and ethical review? How many are not? Have all members been trained? The donation of gametes or embryo used in human embryonic stem cell research should be voluntary, and valid informed consent must be obtained. Now, what is the process of informed consent in those institutes which conduct the research? And what does the written informed consent form look like? Documentation of the provenance of oocytes and stem cell lines and human embryo or human/animal embryo creation/disposition is necessary for oversight. It may include: Origin of the oocytes, the process of informed consent, the origin of the stem cell lines, the date, and the means of human embryo or human/animal embryo creation and disposal etc. All these are not clear for regulators nor for the public. It is necessary to revise and update it to fill the regulatory gaps on the basis of 9 years implementation of *Ethical Guiding Principles*.

We are not only faced with regulatory gaps, but also faced with ethical and governance challenges presented by rapid advancement of science and technology (and stem cell research in particular). The emerging issues we have to address include: induced pluripotent stem cells, hybrids/cybrids and chimeras, parthenogenesis, genetic modification/modification of chromosome DNA, artificial gametes from adult cells, PGD and preimplantation tissue typing, designer babies, the translation of stem cells science from bench to bedside, medical tourism, and oocyte trading over vast internet networks.

Since 2005 in mainland China, there might be about 500 health institutions which offered unproven and unregulated "stem cell therapy" (e.g. inject undifferentiated stem cells into patient's body) and attracted medical tourism to China from all over the world. It is difficult to ascertain whether the stem cell products offered are safe and effective, because no third party has tested them, and the methods and results have yet to be published in reputable journals. They usually recruit patients by advertisements on the internet. In these advertisements, the stem cell therapies are described as 'magic bullets' which can cure any diseases. The patients they enroll are desperate with untreatable, debilitating, and ultimately fatal illnesses usually from rural areas. They do not know how to assess the efficiency of therapy or how to claim their rights when they feel they have been deceived.

One reason is that in mainland China, the therapeutic misconception has tended to be common. Physicians and investigators have deliberately confused clinical trial with medical care, thereby avoiding any ethical review and informed consent process, but with the intention to make money from uninformed patients. And no regulation on the relationship between drugs/biological products/equipment manufacturing companies and physicians and investigators or IRB members has been developed.

Regulations/Guidelines

Although in 2007 MOH has promulgated *Interim Regulations on Ethical Review of Biomedical Research Involving Human Subjects* in which it is stipulated that experimental application of health care technique or product created by biomedical research shall be subject to ethical review, however, all so called "stem cell therapy" evaded ethical review. In view of the chaos caused by unproven and unregulated "stem cell therapy", MOH made efforts to put stem cell therapy under its control. In 2009 MOH promulgated *Regulations on Clinical Application of Medical Technologies* in which the treatment with heterologous stem cells is classified as the third category of medical technology. The third category of medical technology needs to be subject to strict control and regulation under MOH, because its safety and efficacy have not been proved by clinical trials yet. However, the implementation of the *Regulations* failed to restore "stem cell therapy" back to normal. On January 10 2012 MOH decided to suspend all stem cell therapy and clinical trial for 1 year.

In order to promote the successful and ethical clinical translation of stem cell research and protect patients' welfare and rights Ethics Committee of MOH considered it is necessary to draft *Ethical Guidelines on Adult Stem Cell Clinical Trials and Clinical Application* and submitted it to MOH. A team with the head of Professor Hu Chingli from Shanghai Jiaotong University and the former Associate General Director of WHO for drafting *Ethical Guidelines on Human Adult Stem Cell Clinical Trial and Application* was appointed by Ethics Committee of MOH. Ethics Committee of MOH reviewed, revised and approved the *Ethical Guidelines* and submitted it to MOH in September 2010.

In the *Ethical Guidelines* it is stipulated that the ethical principles for clinical trial of adult (including mesenchymal) stem cells should include: Scientific validity, non-maleficence/beneficence, informed consent, justice, public good, and non-commercialization.

The *Ethical Guidelines* emphasize the importance of distinguishing between pre-clinical research, clinical trials and clinical application pointing out that pre-clinical (lab or/animal) research is the premise of clinical trials and clinical trial is the necessary condition for clinical application. Only sufficient evidences of safety and efficacy are obtained during the two steps of scientific research (pre-clinical research and clinical trials), the results are evaluated scientifically and ethically, and the application is approved by health administration, adult stem cells are permitted to be translated into clinical application.

The conditions for clinical application of adult stem cells are: Except those adult stem cells which are not specially treated *in vitro* such as hemopoietic stem cells, or cartilage cells and are routinely used to treat diseases of blood system, cornea injuries or cartilage injuries, the safety and efficacy of all other adult stem cells used for treating diseases have not been proved yet, so before adult stem cells be applied clinically, clinical trial must be conducted in compliance with scientific and ethical principles. It is not permissive for any institution to use adult stem cells in clinics as a routine service until the safety and efficacy are proved by clinical trial.

The *Ethical Guidelines* require that in order to ensure the quality and safety of adult stem cells, medical professionals must do screening for genetic and epidemic diseases to donor, must have strict aseptic techniques and preventive measures to ensure non-contamination and non-pathogenic bacteria transmission, must establish unified criteria of preparation and system of quality management, must ensure minimization of genetic variation *in vitro*, such as genetic recombination, genetic deletion and other genetic abnormal change, or oncogenesis etc., must establish unified normative criteria of transplantation (time, route, number and evaluation index of clinical observation to ensure the safety of human subjects).

The institutional ethical review committee of the health institution conducting clinical trial of human adult stem cells shall review the protocol of clinical trial of adult stem cells. What they review shall include:

1. Are the adult stem cells used for clinical trials provided by qualified institution? There shall be a scientific authentication report to prove whether their biological features provided meet scientific criteria.
2. Provided data, reports and scientific evaluation of safety and efficacy of pre-clinical research on adult stem cells.
3. Do investigators' qualification and experiences meet the requirements of clinical trial?
4. How is the scientific validity of clinical trial? Is the protocol scientific valid?
5. Is the degree of risks exposed to subjects and expected benefits acceptable?
6. During the process of informed consent is the information disclosed to subjects complete and understandable? Is the method for obtaining consent appropriate? Is the consent form appropriate?
7. Is the measure for confidentiality adequate for protecting subjects' personal information?
8. Are the criteria for inclusion and exclusion of subjects appropriate and fair?
9. Are subjects clearly informed of their rights including the right to withdrawal at any time without reason and not being discriminated?
10. Do the subjects get compensations when they are injured or even died for participating in clinical trial?
11. Is there any staff among investigators who is responsible for informed consent and safety of subjects?

12. Are protective measures taken for the risks which are borne by subjects?
13. Is there any conflict of interest between investigators and subjects?

As for the innovative therapy or experimental treatment with adult stem cells the *Ethical Guidelines* stipulates that in the case that the patients with untreatable and fatal diseases insistently request the treatment with adult stem cells, to provide experimental treatment of this unproven therapy for few patients is permissive. But the patients selected shall be those who suffer cancer at late stage or other untreatable and fatal diseases.

In such cases clinicians shall provide written complete plan of such experimental treatment including scientific validity to choosing adult stem cells treatment; data of safety and efficacy of pre-clinical research; qualifications of medical staff; patient's voluntary choice; valid consent form; qualified stem cells technical operation facilities; responses to side-effects, complications and adverse effects; follow-up plans; etc.

Clinicians shall take responsibility for establishing therapeutic results with systematic and objective manner; reporting therapeutic results including negative results and adverse events to scientific community at conference or in journal; and timely turning to clinical trials after obtaining positive results in order to get universalized knowledge.

Based on the *Ethical Guidelines* the MOH now is drafting:

Administrative Measures on Research Base for Stem Cell Clinical Trial
Guiding Principles for Quality Control of the Stem Cell Preparation and Pre-clinical Trial
Administrative Measures on Stem Cell Clinical Trial

Appendix

Ethical Guidelines on Human Adult Stem Cell Clinical Trial and Clinical Application
Ethics Committee, Ministry of Health
September 2010

Chapter 1
General Provisions

Article 1
In order to improve the ethical governance of human adult stem cell research and application for better treating human diseases by use of stem cell technology, improving people's health, and protecting patients' and subjects' rights and interests, the Guidelines are developed according to *Law on Medical Practitioners*, *Regulation on Human Organ Transplantation*, *Ethical Guiding Principles for Human Embryonic Stem Cell Research*, *The Interim Rule for Ethical Review of Biomedical Research Involving with Human Subjects*, *The Rule for Clinical Application of Medical Technologies* and *Norms for Quality Control of Drug Clinical Trials*

with the reference with *Guidelines on Clinical Translation of Stem Cell Research* developed by International Society for Stem Cell Research.

Article 2
Human adult stem cell refers to those pluripotent or multipotent cells which have potential of self-renewing and differentiation and exist in all sorts of tissues or organs of human body (such as bone marrow, skin, fat, etc.), in human embryonic tissues or germ cells (such as amniotic fluid, umbilical cord, umbilical cord blood, etc.). Induced pluripotent stem cell is one kind of adult stem cells. Adult stem cells and their derivatives may have promising prospect for treating patients who suffer from incurable or stubborn diseases such as cancer (leukemia, lymphoma, etc.), Alzheimer disease, Parkinson disease, diabetes, cerebral palsy, paraplegia and so on. This caused the attention from scientists and the public.

Article 3
The transplantation technology of human adult stem cells (including those derived from tissues outside embryo) refers to transplanting patient's autologous or allogenic bioactive adult pluripotent stem cells or those of particular type after induction and differentiation into the patient's body for continuous proliferation so as to repair injured tissue or organ. This technology is classified into three categories:

Category 1: Human primary cells or tissues which are not processed *in vitro* with special technology, such as hemopoietic stem cells, cartilage cells etc. are transplanted for treating diseases of blood system, cornea injuries and cartilage injuries;

Category 2: Adult stem cells which are proliferated, induced and differentiated *in vitro* with special technology, such as neurons, mesenchymal stem cells etc. used for clinical trials or treatment of certain diseases;

Category 3: Adult stem cells with gene modification, such as adult stem cells used in gene therapy or iPS cells used for treatment, or adult stem cells used as vector for non-medical purpose,

Among the categories above Category 1 is routine therapy; Category 2 is at the stage of exploration, there has been no systematic scientific evaluation on their safety and effectiveness, so standard trial and research shall be actively conducted and promoted. Category 3 (such as the use of iPS cells) is at the stage of basic and preclinical research, the condition of clinical trial has not been satisfied with, let alone its clinical use. The adult stem cells transplantation for non -medical use shall be explicitly prohibited.

Article 4
So far the safety and effectiveness of adult stem cells used for clinical treatment have not been proved yet, preclinical study and standard clinical trial shall be first conducted in conformity with scientific and ethical principles Any institution shall be not allowed to use adult stem cells as routine treatment in clinics before the safety and effectiveness are proved by clinical trial and the clinical use is to be approved and licensed.

Article 5
The institution which conducts clinical trial of adult stem cells of the second or part of third category shall submit the application to provincial, municipal or autonomous region's health administration after finishing preclinical studies and obtain empirical evidences of safety and effectiveness.

Article 6
The clinical trial of adult stem cells must be based on scientific literatures and preclinical studies including animal experiments. The trial design, procedures, collection and process of data etc. all should be rigorously scientific and conformed with generally recognized scientific principles, include the principle of research integrity. As for biomedical research involving human subjects any research that is not meet scientific requirement must be a treated as a violation of ethical principle.

Chapter 2
Ethical Principles

Article 7
Principle of Nonmaleficence/Beneficence
Adult stem sell products shall be tested by a qualified third party before the clinical trial of adult stem cells be conducted. If the products fail to reach the standard (the standard shall be set later), the trial shall not be conducted. According to data obtained from preclinical research, the possible risk/benefit to human subjects shall be evaluated and risks shall be minimum. If risks are higher than minimum, it shall be considered that whether the benefits to society are great enough to justify the trial. In clinical trial efforts shall be made to minimize the risks and maximize the benefits. If risks are high, while the benefits to human subjects and society are marginal, the research project shall not be approved

Article 8
Principle of Informed Consent
In clinical trial of adult stem cells adequate information shall be provided to human subjects, possible therapeutic effect, risks and toxic/side effects shall be objectively disclosed to them, and efforts shall be made by investigator to help them understand all the information concerned. Human subjects shall be given enough time to consider it, then voluntarily make their own decision to agree or refuse to participate in the clinical trials. Human subjects shall be permitted to withdraw from the trial at any time with any reason, and shall not be discriminated because of the withdrawal. Personal information of human subjects shall be strictly kept confidential.

Article 9
Principle of Justice
When human subjects are recruited, the criteria for inclusion and exclusion shall be developed and the benefits/burdens shall be justly distributed.

Article 10
Principle of Public Good
When the results of clinical trial of adult stem cells are proved to be safe and effective, the investigator shall sum up the trials realistically and scientifically, and publish the results on professional journals. Investigators, funding units, and policy-making departments shall thoroughly considerate social good of the research and maximize it.

Article 11
Principle of Non-commercialization
Adult stem cells shall be collected with non-compensation and voluntary donation, but appropriate subsidy is permitted. Expenses of adult stem cell clinical trial shall be supported by institute/department concerned or foundation. It shall be not permitted to charge any fee to human subjects.

Chapter 3
Norms of Conduct

Article 12
The borderline between preclinical research, clinical trial and clinical application shall be strictly distinguished and not be confused. Preclinical research and clinical trial are two important phases of clinical scientific trial of medicine. The results of the research can be applied to clinical practice only if those two phases of research are finished, adequate evidences of safety and effectiveness are obtained and scientifically evaluated and ethically reviewed, and approved by the state health care administration. It shall be ethically impermissible to apply adult stem cells into patients' market in commercialized way and it causes patient's physical, mental and economic harm without being proved scientifically and approved by health care administration.

Article 13
The collecting, handling and processing of adult stem cells shall be under strict quality control. In order to ensure the quality and safety of adult stem cells, The screening of genetic and epidemic diseases shall be conducted to the donors; adult stem cells shall be collected, handled and processed under adequately safe and aseptic (germ-free) condition in conformity with GMP; the entire production cycle shall be monitored by standard protocol, during the differentiation and production the batch homogeneity and verification method shall be ensured, the diversity of cell sources shall be reduced, the differentiation and production process shall be regulated, the function and composition of the final product shall be strictly tested. In order to maintain the stability of stem cells, their phenotype, karyotype, genetics and epigenetics shall be analyzed and marked; close attention shall be paid to the tumorigenicity and toxicity of pluripotent stem cells in treatment (including the acute and chronic toxicity of main organs), and blood biochemical changes and immunogenicity of expected parts after cell transplantation; unified standard criteria of transplantation, including timing, route, number and main evaluation indices of clinical observation shall be established in order to ensure the safety of human subjects, and the obtained data being able to determine the effectiveness of the graft.

Article 14
Preclinical research of adult stem cells is the necessary premise of clinical trial. In preclinical research the characteristics of the stem cells that are planned to be used for treatment, the pathway of entering target, the mechanism in human body, the adverse/side effects and tumorigenicity shall be studied systemically and in conformity to norms. Only after the scientific data which prove the safety, effectiveness and controllability of stem cell therapy technology are obtained, clinical trial can be conducted.

Article 15
Clinical trial of adult stem cells is the necessary condition for clinical application (translation). Clinical trial projects shall be strictly reviewed both in scientific and ethical aspects by provincial, municipal or autonomous region ethical review committee. The main contents of ethical review include:

1. Whether the adult stem cells used for clinical trial are provided by qualified institution. There shall be a scientific identification report to prove the biological characteristics of the provided stem cells and their derivatives meet scientific standard.
2. Whether data, reports and scientific evaluation about the safety and effectiveness of preclinical adult stem cell study (including laboratory research and animal research) are provided.
3. Whether the qualification and experiences of the investigator meet the requirements of clinical trials.
4. Whether the scientific grounds of the clinical trial, trial protocol, and the aim and significance of the trial are provided.
5. Whether the ratio of potential risks exposed to subjects and expected benefits is acceptable. Whether protection measures are taken to protect subjects from potential risks.
6. In the informed consent process of, whether the information provided to subjects (or their family member, guardian or legal proxy) is complete and understandable, whether the informed consent is obtained in an appropriate way, and whether the informed consent form is appropriate.
7. Whether confidential measures are taken to protect subjects' data.
8. Whether the criteria for subjects' inclusion and exclusion are appropriate and fair?
9. Whether subjects are explicitly disclosed of the right they enjoy including being able to withdraw at any time during the trial without need to provide any reason and the right to non-discrimination.
10. Whether the subjects get reasonable subsidies due to participating in the trial. If they get injury or even die caused by participating in the trial, whether the medical care or compensation they receive is appropriate.
11. Whether there is any specially assigned member in research team who is responsible for handling the issues of informed consent and subjects' safety.
12. Whether protective measures are taken for potential risks subjects may exposed to.
13. Whether there is conflict of interest between investigator and subjects.

Article 16
Strict licensing system shall be adopted for clinical application of adult stem cells. According to the norms laid down in Rules of Clinical Applications of Medical Technologies promulgated by Ministry of Health, stem cell treatment is classified as the Third Type of Medical Technologies than shall be strictly controlled and governed by health care administration. It is prohibited that adult stem cells are applied into clinics, and operated commercially by any institution without the approval from health care administration.

Article 17
The transplantation technologies of bone marrow hemopoietic stem cells, peripheral hemopoietic stem cells and umbilical cord stem cells that are applied to treat hemopoietic system diseases, hematopoietic injuries caused by radiation or chemotherapy for tumor, autoimmune diseases, radiation diseases, genetic diseases and etc. have been used for several decade, their safety and effectiveness have been scientifically proved. There has already existed mature governance mechanism, and these procedures have become routine treatment, so there is no need for them to conduct clinical trials.

Article 18
Based on the implications of Article 35 Helsinki Declaration, autologous or allogenic stem cells are permissive to provide to individual critical patients with their voluntary consent by medical professionals as experimental treatment or innovative therapy. This kind of experimental treatment is different from clinical trial, nor from routine clinical practice, which shall meet the following requirements:

1. Patients of experimental treatment shall be those with advanced cancer or serious disease, for which doesn't exist any alternative to be selected as better medicine or medical technology;
2. Clinicians shall propose written protocol of experimental treatment including reasonable scientific ground for choosing adult stem cell therapy, scientific data about the safety and effectiveness of preclinical studies, qualification of the clinician, voluntariness of the patient, qualified informed consent form, operation facilities for stem cell technology that meet scientific requirement, measures to deal with toxic/side-effects and plan to handle complication and adverse reaction and follow-up plan.
3. Clinicians shall commit to use the experiences that are obtained from their individual patients to pursuit universal knowledge. For this goal, the written protocol shall also include: determining therapeutic results in a systematic and objective way, reporting results, including negative results and adverse events to medical community at academic conferences or professional journals, and promptly turning into formal clinical trials (according to Article 15) after positive results being obtained from patients.
4. The written protocol shall be reviewed by provincial, municipal or autonomous region ethical review committee, and also approved by health care administration at the same level.
5. The patient shall be charged for experimental treatment according to the cost.

Article 19
Mesenchymal stem cell is a kind of cell with low-grade differentiation which has been in fetus' mesenchymal tissues. In adult tissues there still preserve highly viable mesenchymal cells in initial state which can differentiate into new cells and tissues under certain condition. Using these cells as the source of adult stem cells can reduce rejection. Its preclinical studies shall be strengthened. Any research that meets relevant rules or requirements set in the Article 15 of the Guidelines can apply for clinical trials.

Article 20
When adult stem cells and their derivatives need to be produced in batch and enter into medical market, they shall be classified as medical products of human cells which in turn shall be strictly reviewed in compliance with SFDA's *Norms for Quality Control of Drug Clinical Trials*. Without SFDA's approval, no such product shall be allowed to enter into the medical market.

Article 21
The application of adult stem cells and their derivatives for non-medical purpose, such as enhancing height, IQ or athletic performance shall be strictly prohibited.

Chapter 4
Oversight and Governance

Article 22
After the safety and effectiveness of adult stem cells are proved in clinical trial and before they are translated into clinical application, the application for clinical translation shall be approved by Ministry of Health, and the qualification of licensing shall be obtained, then clinical application shall be permissive.

Article 23
Expert Committee of Stem Cell Research and Application at national level shall be founded and affiliated with Ministry of Health. Strictly licensing system shall be established. Those health care institutions, which are qualified and competent in standard management shall be granted the license for stem cell therapy, and public notice shall be declared timely. Those health care institutions which already obtain the license shall be reviewed and reexamined regularly. Those health care institutions which fail to pass the capacity of clinical applications and ethical assessment shall be prohibited to perform any kind of stem cell therapies.

Article 24
The provincial, municipal or autonomous region health care administration has the responsibility to govern and oversee adult stem cell clinical trial and clinical application in its jurisdiction. For clinical trial and application of adult stem cells and their derivatives, all health care institutions in province, city, People's Liberation Army and Armed Police Force shall observe the Guidelines.

Since the date on which the Guidelines take into effect, any guidelines or rules that are incompatible with the Guidelines shall be overturned.

Article 25
Any health care institution which violates the Guidelines, such as providing stem cell therapy before clinical trials or providing long term stem cell therapy without clinical trials being conducted in the name of experimental treatment, not only shall be ordered to terminate its stem cell therapy immediately, but also shall be banned from applying for the application of the third category of medical technologies specified in *The Rule for Clinical Application of Medical Technologies* and clinical trial of stem cell therapy in coming 5 years.

Article 26
The Guidelines shall go into effect on the date it is promulgated.
Translated by Wang Zhaochen, Zhang Di, Cui Gengshen, & Liu Ran
Reviewed by Zhai Xiaomei & Qiu Renzong
Center for Bioethics
Chinese Academy of Medical Sciences/Peking Union Medical College

International Perspectives[1]

Introduction

The use of stem cells as medicines is a promising and upcoming area of research as they may be able to help the body to regenerate damaged or lost tissue in a host of diseases like Parkinson's, multiple sclerosis, heart disease, liver disease, spinal cord damage, cancer and many more. Translating basic stem cell research into routine therapies is a complex multi-step process which entails the challenge related to managing the expected therapeutic benefits with the potential risks while complying with the existing regulations and guidelines. While in the United States (US) and European Union (EU) regulations are in place, in India, there does not exist a well-defined regulatory framework for "stem cell based products (SCBP)". There are several areas that need to be addressed as it is quite different from that of pharmaceuticals. These range from establishing batch consistency, product stability to product safety and efficacy through pre-clinical, clinical studies and marketing authorization [22].

Stem cells have a unique ability to differentiate into the specific cells required for repairing damaged or defective tissues or cells. Stem cell based therapies, encompassing collection, purification, manipulation, characterization delivery of cells for therapeutic purposes, have existed since the first successful bone marrow transplantation in 1968 [20]. Presently, human embryonic stem cells (hESCs) are used in 13 % of cell therapy procedures, while fetal stem cells are used in 2 %, umbilical cord stem cells in 10 %, and adult stem cells in majority (75 %) of treatments [1]. Among adult stem cells one of the cell types most widely used to date in cell therapy

[1] This part is compiled and pieced together on the basis of published or online writings and originally intended to be provided as internal reference.

are mesenchymal stem cells (MSCs), which are of a mesodermal origin and have been isolated from bone marrow, umbilical cord blood, muscle, bone, cartilage, and adipose tissue. Recent studies have shown that MSCs also represent an optimum tool in cell therapy because of their easy *in vitro* isolation and expansion and their high capacity to accumulate in sites of tissue damage, inflammation, and neoplasia. MSCs are therefore useful in regenerative therapy, in graft-versus-host disease and in Crohn's disease, or in cancer therapy. The development in the future of an optimum methodology for genetic manipulation of MSCs may even increase their relevant role in cell and gene therapy. So the multipotent mesenchymal stem cells (MSCs) have been suggested as a suitable cell source for cell-based treatments for diseases such as osteoarthritis due to their ability to differentiate towards chondrogenic and osteogenic lineages. MSCs can be obtained from a variety of tissue sources, are scalable for mass-production and immuno-privileged enabling their use for allogeneic cell therapy [31, 2].

The term "stem cell based products (SCBP)" is used to refer to products intended to be administered to a patient and that contain or are derived from stem cells [23]. Commercial clinics worldwide are currently advertising so-called stem cell "therapies" for a host of diseases. Most of the clinics providing stem cell based interventions do not operate within the context of a formal clinical trial (CT). Whether the motive is outright profiteering or an attempt to help needy patients, the risks to patients of physical harm and financial exploitation remain extremely high. Globally, many pharmaceutical companies, including the big ones, are reluctant to enter this segment because of the great investment required and the uncertainties associated with it which include the regulatory framework. While some have regulations in place, others do not even have their own national guidelines to follow. Appropriate regulation of SCBP is essential to ensure public safety and trust while minimizing unnecessary barriers to product development, but presents numerous regulatory challenges [22].

As the pace of translational stem cell research accelerates, researchers and governing bodies must work together to develop and implement rigorous ethical standards to guide the transition into the clinical sphere. The field of stem cell research has entered an invigorating translational phase aiming to yield discoveries that will pave the way for regenerative medicine to cure diseases for which traditional methods have failed. Government agencies and charities are increasingly directing funding towards translational research in both the US and Europe. In the UK, the Medical Research Council has set up a translational stem cell research committee to fund research proposals with clear translational goals. In the US, the portfolio of the Californian Institute of Regenerative Medicine includes more than 40 translational projects. The focus on applied stem cell research and the transition into stem-cell-based therapy in the clinic must be accompanied by the development of regulatory oversight of basic research with translational potential [14].

The capacity to reprogram human adult somatic cells into a pluripotent state or a different cell type has opened the door to the development and study of patient-specific cells. These cells not only provide a tool for researchers to understand more about the mechanistic basis of disease, but also offer the possibility of drug testing in a dish. Fundamentally, it is the responsibility of researchers to ensure that appropriate consent has been obtained from patients for the initial study and follow-up research, especially in cases involving genomic analysis that could potentially disclose sensitive information. Although clinical trials using reprogrammed patient-specific cells are a long way off, other areas of stem cell research are closer to clinical application. The International Society for Stem Cell Research (ISSCR) has taken a strong lead in promoting stringent guidelines for translational stem cell research. In 2008, they produced a booklet for researchers and clinicians who are moving their research to the clinical phase, in which they called for rigorous standards and evaluation, a thorough informed consent process for patients involved in clinical trials, and transparency [14].

Few stem-cell-based treatments, such as bone marrow transplantations to treat blood-related disorders, have proven beneficial for patients in rigorous clinical trials and are now offered as treatments. Most other such therapies are in an experimental phase, and only a handful of clinical trial results have been published so far. In 2010, de Luca, Pellegrini and co-workers presented the results of their 10 year study using the human cornea, demonstrating the transplantation of limbal adult stem cells to restore retinal epithelium destroyed by burn (*New Engl. J. Med.* 363, 147–55; 2010). Earlier this year, Lanza and colleagues reported no adverse effects when they transplanted retinal pigment epithelium cells derived from human embryonic stem cells into patients suffering from advanced stages of macular degeneration (*Lancet* 379,713–720; 2012) [14].

Unfortunately, beyond such strictly regulated clinical trials, many treatments proposing to use stem cells (in particular, adult mesenchymal stem cells) to cure a range of ailments are being offered to patients around the world, with no clinical trial results to support their claims. The challenge for the field is to develop stringent rules in conjunction with government authorities, so that clinical trials are appropriately identified and regulated, and to remain vigilant about informing the public and the authorities in cases of non-compliance. The ISSCR has taken significant steps in this direction by developing a comprehensive resource for patients considering stem-cell-based therapies by encouraging their members to promote the dissemination of this information [14].

Governments have also taken note of this need. In January, China halted unapproved stem cell treatments and placed applications for new trials on hold until July 2012. In a press release from the ISSCR, Chinese stem cell researchers welcomed this measure and noted that it demonstrates that governing bodies are taking steps to put in place much-needed regulation at the same time as increasing their investment in stem cell research. Last year, the X-Cell Center in Germany closed after 4 years of proposing expensive treatments involving the injection of stem cells (derived from bone marrow) into various affected body parts of patients, a method radically different from using bone marrow transplantation to treat blood disorders. The closure of this clinic was a result of a change in European laws requiring that hospital doctors apply for European-Union-wide licenses to use innovative therapies such as stem-cell-based treatments, and of the active lobbying of German stem cell researchers belonging to the North Rhine Westphalia Stem Cells Network. The latter wrote an open letter against the activities of this clinic and directed patients to the ISSCR handbook on stem cell therapies [14].

However, some physicians seem to believe that full approval of a possible treatment by national organisations such as the Food and Drug Administration could take too long, or would not be necessary if patients are made aware of the risks involved and if a review panel has estimated their safety. Last year, the governor of Texas underwent an experimental stem cell treatment. In parallel, a change was proposed in Texas regulation that could allow experimental stem cell therapies to be made commercially available to patients. This triggered the reaction of ISSCR members, who stated that such changes would breach their guidelines for clinical trials involving stem cells [14].

Unapproved stem-cell-based therapies represent a danger for the patients and will ultimately be detrimental for the development of regenerative medicine. By taking responsibility and implementing regulatory oversight, researchers can enhance the move towards safe and effective translational applications [14].

Regulatory Challenges

There are several regulatory issues that relate to the safety, efficacy, and quality of SCBPs to be considered while preparing a cell- and tissue-based therapy for clinical and commercial use. Initially, safety testing is critical, including assays for potential microbial, fungal, endotoxin, mycoplasma, and viral contamination; karyotype testing; and enrichment for the required cell population. Once safety has been established, the product must pass *in vitro* functional assays designed to act as surrogate measures for clinical effectiveness [8, 34]. These potency assays must be fully validated to meet regulatory requirements, including appropriate standards and controls. The product has to be made to a certain set of specifications, ensuring high quality. Another aspect is the scarce availability of classical toxicology studies from the pre-clinical development. All animal models have inherent limitations, like, for example, the application of human cells in a xenogenic milieu [5]. This requires the use of severely immuno-compromised small animals. Furthermore, for a variety of diseases, for example, in orthopedics, small animals are not capable of modeling the disease. Selection of the most appropriate and sensitive model for conducting tumorigenicity studies should take into account the biological characteristics, conditions of *in vitro* manipulation, persistence of cells, route of administration and the intended clinical use of the SCBP. In the presence of reduced pre-clinical data, it is required that the CTs should be performed, with the highest attention being paid to the safety and ethical issue involved [22].

Cell therapy is one of the advanced therapy products (ATPs), together with gene therapy and tissue engineering in Europe. A regulatory framework is required for ATPs to ensure patient accessibility to products and governmental assistance for their regulation and control. Certainty, scientific reality and objectivity, and flexibility to keep pace with scientific and technological evolution are the characteristics defining an effective regulation [31].

Aspects to be regulated mainly include control of development, manufacturing, and quality using release and stability tests; non-clinical aspects such as the need for studies on biodistribution, cell viability and proliferation, differentiation levels and rates, and duration of *in vivo* function; and clinical aspects such as special dose characteristics, stratification risk, and specific pharmacovigilance and traceability issues [31].

Guidelines – ranging from total prohibition to controlled permissiveness – defining what may be permitted in research with pluripotent stem cells have been issued in countries all over the world. All such guidelines reflect the different views about when life starts during the human embryonic development, as well as regulation of measures to protect oocyte donors and to reduce the probability of human embryo destruction. There is general international agreement in that the results of stem cell research should not be applied in humans without prior ethical scrutiny. For this purpose, 42 European countries have national ethics committees since 2006, and a President's Council on Bioethics with an advisory role in bioethical matters was created in the US in 2001. The European Commission currently has the *Group on Ethics in Science and New Technologies*, an advisory, independent, and plural multidisciplinary body, and in other countries, such as the United Kingdom, legislation on action and bioethics is clearly established since several years ago [31].

Regulatory Framework in the US

In the United States of America, restrictions are limited to research with federal funds. No limitations exist for research with human embryonic stem cells provided the funds come from private investors or specific states. In countries such as Australia, China, India, Israel, Japan, Singapore, and South Korea, therapeutic cloning is permitted [31].

Stem-cell–based therapies have existed since the first successful bone marrow transplantations in 1968. The FDA has developed a regulatory framework that controls both cell- and tissue-based products, and recently issued updates to previous regulations referring to human cells, tissues, and all derived products [42], 2006]. This regulation provides an adequate regulatory structure for the wide range of stem cell-based products which may be developed to replace or repair damaged tissue, as both basic and clinical researchers and those working in biotechnological and pharmaceutical companies which need greater understanding and information to answer many questions before submitting a stem cell-based product for clinical use [31].

It should be reminded that, unlike conventional medicinal products, many stem cell-derived products are developed at universities and basic research institutions, where preclinical studies are also conducted, and that researchers there may not be familiar with the applicable regulations in this field. The FDA also provides specific recommendations on how scientists should address the safety and efficacy issues related to this type of therapies [23].

Any product based on stem cells or tissues undergoes significant processing, and it should therefore be fully verified that they retain their normal physiological function, either combined or not with other non-tissue components, because they will generally be used for metabolic purposes. This is why many such products, if not all, must also comply with the Public Health Services Act, Section 351 [39], governing the granting of licenses for biological products, which requires FDA submission and application for investigational protocols of new drugs before conducting clinical trials in humans [31.]

The key points of the current FDA regulation for cell therapy products [23] include: (i) demonstration of preclinical safety and efficacy; (ii) no risk for donors of transmission of infectious or genetic diseases; (iii) no risk for recipients of contamination or other adverse effects of cells or sample processing; (iv) specific and detailed determination of the type of cells forming the product and what are their exact purity and potency; (v) *in vivo* safety and efficacy of the product [31].

In the US, use of cell therapy products is codified within the Code of Federal Regulations in the following sections: Investigational New Drug Regulations (21 CFR 312), biologics regulations (21 CFR 600) and cGMP (21 CFR 211).[2] In particular, US federal regulation on cellular therapy is divided into two sections of the Public Health Service Act (PHSA), referred as "361 products" [35] and "351 products" [39]. Traditional blood and bone marrow progenitor cells as well as other tissues for transplantation fall into 361 products definition. The Food and Drug Administration (FDA) has established that cells or tissues used for therapeutic purposes and the regulation that pertains to processing of 361 products are codified under the Good Tissue Practice (GTP). CFR, Part 1271 provides US regulations on Cells, Tissues, and Cellular and Tissue-Based Products (HCT/Ps) [41]. This became effective in 2005 as rules for HCT/Ps. The FDA has also issued guidance documents about how the drug, biologic, and device regulations apply to cellular and genetic therapies [22, 40].

Classification of stem cell based therapies is based on indication to be treated. Restrictions are limited to research with federal funds. No limitations exist for research with hESCs, provided the funds come from private investors or specific states. The FDA has developed a regulatory framework that controls both cell- and tissue-based products, based on three general areas [31]:

- Prevention of use of contaminated tissues or cells (e.g. AIDS or hepatitis);
- Prevention of inadequate handling or processing that may damage or contaminate those tissues or cells; and
- Clinical safety of all tissues or cells that may be processed, used for functions other than normal functions, combined with components other than tissues, or used for metabolic purposes.

The Center for Biologics Evaluation and Research (CBER), the division of US FDA that regulates stem cell based therapies, has so far approved ApliGraf®, Carticel® and Epicel®. Those cell-based therapeutics "that are, minimally manipulated, labeled or advertised for homologous use only, and not combined with a drug or device" do not require FDA approval [33]. In contrast, manipulated autologous cells for structural use meet the definition of somatic cell therapy products and

[2]Please see Legal Information Institute, Cornell University Law School, http://www.law.cornell.edu/cfr/text/21/312.

require an "investigational new drug" (IND) exemption or the FDA license approval. In 2007, the "Guidance for Industry: Regulation of HCT/Ps – Small Entity Compliance Guide" and in 2009, the "Guidance for Industry on Current Good Tissue Practice (cGTP) and Additional Requirements for Manufacturers of HCT/Ps" (http://www.fda.gov) had been released [6]. Clinical studies employing mescenchymal stem cells (MSCs) underlie the IND mechanism. Accordingly, the investigators have to make an IND application, which necessitates detailed study protocols describing the clinical plan as well as the preparation and testing of the therapeutic cell product [21, 22].

Under the current FDA policies, there are at least two ways in which physicians may administer more than minimally manipulated stem cell products to patients. The first is under the FDA's program for expanded access to investigational drugs and biological products for treatment use (what is sometimes referred to as "compassionate use") as long as these products are currently being tested elsewhere in a CT and only if expanded access will not interfere with the conduct of clinical investigations. FDA allows clinicians to charge for direct cost recovery and administrative costs associated with expanded access use [24]. The second is the off-label prescribing of FDA-approved stem cell products. Off-label prescribing is premised on the position that the FDA does not have the authority to regulate medical practice and the assumption that physicians can be trusted to use their professional judgment in deciding how to treat their patients [13, 22].

There is still much to be learned about the procedures to establish the safety and efficacy of cell therapy products. The greater the understanding of the biology of stem cell self-renewal and differentiation, the more precise the evaluation and prediction of potential risks. Development of techniques for cell identification within a mixed cell culture population and for follow-up of transplanted cells will also be essential to ascertain the potential *in vivo* invasive processes and to ensure safety [31].

Since new stem cell-based therapies develop very fast, the regulatory framework must be adapted and evolve to keep pace with such progress, although it may be expected to change more slowly. Meanwhile, the current regulations must provide the framework for ensuring the safety and efficacy of the next generations of stem cell-based therapeutic products [31].

In July 2012 a U.S. District Court issued an injunction against use of the "Regenexx™ Procedure," which purports to treat joint, muscle, tendon or bone pain due to injury or other conditions with culture expanded autologous adult stem cells. The court agreed with the FDA that the cell product used in the procedure is both a drug and a "biological product" subject to FDA regulation. The FDA notified Regenerative Sciences in 2008 that the FDA believed the cell product used in the Regenexx™ Procedure constituted a drug under the Federal Food Drug and Cosmetics Act (FFDCA) and a biological product under the Public Health Service Act (PHSA). In 2009 and 2010, the FDA investigated a Regenerative Sciences laboratory and determined that it did not operate in conformity with current good manufacturing practices. In 2010, the FDA filed suit to enjoin further use of the Regenexx™ Procedure claiming that it constitutes the manufacturing and holding for sale an unapproved biological drug product [3, 4, 29].

In October 2012 FDA approved Stemedica phase II clinical trial for acute myocardial infarction with ischemia tolerant mesenchymal stem cells. Stemedica Cell Technologies, Inc., a leader in adult allogeneic stem cell manufacturing, research and development, announced today that the U.S. Food and Drug Administration (FDA) approved its application for an Investigational New Drug (IND) to assess the clinical effects of Stemedyne-MSC (Stemedica's human bone marrow-derived ischemia tolerant mesenchymal cells) in subjects with a myocardial infarct. The Phase IIa double-blinded randomized clinical trial will study approximately forty (40) patients. All patients will initially receive standard care including percutaneous transluminal coronary angioplasty (PTCA) and stenting and, upon completion, will be randomized to receive Stemedyne-MSC intravenously or placebo [36].

Regulatory Framework in EU

European countries may be classified into three groups based on their different positions regarding research with embryonic stem cells of human origin. (i) Countries with a restrictive political model (Iceland, Lithuania, Denmark, Slovenia, Germany, Ireland, Austria, Italy, Norway, and Poland); (ii) Countries with a liberal political model (Sweden, Belgium, United Kingdom, and Spain); and (iii) Countries with an intermediate model (Latvia, Estonia, Finland, France, Greece, Hungary, Switzerland, the Netherlands, Bulgaria, Cyprus, Portugal, Turkey, Ukraine, Georgia, Moldavia, Romania, and Slovakia) [31].

The *Seventh Framework Program for Research of the European Union*, coordinated by the European Medicines Agency, was approved on July 2006. This Seventh Framework Program provides for funding of research projects with embryonic stem cells in countries where this type of research is legally accepted, and the projects involving destruction of human embryos will not be financed with European funds. Guidelines on therapeutic products based on human cells are also established [15].

This regulation replaces the points in the prior 1998 regulation (CPMP/BWP/41450/98) referring to the manufacture and quality control of therapy with drugs based on human somatic cells, adapting them to the applicable law and to the heterogeneity of products, including combination products. Guidance is provided about the criteria and tests for all starting materials, manufacturing process design and validation, characterization of cell-base medicinal products, quality control aspects of the development program, traceability and vigilance, and comparison. Is also provides specific guidance of matrixes and stabilizing and structural devices or products as combination components [31].

The directive recognizes that conventional non-clinical pharmacology and toxicological studies may be different for cell-based drugs, but should be strictly necessary for predicting response in humans. It also establishes the guidelines for clinical trials as regards pharmacodynamic and pharmacokinetic studies, defining the clinically effective safe doses. The guideline describes the special consideration to be given to pharmacovigilance issues and the risk management plan for these products [31].

The guideline has therefore a multidisciplinary nature and addresses development, manufacture, and quality control, as well as preclinical and clinical development of medicinal products based on somatic cells [10] and tissue engineering products [19]. Includes autologous or allogeneic (but not xenogeneic) protocols based on cells either isolated or combined with non-cell components, or genetically modified. However, the document does not address non-viable cells or fragments from human cells [31].

Legislation on cell therapy in Europe is based on three directives [31]:

- Directive 2003/63/EC (amending Directive 2001/83/EC), which defines cell therapy products as clinical products and includes their specific requirements [10, 11].
- Directive 2001/20/EC, which emphasizes that CTs are mandatory for such cell therapy products and describes the special requirements for approval of such trials [9].
- Directive 2004/23/EC, which establishes the standard quality, donation safety, harvesting, tests, processing, preservation, storage, and distribution of human tissues and cells [12].

The EU directives recognize that conventional nonclinical pharmacology and toxicological studies may be different for cell-based drugs, but should be strictly necessary for predicting response in humans. The EU regulation (1394/2007) on Advanced Therapy Medicinal Products (ATMPs) became effective from December 2008 and is binding in its entirety and directly applicable in all Member States of the European Parliament and of the council [19]. ATMPs include gene therapy medicinal products, somatic cell therapy products (as defined in Directive 2001/83/EC) [10], and tissue engineered products. Cells fall under this regulation, in case they have been subjected to substantial manipulation, resulting in a change of their biological characteristics, physiological functions or structural properties relevant for the intended therapeutic application. The Committee for Advanced Therapies (CAT) within European Medicines Agency (EMEA) is responsible, among other tasks, for preparing a draft opinion on the quality, safety, and efficacy of ATMPs that follow the centralized marketing authorization (MA) procedure. Yet, no MA has been granted for any stem cell based medical product (SCBPM) in the EU [17].

EMA has very recently released a “Reflection Paper” [16] which covers specific aspects related to SCBPs with an intention for MA application. This reflection paper is relevant to all medicinal products using stem cells as starting material regardless of their differentiation status at the time of administration. SCBPs intended for clinical use should be produced via a robust manufacturing process governed by quality control sufficient to ensure consistent and reproducible final product. EMA suggests a risk-based approach according to Annex I, part IV of Directive 2001/83/EC for SCBPs [10, 22].

Generally, the clinical development plan should follow corresponding EU guidance on medicinal products and specific relevant guidance for the diseases to be treated. CTs should be designed to demonstrate safety and efficacy as well as provide evidence to substantiate the mode of action identified during the CT.

For first-in-man studies, the principles of the guideline on strategies to identify and mitigate risks for first-in-human CTs with investigational medicinal products (EMEA/CHMP/SWP/28367/07) [18] should be considered. In first-in-man studies, specific safety endpoints may need to be defined based on theoretical considerations and in order to detect early any toxicity arising from potential contaminants in the final product. In those cases where sufficient proof-of-concept and safety cannot be established in the nonclinical studies, for example, due to justified difficulties in finding an appropriate animal model, the evidence should be generated in CTs by including additional endpoints for efficacy and safety, respectively. Clinically meaningful endpoints related to the pharmacodynamic effect of the product should be used for efficacy assessment in the target indication. The effective range of stem cells and/or stem-cell derived cells administered should be defined during dose finding studies, unless justified. A safe and effective treatment dose should be identified, and where possible, the minimally effective dose should be determined. The selected biomarkers should permit delineation of the differentiation status of the SCBP at time of patient administration as well as facilitate *in vivo* monitoring once administered. The presence of the administered stem cells in places other than those intended should be investigated. It is important to evaluate the time to achieve the clinical outcome and, where relevant, the time to engraftment in order to correctly define the cell population required for such an *in vivo* effect. The need for and duration of post-authorization long-term efficacy follow-up should be identified during the CTs, taking into consideration results from nonclinical studies and the intended therapeutic effect [22].

In the UK Code of Practice for the Use of Human Stem Cell Lines [38] it is stipulated:

In 3.4 Medicines for Human Use (Clinical Trials) Regulations 2004: Stem cell (gamete derived) cell based products that involve the destruction of a human embryo in their formulation are initially licensed by the HFEA. At the point where the embryo has been destroyed and cells are harvested these human cells would fall under the remit of the HTA. The development of a product using these cells is under the remit of the HTA until such time as the MHRA classifies the product as an Investigational Medicinal Product (IMP) or the product is classified as an Advanced Therapy Medicinal Product (ATMP). Once this classification has been confirmed the Manufacture, Clinical Trial Approval and Marketing approval (for IMPs) are under the remit of the MHRA and not the HTA. Trials of IMPs in the UK are authorised and regulated by the Medicines and Healthcare products Regulatory Agency (MHRA). Stem cell lines that fall within the EU definition of medicinal product which are used in clinical trials to assess safety or efficacy in humans will be IMPs and such trials must be authorised by the MHRA.

In 3.5 EC Regulation on advanced therapy medicinal products 2007:

It is obligatory under the Regulation that all advanced therapy medicinal products (ATMPs) which may include stem cells therapies (regardless of derivation) are subject to the European centralised marketing authorisation procedure which is coordinated on behalf of the European Commission by the European Medicines Agency (EMEA).

In 3.6 Clinical trials involving the use of human stem cell lines the role of the Gene Therapy Advisory Committee (GTAC):

GTAC has UKwide responsibility for the ethical oversight of proposals to conduct clinical trials involving gene therapy or stem cell therapies derived from stem cell lines. It is both a Government Scientific Advisory Committee and a Research Ethics Committee formally recognised in statute. Its terms of reference are:

- To consider and advise on the acceptability of proposals for gene therapy research on human subjects, on ethical grounds, taking account of the scientific merits of the proposals and the potential benefits and risks.
- To consider and advise on the acceptability of proposals for research on human subjects using cells derived from stem cell lines, based on ethical grounds, taking account of the scientific merits of the proposals and the potential benefits and risks.
- To provide ethical advice on the use of unlicensed gene therapy and stem cell line derived therapies in humans.
- To work with other agencies which have responsibilities in this field, including research ethics committees, and agencies with statutory responsibilities the Medicines and Healthcare products Regulatory Agency, the Human Tissue Authority, the Health and Safety Executive and the Department for Environment Food and Rural Affairs.
- To provide advice to United Kingdom Health Ministers on the above matters.

Researchers wishing to conduct clinical trials using products derived from stem cell lines should contact the GTAC Secretariat for initial discussions (gtac@dh.gsi.gov.uk).

In 3.7 Overview of regulatory requirements:

Research involving human stem cells, and in particular the development of stem cell therapies, may involve many regulatory approvals. This is due to the nature of UK and EU legislation by which embryos, cells, tissue, clinical trials and licensing of therapies fall under separate legislation. An interactive resource explaining all the UK regulatory requirements, information and points of contact within the relevant organisations is provided through the Department of Health/Medical Research Council (MRC) "UK Stem Cell Tool Kit". This resource allows researchers to build a customised 'map' outlining all of the regulatory steps necessary to undertake research involving human stem cells and to translate ideas for a new treatment from the laboratory to patients [38].

Scenario in India

The "Ethical Guidelines for Biomedical Research on Human subjects" released by Indian Council of Medical Research (ICMR) in 2006 [26] has provided under Section V, the requirements for carrying out "stem cell research and therapy". These guidelines have categorized research on stem cells into mainly three areas, namely, permissible, restrictive and prohibited areas. Under permissible category, CT with clinical grade stem cells, following ICMR Guidelines for Biomedical Research and GCP guidelines of the Government of India (GOI), may be carried out with prior approval of Institutional Committee for Stem Cell Research and Therapy (IC-SCRT), Institutional Ethics Committee (IEC) and Drug Controller General of India (DCGI). Clinical grade stem cells are required to be produced under international GMP/GTP conditions. The headings under which the CT protocols should be written need to be as per Annexure III of the guideline. All CTs on stem cells shall be registered with National Apex Committee for Stem Cell Research and Therapy (NAC-SCRT) through IC-SCRT. Restricted category includes CTs sponsored by multinationals, involving stem cell products imported from abroad. Such collaboration shall require prior approval of the NAC-SCRT through IC-SCRT, IEC, DCGI and respective funding agency as per its procedure/Health Ministry's Screening Committee (HMSC). Each institution shall constitute an IC-SCRT as provided in these guidelines and provide adequate support for its functioning.

ICMR and the Department of Biotechnology (DBT) have together laid down "Guidelines for Stem Cell Research and Therapy" in November 2007 [27]. The guideline has many commonalities with the ICMR, 2006 guidelines. The guideline has emphasized on mechanism for review and monitoring research and therapy in the field of human stem cells, one at the National level (the NAC-SCRT) and the other at the institutional level (the IC-SCRT). All established human stem cell lines from any source, imported or created in India, should be registered with IC-SCRT and NAC-SCRT. The investigators should ensure that the cell lines have been established in accordance with the existing guidelines of the country. An appropriate Material Transfer agreement (MTA) should be adopted for the purpose. The investigators and the institutions where the stem cell research is being conducted need to bear the ultimate responsibility of ensuring that research activities are in accordance with laid down standards and integrity. CTs with cells processed as per National GTP/GMP guidelines (minimally manipulated or manipulated with alteration in functionality or genetic characteristics) may be carried out with prior approval of IC-SCRT/IEC/DCGI, as applicable. The informed consent process for participation in CTs for SCBP encompasses many more details and conditions than those for other type of products. All records pertaining to adult stem cell research must be maintained for at least 5 years and those related to hES cell research must be maintained for 10 years.

Clinical use of stem cells is not permitted until the

- Efficacy and safety of the procedure is established;
- Origin, safety and composition of the product is adequately defined and labeled; and
- Conditions for storage and use are given in detail.

Our Central Drugs Standards Control Organization (CDSCO) has released guidance document on submission requirements for new drug approvals for Biotechnological/Biological products in Dec 2008 [25] along the lines of the

CTD format. However, the same format cannot be directly applied for SCBP due to inherent differences. Again, under the Drugs and Cosmetics Act and Rules, there is no specific "Form" applicable either to apply for grant or renewal of manufacturing licence for SCBPs. While there has been a subtle growth in the number of private hospitals and clinics providing stem cell therapies across India, the Indian industry is at crossroads in deciding how to take their SCBPs, for which they have gone through the CTs (after taking due approvals from DCGI, IEC and ISCRT), to a commercially licensed product within India! In the absence of laws/regulations specifying the requirements, it is difficult to enforce the existing guidelines in India. Also, NAC-SCRT is yet to become functional. Once regulations are laid down, one can be either in compliance or out of compliance, and automatically an enforcement mechanism would get built-in against non-compliance. Indian government has taken steps in this direction. A new central committee, viz. Cell Biology Based Therapeutic Drug Evaluation Committee (CBBTDEC) has been set up, under the chairmanship of DG ICMR, with the mandate to advice on regulatory pathway for CT marketing approval for therapeutic products derived from stem cells, human gene manipulation and xenotransplant technology. CBBTDEC had its first meeting to discuss various proposals put up to DCGI by the sponsors/CROs on March 9, 2011. Formal recommendations have been communicated in May 2011 [22].

Toward Harmonization

Though the International Conference on Harmonisation of Technical Requirements for Registration of Pharmaceuticals for Human Use (ICH) has not yet formulated any guidelines specific to SCBPs, some of its guidelines on biotechnology products are relevant to this area [7]. Various non-binding codes of practice and guidelines to cover stem cell research have also been published by international bodies such as the International Society for Stem Cell Research [28] and the Hinxton Group [37]. "The Guidelines for the Clinical Translation of Stem Cells" drafted by ISSCR emphasize the CTs approach in the majority of translational stem cell studies. Fundamental principles in the responsible clinical application of stem cells are the following:

- Only quality-controlled cells with known biological characteristics are used;
- Efficacy and safety after delivery of the cells have been demonstrated in appropriate animal models;
- Stem cell specific expertise is involved in the peer review of the clinical protocols and the underlying pre-clinical research; and
- Voluntary informed consent is obtained prior to a CT to ensure that recipients are aware of the risks of tumor formation and lack of proof of clinical benefits.

But, as guidelines, the ISSCR's recommendations are essentially an unenforced code of professional conduct! Both the regulatory frameworks in the EU and USA are structured to assure safety and thus they require a thorough analysis of all critical steps and aspects in advance. Although there are still differences, the authorities are in contact to further harmonize them [32].

ISSCR's Recommendations

Recommendation 1: Institutions where preclinical or clinical research involving stem cells or their direct derivatives is performed should take efforts to ensure that investigators are aware of these Guidelines and other relevant policies and regulations and put them into practice.

Recommendation 2: Human subjects review committees must review clinical research involving (a) products from human embryonic or other pluripotent stem cells; (b) novel applications of fetal or somatic (adult) stem cells; and (c) hematopoietic or other stem cells used for applications outside established standards of care. The human subjects review of stem cell-based clinical protocols must enlist stem cell-specific scientific and ethical expertise. The ISSCR does not anticipate that stem cell research oversight committees will be required to conduct a separate review, although some members of stem cell research oversight committees may be used as consultants to the human subjects review process.

Recommendation 3: In the case of donation for allogeneic use, the donor should give written informed consent that covers, where applicable, the following issues:

(a) that cells and/or cell lines may be subject to storage. If possible, duration of storage should be specified;
(b) that the donor may (or may not) be approached in the future to seek additional consent for new uses, or to request additional material (blood or other clinical samples) or information;
(c) that the donor will be screened for infectious and possibly genetic diseases;
(d) that the donated cells may be subject to genetic modification by the investigator;
(e) that with the exception of directed altruistic donation, the donation is made without restrictions regarding the choice of the recipient of the transplanted cells;
(f) disclosure of medical and other relevant information that will be retained, and the specific steps that will be taken to protect donor privacy and confidentiality of retained information, including the date at which donor information will be destroyed, if applicable;
(g) explanation of what types of genomic analyses (if any) will be performed and how genomic information will be handled; and
(h) disclosure that any resulting cells, lines or other stem cell-derived products may have commercial potential, and whether any commercial and intellectual property rights will reside with the institution conducting the research.

Recommendation 4: Donors must be screened for infectious diseases, as is done for blood and solid organ donation, and for genetic diseases as appropriate.

Recommendation 5: In the course of development of stem cell-based products, it is imperative to validate surrogate markers of the identity and potency of cell products.

Recommendation 6: Where possible, components of animal origin used in the culture or preservation of cells should be replaced with human components or with chemically defined components to reduce the risk of accidental transfer to patients of unwanted chemical or biological material or pathogens.

Recommendation 7: Acknowledging the limitations in current assays, scientists and regulators must work together to develop common reference standards for minimally acceptable changes during cell culture, to ensure quality and safety of cell therapy, and to facilitate comparisons across studies.

Recommendation 8: The level of regulation and oversight should be proportional to the degree of risk raised by the particular cell product and intended use (autologous versus allogeneic use, minimally versus highly manipulated cell products, use for homologous versus non-homologous functions).

Recommendation 9: To facilitate international collaboration and universal access to stem cell-based treatments (both during clinical trials and when established as standards of clinical care), there is a need to develop appropriate quality management systems for donation, procurement, testing, coding, processing, preservation of stem cell potency, storage, and distribution of the cells. For extensively manipulated stem cells (either autologous or allogeneic) destined to clinical application, the ISSCR recommends adherence to GMP procedures, which includes minimizing risks to patients from unwanted cell products.

Recommendation 10: Cellular therapeutics that incorporate gene repair or genetic modification must adhere to regulatory guidelines set forth for both gene therapy and cell therapy.

Recommendation 11: Sufficient preclinical studies in relevant animal models – whenever possible for the clinical condition and the tissue physiology to be studied – are necessary to make proposed stem cell-based clinical research ethical, unless approved, controlled, and conclusive humans studies are already available with the same cell source. Investigators should develop preclinical cell therapy protocols in small animal models, as well as in large animal models when deemed necessary by independent peer review or regulatory review.

Recommendation 12: Because new and unforeseen safety concerns may arise with clinical translation, frequent interaction between preclinical and clinical investigators is strongly encouraged.

Recommendation 13: Small animal models should be used to test the transplantation of wild-type and/or diseased and genetically-corrected stem cells, to assess the morphological and functional recovery caused by cell therapy, and to investigate the biological mechanisms of tissue restoration or repair. Small animal studies should also assess the dosage and route of administration of potential cell therapies, the optimal age and disease stage for therapeutic efficacy, and the cellular distribution, survival, and tissue integration.

Recommendation 14: Large animal models should be used for stem cell research related to diseases that cannot be sufficiently addressed using small animal models or where structural tissue such as bone, cartilage, or tendon need to be tested in a load-bearing model. The selected large animal model must offer an appropriate context for studying the human disease and conditions of specific interest.

Recommendation 15: The need for studies in non-human primates should be evaluated on a case-by-case basis, and performed only if the studies promise to provide necessary and otherwise unobtainable information for experimental therapeutic application of stem cells or their progeny in patients. All studies involving the use of non-human primates must be conducted under the close supervision of qualified veterinary personnel with expertise in their care and their unique environmental needs.

Recommendation 16: Cells to be employed in clinical trials must first be rigorously characterized to assess potential toxicities through *in vitro* studies and (where possible for the clinical condition and tissue physiology to be examined) in animal studies.

Recommendation 17: Criteria for release of cells for transfer to patients must be designed to minimize risk from culture-acquired abnormalities.

Recommendation 18: Risks for tumorigenicity must be assessed for any stem cell-based product, especially when extensively manipulated in culture or when genetically modified. A clear plan to assess the risks of tumorigenicity for any cell product must be implemented under the direction of an independent review body prior to approval for human clinical use.

Recommendation 19: Cell cultures and animal models should be used to test the interaction of cells with drugs to which recipients will be exposed. These include the immunosuppressants planned for recipients, as well as other drugs that might be used to treat their underlying disease process.

Recommendation 20: Stem cell-based clinical researchers should:

(a) cooperate with and share scientific expertise to assist other investigators and human subjects research review committees in assessing:
   (i) the biological characteristics of the cells to be used in clinical trials;
   (ii) whether these cells have been developed with appropriate manufacturing standards;
   (iii) preclinical data on their use in animal and/or other models for evaluating their safety and efficacy; and
   (iv) any early clinical data, if available, which address safety issues in the short and medium term and continued observation for long term effects;
(b) address the risks of stem cell-based interventions including, for example, cell proliferation and/or tumor development, exposure to animal source materials, risks associated with viral vectors, and risks as yet unknown;
(c) provide the utmost clarity regarding the potential benefits of participating in the trial with stem cells, since patients may have recourse to reasonable therapeutic

alternatives; the informed consent process must emphasize the novel and experimental aspects of cell based interventions. It is important to minimize misconceptions patients may have about the potential for therapeutic efficacy;
(d) disclose any financial and non-financial conflicts of interest among the investigators, sponsors, and institutions in which the stem cell research is being conducted;
(e) monitor research subjects for long-term health effects and protection of the confidentiality of their health data;
(f) provide a clear, timely, and effective plan for adverse event reporting;
(g) offer a clinical plan to provide treatment for toxicity, including treatment of tumors that might arise. This plan might include compensation for research-related injuries; and
(h) ensure that insurance coverage or other appropriate financial or medical resources are available to patients to cover potential complications arising from their research participation.

Recommendation 21: All studies involving clinical applications of stem cells, whether publicly or privately sponsored, must be subject to independent review, approval, and ongoing monitoring by human subjects research oversight bodies with supplemental appropriate expertise to evaluate the unique aspects of stem cell research and its application in a variety of clinical disciplines. This review and oversight process must be independent of the investigators regardless of whether it occurs at the institutional, regional, or national level, and regardless of whether investigators employ the services of a contract research organization.

Recommendation 22: In countries where there is no official national regulatory body, the ISSCR strongly encourages governments to develop a regulatory competence at the national, regional, or local level to monitor clinical interventions with stem cell-based products. The ISSCR will strive to provide professional advice to those governing bodies interested in building their own capacities for regulatory oversight.

Recommendation 23: The peer review process for stem cell-based clinical trials should have appropriate expertise to evaluate (a) the *in vitro* and *in vivo* preclinical studies that form the basis for proceeding to a clinical trial and (b) the scientific underpinnings of the trial protocol, the adequacy of planned end-points of analysis, statistical considerations, and disease-specific issues related to human subject protection.

Recommendation 24: Risks should be identified and reduced, and potential benefits to subjects must be realistically delineated but not overemphasized. Subject selection can affect the risks and benefits of the study and subjects should be selected to minimize risks, maximize the ability to analyze results, and enhance the benefits to individual subjects and society.

Recommendation 25: As a general principle, a stem cell-based approach must aim at being clinically competitive or superior to existing therapies. If an efficacious therapy already exists, the risks associated with a stem cell-based approach must be low and the stem cell-based approach must offer a potential advantage (for example, better functional outcome; single procedure (cell administration) versus life-long drug therapy with associated side effects; reduction in long-term cost). If an efficacious therapy is not available, then the severity of the disease, especially if the disease to be treated is severely disabling and life-threatening, might justify the risks of a stem cell-based experimental intervention in patients. Maximum effort should be made to minimize the risks for all possible adverse events associated with stem cell-based approaches. Care must also be taken to not take advantage of the hopes of patients with poor short-term prognoses.

Recommendation 26: Clinical research should compare new stem cell-based therapies against the best medical therapy currently available to the local population.

Recommendation 27: As far as possible, groups or individuals who participate in clinical stem cell research should be in a position to benefit from the results of this research. Groups or individuals must not be excluded from the opportunity to participate in clinical stem cell research without rational justification.

Recommendation 28: Informed consent is particularly challenging for clinical trials involving highly innovative interventions.

(a) Patients need to be informed when novel stem cell-derived products have never been tested before in humans and that researchers do not know whether they will work as hoped.
(b) Cell-based interventions, unlike many pharmacological products or even many implantable medical devices, may not leave the body and may continue to generate adverse effects for the lifetime of the patient. The possible irreversibility of a cellular transplant should be explained clearly.
(c) Subjects should be informed about the source of the cells so that their values are respected.
(d) Ensuring subject comprehension must be done at each phase of the clinical trials process. Ideally, the subject's comprehension of information should be assessed through a written test or an oral quiz during the time of obtaining consent.
(e) Human subjects research committees should ensure that informed consent documents accurately portray these uncertainties and potential risks, and clearly explain the experimental nature of the clinical study.

Recommendation 29: A data monitoring plan, which may involve an independent data safety and monitoring process, is required for all clinical studies, and aggregate updates should be provided to peer review committees on demand, complete with adverse event reporting and ongoing statistical analysis.

Recommendation 30: Subject withdrawal from the research should be done in an orderly fashion to promote physical and psychological safety. Given the potential for transplanted cellular products to persist long-term, and depending on the nature of the experimental stem cell-based intervention, patients may have to undergo long-term health monitoring, and additional safeguards for ongoing patient privacy should be provided.

Recommendation 31: To advance scientific understanding, research subjects should be asked, in the event of death, for consent to the performance of a partial or complete autopsy to obtain information about the extent of cellular implantation and its morphological and functional consequences. Any request for an autopsy must consider cultural and familial sensitivities.

Recommendation 32: Researchers should facilitate the gathering of empirical data about socio-demographic characteristics of participants in clinical trials, financial compensation levels (if applicable), and the nature and extent of any benefit and harm resulting from research participation. Such data are crucial for health services researchers and policy-makers to improve the conduct of future clinical trials and to assess the utility of the information obtained in these trials for informing policy decisions such as approval and insurance coverage for cell-based interventions.

Recommendation 33: Researchers should publish both positive and negative results and adverse events. To ensure the integrity of scientific information and to promote the highest standards of professional conduct, researchers should present their results at professional scientific conferences or in peer-reviewed scientific journals before reporting their research to the lay media or to patient advocacy groups and associations.

Recommendation 34: Clinician-scientists may provide unproven stem cell-based interventions to at most a very small number of patients outside the context of a formal clinical trial, provided that:

(a) there is a written plan for the procedure that includes:

  (i) scientific rationale and justification explaining why the procedure has a reasonable chance of success, including any preclinical evidence of proof-of- principle for efficacy and safety;
  (ii) explanation of why the proposed stem cell-based intervention should be attempted compared to existing treatments;
  (iii) full characterization of the types of cells being transplanted and their characteristics as discussed in Section 4, Cell Processing and Manufacture;
  (iv) description of how the cells will be administered, including adjuvant drugs, agents, and surgical procedures; and
  (v) plan for clinical follow-up and data collection to assess the effectiveness and adverse effects of the cell therapy;

(b) the written plan is approved through a peer review process by appropriate experts who have no vested interest in the proposed procedure;
(c) the clinical and administrative leadership supports the decision to attempt the medical innovation and the institution is held accountable for the innovative procedure;
(d) all personnel have appropriate qualifications and the institution where the procedure will be carried out has appropriate facilities and processes of peer review and clinical quality control monitoring;
(e) voluntary informed consent is provided by patients who appreciate that the intervention is unproven and who demonstrate their understanding of the risks and benefits of the procedure;

(f) there is an action plan for adverse events that includes timely and adequate medical care and if necessary psychological support services;
(g) insurance coverage or other appropriate financial or medical resources are available to patients to cover any complications arising from the procedure; and
(h) there is a commitment by clinician-scientists to use their experience with individual patients to contribute to generalizable knowledge. This includes:
    (i) ascertaining outcomes in a systematic and objective manner;
    (ii) a plan for communicating outcomes, including negative outcomes and adverse events, to the scientific community to enable critical review (for example, as abstracts to professional meetings or publications in peer-reviewed journals); and
    (iii) moving to a formal clinical trial in a timely manner after experience with at most a few patients.

<u>Recommendation 35</u>: Regulatory and oversight agencies, (local, national and international) must explicitly include the consideration of social justice principles into their evaluations. Mechanisms include (a) involvement of community and patient advocates in public discussions, committee representation, and oversight board evaluation procedures; (b) opportunity for open discussions about ethical issues; (c) enforcement of social justice considerations by appropriate agencies.

<u>Recommendation 36</u>: Reporting on stem cell research must be based in scientifically-grounded research. Frank disclosure of failures in research, adverse incidents, and lack of significant change in the status of treated patients will need to be made. Patient advocates must follow the same standards of discourse.

<u>Recommendation 37</u>: There should be public engagement in the policy making of individual governmental agencies. Such consultation should aim to be inclusive and interactive.

<u>Recommendation 38</u>: The ISSCR seeks to maximize social good, which leads to the following considerations:

(a) Stem cell collections with genetically diverse sources of cell lines should be established.
(b) Collaborations among researchers and institutions should be structured to maximize the fairness of the parties' roles, and to increase joint capacity and social benefit.
(c) Fair access is important. Access will depend on financial terms and business models that are perceived as fair by all stakeholders, including patients, providers, payers, companies, and governments. The ISSCR therefore:
    (i) encourages open stakeholder discussion to identify and evaluate alternative models and terms; and
    (ii) encourages development and assessment of alternative models of intellectual property, licensing, product development, and public funding to promote fair and broad access to stem cell-based diagnostics and therapies.

Recommendation 39: As an aspirational ethical goal – provided that a stem cell-based therapy is proven to offer a major therapeutic benefit – commercial companies, subject to their financial capability, should offer affordable therapeutic interventions to persons living in resource-poor countries who would otherwise be wholly excluded from benefiting from that stem cell-based therapy. Academic and other institutions that are licensing stem cell therapeutics and diagnostic inventions should incorporate this requirement in their intellectual property license.

Recommendation 40: These guidelines will be reviewed and revised as needed to accommodate new scientific advances and to address specific translational research issues.

Way Forward

There is still a significant gap between promising laboratory-based research and approved SCBPs in this fast emerging field. Legislation in this field must seek to both regulate and enable scientific progress without being confusing, difficult to interpret or unnecessarily onerous. In addition, the public must have confidence that its interests are protected [30]. Few of the measures which could help to speed up the translation of SCBP from bench to bedside while still ensuring patient safety include the following.

- Compliance with the existing regulations and guidelines to ensure that the product is safe, pure, and potent meeting GTP, GMP and GCP requirements.
- Nonclinical evidence on the proof-of-principle and safety in a relevant animal model should be tried before administration to humans.
- Encourage companies to develop and validate new non-invasive methods for biodistribution studies in humans to follow the cells during the CTs. Possible markers/tracers should be evaluated and justified.
- A risk-based approach to be applied while giving regulatory approvals. Conditional marketing autho-rization could be a possible approach without compromising on patient safety [22].

References

1. Ahrlund RL, De Luca M, Marshak DR, Munsie M, Veiga A, Rao M. Isolation and production of cells suitable for human therapy: challenges ahead. Cell Stem Cell. 2009;4:20–6.
2. Ansboro S et al. Strategies for improved targeting of therapeutic cells: implications for tissue repair. Eur Cell Mater. 2012;23:310–9.
3. Bellamy J. Stem cell therapy regulation plays catch up. Science-Based Medicine, August 9, 2012.
4. Bersenev A. Regulation of autologous adult stem cells – the case of regulatory regress. Cell Trials, October 18. http://celltrials.info/2012/10/18/regulation-autologous-adult-stem-cells-case-regulatory-regress/ (2012).

5. Bianco P, Robey PG, Robey PJ. Mesenchymal stem cells: revisiting history, concepts, and assays. Cell Stem Cell. 2008;2:313–9.
6. Bieback K, Kinzebach S, Karagianni M. Translating research into clinical scale manufacturing of mesenchymal stromal cells. Stem Cells Int. 2010;25:1–11.
7. Catalano J. The international conference on harmonization (ICH) and its relevance to cell therapy. In: ISCT 6th annual somatic cell therapy symposium. 2. Available from: http://www.fda.gov/cber/genetherapy/isct092506jc.htm (2006).
8. Collins NH. Product review, release, and administration. In: Gee A, editor. Cell therapy: cGMP facilities and manufacturing. New York: Springer; 2009. p. 215–28.
9. EU Directive 2001/20/EC. Relating to the implementation of good clinical practice in the conduct of clinical trials on medicinal products for human use. Official J Eur Commun, 1 May 2001.
10. EU Directive 2001/83/EC. The community code relating to medicinal products for human use. http://www.emea.europa.eu/docs/en_GB/document_library/Regulatory_and_procedural_guideline/2009/10/WC500004481.pdf
11. EU Directive 2003/63/EC. Amending Directive 2001/83/EC on the Community code relating to medicinal products for human use. Official J Eur Union, 27 June 2003.
12. EU Directive 2004/23/EC. Setting standards of quality and safety for the donation, procurement, testing, processing, preservation, storage and distribution of human tissues and cells. http://eur-lex.europa.eu/LexUriServ/LexUriServ.do?uri=OJ:L:2004:102:0048:0058:en:PDF
13. Dresser R, Frader J. Off-label prescribing: a call for heightened professional and government oversight. J Law Med Ethics. 2009;37:476–86.
14. Editorial 2012. Stem cell research: regulating translational application. Nat Cell Biol. 14557, published online 30 May 2012.
15. EU EMEA. Guideline on human cell-based medicinal products. http://www.emea.europa.eu/docs/en_GB/document_library/Scientific_guideline/2009/09/WC500003894.pdf (2008).
16. EU CAT. Reflection paper on stem cell-based medicinal products. EMA/CAT/571134/2009, 2011;1–14. http://www.ema.europa.eu/docs/en_GB/document_library/Scientific_guideline/2011/02/WC500101692.pdf.
17. EU Committee for Advanced Therapies (CAT). Use of unregulated stem-cell based medicinal products. Lancet. 2010;376:514.
18. EU European Medicines Agency (EMEA). Guideline on strategies to identify and mitigate risks for first-in-human clinical trials with investigational medicinal products, Doc. Ref. EMEA/CHMP/SWP/28367/07. http://www.emea.europa.eu/docs/en_GB/document_library/Scientific_guideline/2009/09/WC500002988.pdf (2007).
19. EU Regulation 1394/2007/EC. On advanced therapy medicinal products and amending Directive 2001/83/EC and Regulation (EC) No 726/2004. Official J Eur Comm, 10 Dec 2007.
20. Gatti RA, Meuwissen HJ, Allen HD, Hong R, Good RA. Immunological reconstitution of sex-linked lymphopenic immunological deficiency. Lancet. 1968;2:1366–9.
21. Gee A. Mesenchymal stem-cell therapy in a regulated environment. Cytotherapy. 2001;3:397–8.
22. George B. Regulations and guidelines governing stem cell based products: clinical considerations. Perspect Clin Res. 2011;2(3):94–9, Jul–Sep.
23. Halme DG, Kessler DA. FDA regulation of stem-cell based therapies. N Engl J Med. 2006;355:1730–5.
24. Hyun I. Allowing innovative stem cell-based therapies outside of clinical trials: ethical and policy challenges. J Law Med Ethics. 2010;38:277–85.
25. Indian CDSCO. Guidance for industry on submission of clinical trial application for evaluating safety and efficacy. Doc No. CT/71108, Version 1.1, 2008. Available from: http://www.cdsco.nic.in (2008).
26. Indian Council of Medical Research (ICMR). Ethical guidelines for biomedical research on human subjects. http://icmr.nic.in/ethical_guidelines.pdf (2006).
27. Indian Council of Medical Research. National guidelines on stem cell research and therapy. http://www.inclentrust.org/uploadedbyfck/file/2007.pdf (2007).

28. International Society for Stem Cell Research (ISSCR). Guidelines for the clinical translation of stem cells. Available from: http://www.isscr.org/clinical_trans/index.cfm (2008).
29. Koleva G. Stem cells, FDA, and the edge of science: three expert viewpoints, Forbes 2/19/2012. http://www.forbes.com/sites/gerganakoleva/2012/02/19/stem-cells-fda-and-the-edge-of-science-three-expert-viewpoints/ (2012).
30. Leanne B, Sarah D. Gaps and overlaps: improving the current regulation of stem cells in the UK. J Med Ethics. 2007;33:621–2.
31. Lelas A. Future research and therapeutic applications of human stem cells: general, regulatory, and bioethical aspects. J Transl Med. 2010;8:131.
32. Martell K, Trounson A, Baum E. Stem cell therapies in clinical trials: workshop on best practices and the need for harmonization. Cell Stem Cell. 2010;7:451–4.
33. Parson A. The long journey from stem cells to medical product. Cell. 2006;125:9–11.
34. Rayment EA, Williams DJ. Concise review: mind the gap: challenges in characterizing and quantifying cell- and tissue-based therapies for clinical translation. Stem Cells. 2010;28:996–1004.
35. Shapiro J, et al. FDA's regulatory scheme for human tissue: a brief overview. Hum Tissue Regul, Update, p. 10 with permission from FDLI, www.fdli.com (2007).
36. Stemedica. Current News Release, October 2. http://www.stemedica.com/info/allogeneic-adult-stem-cells/stem-cell-clinical-trials/2012-1002-FDA-Approves-Stemedica-Phase-II-Clinical-Trial-Acute-Myocardial-Infarction-Ischemia-Tolerant-Mesenchymal-Stem-Cells.asp (2012).
37. The Hinxton Group. An international consortium on stem cells, ethics and law. Law Hum Genome Rev. 2006;24:251–5.
38. UK Medical Research Council (MRC). Code of practice for the use of human stem cell lines, April 2010. http://www.mrc.ac.uk/Utilities/Documentrecord/index.htm?d=MRC003132 (2010).
39. US FDA. Biological products regulated under Section 351 of the Public Health Services Act; implementation of biologics license; elimination of establishment license and product license; correction – FDA. Proposed rule; correction. Fed Regist. Sept 2, 1998a;63(170):46718.
40. US FDA. Guidance for human somatic cell therapy and gene therapy, FDA Center for Biologics Evaluation and Research. http://www.fda.gov/biologicsbloodvaccines/guidancecompliance-regulatoryinformation/guidances/cellularandgenetherapy/ucm072987.htm (1998b).
41. US FDA Human cells, tissues and cellular and tissue-based products, Code of Federal Regulations Title 21 Part 1271. http://www.accessdata.fda.gov/scripts/cdrh/cfdocs/cfcfr/CFRSearch.cfm?CFRPart=1271 (2006).
42. US Food and Drug Administration (FDA). Proposed approach to regulation of cellular and tissue-based products. The Food and Drug Administration. http://www.fda.gov/downloads/BiologicsBloodVaccines/GuidanceComplianceRegulatoryInformation/Guidances/Tissue/UCM062601.pdf (1997).